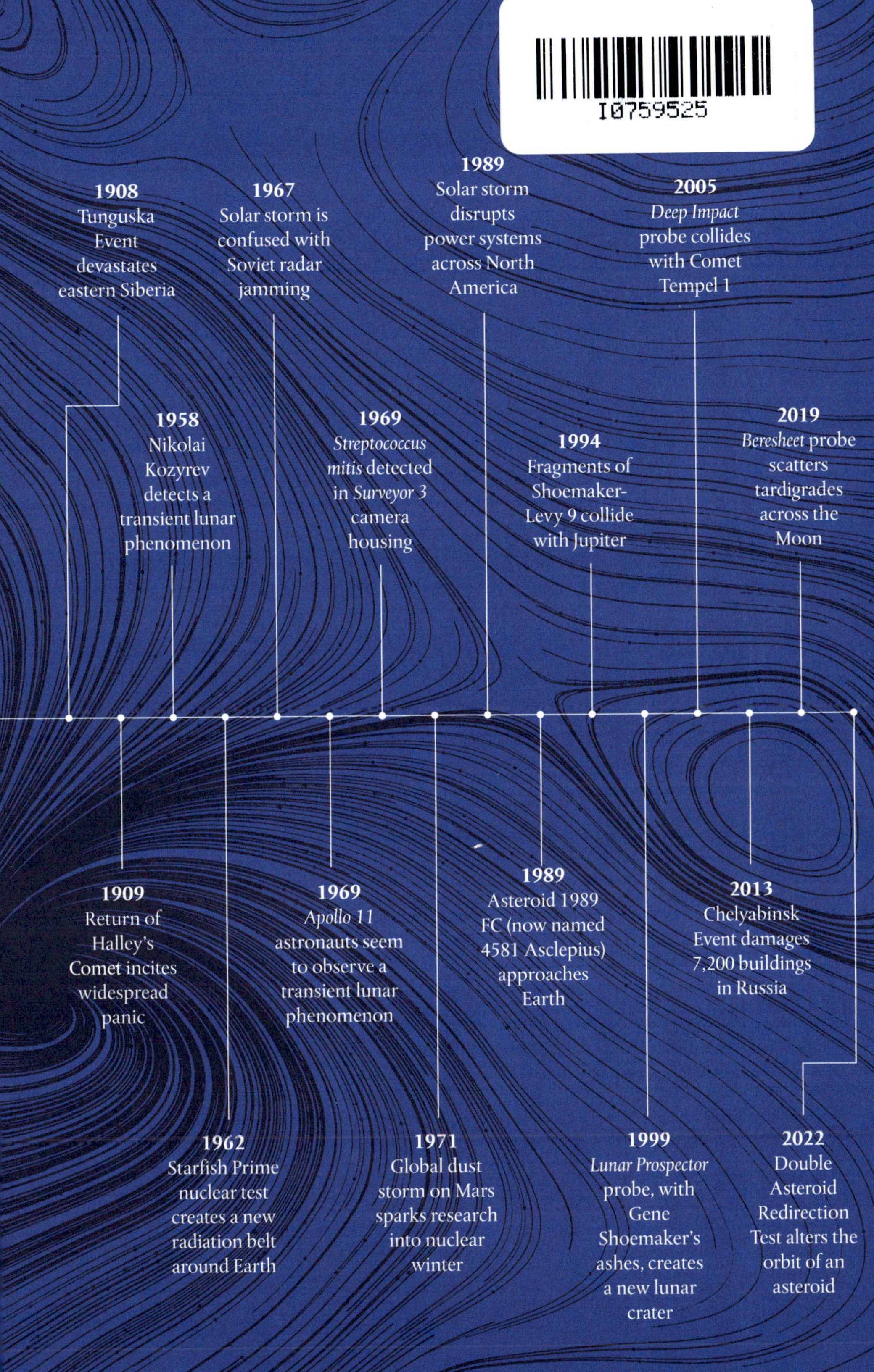

1908
Tunguska Event devastates eastern Siberia
1909
Return of Halley's Comet incites widespread panic
1958
Nikolai Kozyrev detects a transient lunar phenomenon
1962
Starfish Prime nuclear test creates a new radiation belt around Earth
1967
Solar storm is confused with Soviet radar jamming
1969
Apollo 11 astronauts seem to observe a transient lunar phenomenon
1969
Streptococcus mitis detected in Surveyor 3 camera housing
1971
Global dust storm on Mars sparks research into nuclear winter
1989
Solar storm disrupts power systems across North America
1989
Asteroid 1989 FC (now named 4581 Asclepius) approaches Earth
1994
Fragments of Shoemaker-Levy 9 collide with Jupiter
1999
Lunar Prospector probe, with Gene Shoemaker's ashes, creates a new lunar crater
2005
Deep Impact probe collides with Comet Tempel 1
2013
Chelyabinsk Event damages 7,200 buildings in Russia
2019
Beresheet probe scatters tardigrades across the Moon
2022
Double Asteroid Redirection Test alters the orbit of an asteroid

RIPPLES ON THE COSMIC OCEAN

Ripples on the Cosmic Ocean

An Environmental History of Our Place in the Solar System

Dagomar Degroot

THE BELKNAP PRESS OF HARVARD UNIVERSITY PRESS
CAMBRIDGE, MASSACHUSETTS
2025

Printed in the United States of America
First printing

EU GPSR Authorised Representative
LOGOS EUROPE, 9 rue Nicolas Poussin, 17000, LA ROCHELLE, France
E-mail: Contact@logoseurope.eu

Library of Congress Cataloging-in-Publication Data
Names: Degroot, Dagomar, author.
Title: Ripples on the cosmic ocean : an environmental history
of our place on the solar system / Dagomar Degroot.
Description: Cambridge, Massachusetts : The Belknap Press of
Harvard University Press, 2025. | Includes bibliographical references and index.
Identifiers: LCCN 2025007566 (print) | LCCN 2025007567 (ebook) |
ISBN 9780674986503 (cloth) | ISBN 9780674301818 (epub) | ISBN 9780674301825 (pdf)
Subjects: LCSH: Planetary science—History. | Solar system—Environmental
conditions. | Earth (Planet)—Environmental conditions.
Classification: LCC QB501 .D45 2025 (print) | LCC QB501 (ebook) |
DDC 525—dc23/eng/20250425
LC record available at https://lccn.loc.gov/2025007566
LC ebook record available at https://lccn.loc.gov/2025007567

Dedicated to my son, James, who wants to be an astronaut
and to my daughter, Elowyn, who would prefer to be a gardener

Contents

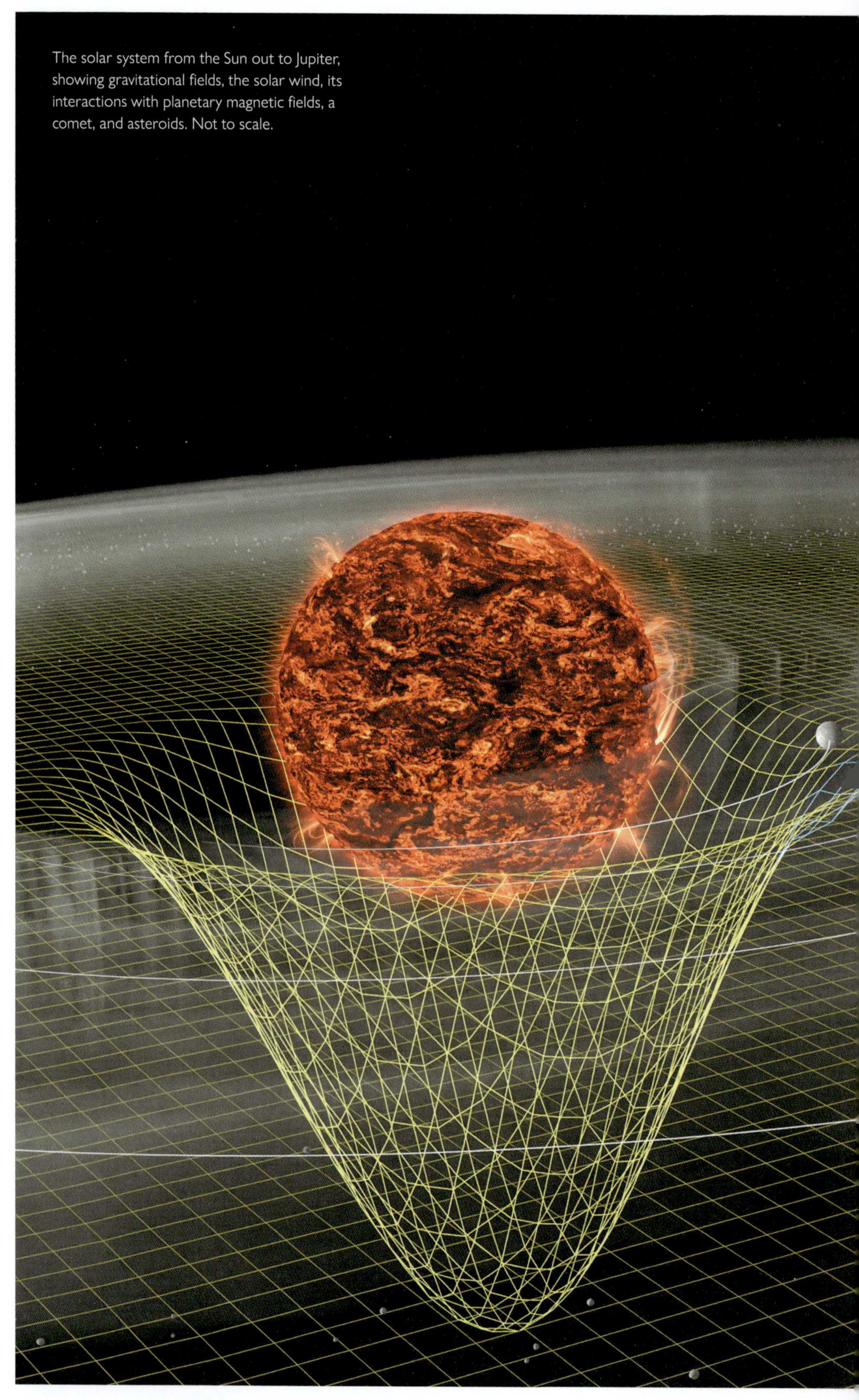

The solar system from the Sun out to Jupiter, showing gravitational fields, the solar wind, its interactions with planetary magnetic fields, a comet, and asteroids. Not to scale.

RIPPLES ON THE COSMIC OCEAN

Introduction
Our Cosmic Environment

A drawing depicting the Milky Way Galaxy.

"When there is any change in the wonted order,
then all eyes are turned to the sky."
—Lucius Annaeus Seneca, *Natural Questions* (64 CE)

"A change always seems to the inquisitive intellect of man
like a breach in the defences of Nature's secrets,
through which it may hope to make its way to the citadel."
—Agnes Clerke, *A Popular History of Astronomy
during the Nineteenth Century* (1885)

Nothing else created on Earth had ever been as lonely as a particular car-sized robot was on February 14, 1990. *Voyager 1* was then nearly six billion kilometers from Earth. It would take almost seven thousand years to drive that far at highway speeds. Across the unimaginable expanse, its caretakers on Earth sent *Voyager 1* a simple command: Take a picture of us. Hurtling through space at over sixty thousand kilometers per hour, the robot complied.[1]

The command was the brainchild of the planetary scientist Carl Sagan, and it took over five hours for the photo to reach Earth. When it did, it revealed a startling new view of our planet: a "pale blue dot," Sagan called it, less than a pixel across. A few years later Sagan reflected on what existed on that dot:

> On it everyone you love, everyone you know, everyone you ever heard of, every human being who ever was, lived out their lives. The aggregate of our joy and suffering, thousands of confident religions, ideologies, and economic doctrines, every hunter and forager, every hero and coward, every creator and destroyer of civilization, every king and peasant, every young couple in love, every mother and father, hopeful child, inventor and explorer, every teacher of morals, every corrupt politician, every "superstar," every "supreme leader," every saint and sinner in the history of our species lived there—on a mote of dust suspended in a sunbeam.[2]

Most historians devote themselves to studying how people have influenced one another across scraps of that mote. Small wonder, since around it there seems to be nothing. It is easy to feel an echo of *Voyager*'s loneliness, for "in all this vastness there is no hint that help will come from elsewhere to save us from ourselves." There is only us and the expanse. Our past is limited by the

A pale blue dot. Look carefully: Earth is just right of center. Photograph of Earth taken by the *Voyager 1* spacecraft from a distance of six billion kilometers.

boundaries of our "tiny stage." It is natural to believe that the *study* of our past should be, too.[3]

But look again at the picture. Is Earth truly alone? The sunbeam in which it is suspended gives it life. How could we consider this dot without recognizing the Sun that nourishes it? Solar light also touches other *worlds,* celestial bodies with complex geological, chemical, and in some cases atmospheric and even biological properties, each shaped by unique evolutionary histories. When sunlight reflects off the solar system's worlds toward Earth, it confuses, intrigues, and occasionally frightens us. Waves of charged particles stream continuously from the Sun, enveloping Earth's magnetic cocoon, with luminous tendrils reaching down from Earth's poles and even extending toward the equator. Gravity bends the fabric of space, warping it around concentrations of mass in the solar system, including the planets and their moons, causing them to tug at one another as they whirl around the deep curvature in space-time created by the Sun. Drawn by the gravitational interplay of planets and the Sun, countless tiny worlds—comets and asteroids—tumble toward Earth along their own curved paths through the rippling tapestry of space. The solar system appears empty to us only because our distant ancestors had no reason to evolve sensory organs capable of perceiving its ceaseless flows of energy and matter. Our dot is merely one particularly comfortable corner of a vast environment created by these flows, all ultimately dependent on the Sun. The solar system is the immense arena surrounding Earth's stage, and though it has too often escaped the notice of historians, it has sustained, inspired, and threatened "every human being who ever was."[4]

Threatened, because the arena is far from stable. The Sun's activity varies with time, and so does Earth's atmosphere. The amount of solar heat retained by Earth waxes and wanes across decades, centuries, even millennia. Bubbles of charged plasma from the Sun hurl toward Earth's magnetic defenses and can transform our skies and scramble our technology. Asteroids and comets zip by Earth and other worlds, occasionally colliding at unimaginable speeds, with catastrophic results. Relentless environmental changes on other planets have sometimes deceived us into believing we have alien company or alerted us to changes in our own world.

In this book I explore the history of these changes—these "ripples on the cosmic ocean"—and imagine them as reverberations in a vast environment that shapes and has shaped human affairs.

Environment is a word with many meanings. We use it broadly to denote our surroundings, which can be physical, social, or metaphorical. The expression *the environment* can also refer to Earth's biosphere, the part of our planet composed of living organisms and their remains. I use the term in a way you might not have encountered before: environment as a network of physical systems that may or may not involve the added complexity of life. I imagine the solar system as a mosaic of environments, organized and entangled by the forces of electromagnetism and gravity. Owing above all to the overwhelming influence of the

Sun, these environments change in ways that reverberate across the solar system, including across Earth, our little corner of the cosmic mosaic.[5]

I argue in this book that these ripples sparked new conversations about the emergence, significance, and ultimate fate of life, especially on Earth. Beginning in the seventeenth century, scholars came to closely associate environmental variability with habitability. When they noticed real or apparent changes in the color, brightness, or even shape of other worlds, they interpreted them as the movement of clouds, the flow of water, and the response of vegetation to autumnal cooling. These changes appeared to reveal that the solar system abounded with analogues of Earth.

Many intellectuals began to imagine how the inhabitants of other worlds would be able to perceive, from afar, how human transformations of Earth differed from our planet's natural variability. Some used new methods of distinguishing purposeful from natural change to search for evidence of alien infrastructure and farming on the other planets in our solar system, and even the Sun and Moon. Scholars repeatedly misinterpreted natural changes in solar system environments as clear signs of alien life, and their "discoveries" sparked widespread fear and excitement. As the nature of environments on the Moon, Mars, and Venus slowly came into sharper focus between the late eighteenth century and early twentieth century, changes in these environments took on a new, frightening meaning for many observers on Earth. Bright flashes and shifting dark patterns on Mars, for instance, appeared to reveal the machinery of a civilization clinging to life on a dying world. Habitability, it seemed, was an unstable condition that would inevitably deteriorate over time.

Because cosmic changes were widely believed to reveal that other planets were inhabited, or at least habitable, they had profound consequences on Earth. They inspired scholars and clerics, artisans and entrepreneurs, writers and artists to develop the principles and technologies for communicating between planets. They spurred the creation of novel machines and practices for observing the heavens, the emergence of innovative literary genres, the invention and reinvention of mass media, and even the basic ideas underpinning what we now call the Anthropocene.

Perhaps most importantly, by the nineteenth century, changes in environments beyond Earth began to hint at a new kind of *risk,* meaning exposure to danger, that threatened all humans, everywhere on Earth. Solar variability, asteroid and comet impacts, shifts in atmospheric chemistry, and the invasion of sapient or microbial aliens all seemed to pose *existential* risks to humanity, threatening to either exterminate our species or imperil our long-term flourishing. The ever-closer union of science with government and industry in the twentieth century eventually enabled governments to imitate the effects of these destructive forces on Earth, and the study of cosmic changes helped reveal precisely how dangerous that could be.

Yet in the twentieth century, changes in cosmic environments, combined with technological, social, and political developments on Earth, also encouraged unprecedented attempts to mitigate existential risks. Efforts to identify and minimize threats caused by deadly microorganisms, deteriorating atmospheric ozone, runaway climate change, solar storms, and asteroid or comet impacts all met with varying degrees of success. But together they created a new era in which powerful governments and large corporations monitored, and to some degree controlled, the relationship between Earth, our stage, and its surrounding arena, the solar system.[6]

This new era involved more than simply negotiating the risks posed by natural changes to cosmic environments. It also involved altering those environments artificially—by exporting terrestrial microorganisms to the surface of other worlds, for example, or by creating and moving swarms of radioactive particles in the space around Earth. Spurred in part by the distant echo of ideas originally inspired by cosmic forces, scientists, engineers, entrepreneurs, and military officers drafted plans for even more dramatic transformations. Some, such as a plan to entirely reshape the environment of Venus, might never leave the drawing board. Others, including the alteration of the Moon's tenuous atmosphere, seem imminent. In this century, settlements with artificial, Earth-like environments could proliferate across the solar system. If they do, some of us could forever breach the narrow boundaries of our pale blue dot, moving from the stage into the arena.

Of course, cosmic changes did not, by themselves, upend human affairs. Many went entirely unnoticed. Some were seen but scarcely mattered. Yet when social, terrestrial, and cosmic sources of change aligned—when the context was just right—a conduit opened through which what happened in space, or seemed to happen in space, truly mattered on Earth. Often, it was only then that human collectives could make their own changes to environments beyond Earth.

Seeing Change in Cosmic Environments

From our perspective here on Earth, every environment in our solar system changes its appearance all the time. Many of these changes follow recurring patterns, as bundles of mass—including Earth—rush along winding paths laid out by gravity. You move with them. While reading this sentence, you have traveled nearly 5 kilometers around Earth's center, 300 kilometers around the Sun, and 2,300 kilometers around the heart of the Milky Way. As a result of all this motion, the core of the Milky Way rises above the horizon after sunset during summers in the Northern Hemisphere (or winters in the Southern Hemisphere), while the outskirts of our galaxy appear in the evenings of Northern Hemisphere winters (or Southern Hemisphere summers). The Moon waxes and wanes through monthly phases—but also appears to grow, shrink, and wobble in the night sky as it revolves in an elliptical orbit around Earth.[7]

Because each successive planet is farther from the Sun and therefore takes longer to complete its orbit, the position of the planets also changes in the night sky. The "inferior" planets—Mercury and Venus—switch from evening to morning stars when they move behind or in front of the Sun, from our perspective, while the "superior" planets—Mars, Jupiter, Saturn, Uranus, and Neptune—appear to move backward, or retrograde, as Earth passes them in its orbit. The inferior planets brighten as they begin to catch up to Earth in their orbits, and the superior do the same as Earth starts to catch up to them. They all dim as we roll away from them, or they roll away from us.[8]

It would have been obvious to our distant ancestors that human lives unfold in a dynamic cosmos. For some three hundred thousand years, a kaleidoscope of brilliant stars and planets, glittering like gem-dust on velvet black, revolved above communities that could not pollute the heavens with artificial lighting. Everyday existence depended on patterns of plant and animal life that seemed to correspond to changes in the position of the stars, the appearance of the Moon, and the movement of the planets. Celestial cycles inspired mythology, guided navigation, influenced architecture, and organized how communities hunted, gathered, sowed, and harvested.[9]

Yet sometimes the heavens changed in unexpected ways. The occasional appearance of a comet or sunspot, for example, revealed another, more erratic kind of variability. In this book I trace how history has been influenced by cosmic changes, real or perceived, natural or artificial, that seemed to follow no cycle, and upended, rather than reaffirmed, age-old ways of organizing human culture. While astronomers in many cultures had, for millennia, observed the heavens, it was in the seventeenth century that scholars in cities scattered across western Europe began to identify or imagine more and more of these surprising changes. Many of this book's chapters accordingly start at that time, and in those places. The chapters focus on Europe and North America not because changes in space environments only influenced people in those places, but because I only know so many languages, and I am keen not to sacrifice too much depth for breadth.

Why were arrhythmic changes in cosmic environments, unobserved for hundreds of thousands of years of human history, suddenly visible in the seventeenth century? The easy answer is that artisans had learned to melt sand into glass, shape the glass into convex and concave discs, and position those discs at just the right distance from one another. In such an arrangement, the first glass disc—the first lens—gathers more light from a distant object than the comparatively narrow human eye does. Its convex shape then directs—refracts—that light, that image of the object, toward the second lens. This lens intercepts the converging light and causes it to diverge just a little, which enlarges, or magnifies, the object's image. Together, the lenses make visible what was otherwise too small or faint to see. When they were arranged in the first refracting *telescope* (Greek for far-seeing), they opened a portal to another reality: the infinite, evolving universe that had always surrounded and shaped our little world.[10]

Plans for a telescope surfaced in high medieval Europe, but in all likelihood no one tried to build one until the late sixteenth century, in the coastal cities of the Dutch Republic. These cities had become hotbeds of technical innovation, owing in part to incessant war with the powerful Spanish Empire. New ideas and new people, mostly refugees from the war-torn southern Low Countries, not only catalyzed remarkable economic and demographic growth but also nourished a dynamic intellectual culture. Unprecedented wealth, comparative religious tolerance, and relative freedom from foreign or noble rule allowed for widespread artistic, artisanal, and philosophical experimentation. It is easy to imagine how this intellectual context encouraged Dutch opticians to make the breakthroughs that enabled the creation of the telescope. A device that allowed users to see up close what was otherwise impossibly far away had obvious and immediate applications for the soldiers, sailors, and surveyors who sustained the Dutch Republic. Yet the most transformative uses of the telescope were less practical. In essence, a leading maritime power had launched a new kind of vessel: one that could transport the human gaze beyond the confines of Earth.[11]

The implications would take years to realize. Only gradually did scholars agree that the telescope provided close-up images of real cosmic environments. Even then, the debut of the telescope did not by itself inspire a new astronomy that sought evidence for change in these environments. Nor did the telescope ensure that environmental changes in the solar system would interest anyone beyond scholars of natural philosophy—interest that would have allowed those changes to shape social, cultural, or political history. Rather, the seventeenth-century breakthrough in the historical influence of cosmic changes depended as much on the telescope as a new relationship that was in the progress of forming between Europe and the rest of the world.

It may have been the introduction of bubonic plague to Europe that encouraged the crucial first steps in this reshaping of the continent. The movement of people and ideas, provoked by the spread of plague, likely encouraged the mingling of shipbuilding traditions from southern and northern Europe, leading in time to the construction of more-seaworthy vessels. Spurred by religious fervor, greed, and simple curiosity, mariners used these vessels to string together continent-straddling networks of trade. They also ferried the organisms of Afro-Eurasia, everything from bacteria to livestock, across oceans that previously had barred their global circulation. In many places, Indigenous peoples, weakened by the introduction of pathogens, pests, and addictive commodities, were conquered, dispossessed, and murdered by European settlers on a staggering scale.[12]

Together, commerce and conquest funneled a torrent of new ideas, observations, organisms, and commodities through the commercial cities of western

Europe. Meanwhile, the recovery of more and more diverse classical texts, many of which had been preserved in the dynamic intellectual culture of the Middle East, encouraged unprecedented efforts at intellectual experimentation that departed from the European canon of ancient wisdom. New ideas and new combinations of older ideas surged through Europe after Johannes Gutenberg mechanized and commercialized movable type printing, a technology originally invented in China. The circulation of ideas permitted by the new printing press helped inspire, among other things, the development of humanistic reforms in education and civic life. By the seventeenth century, intellectuals across the growing colonial world shared novel ideas, observations, and artifacts in a Republic of Letters that functioned like an early analogue to today's internet, linking the learned across national boundaries.[13]

Printing helped create the precondition for more dramatic challenges to authority. Information could now be easily circulated in vernacular languages accessible to ordinary people instead of in Latin, which was exclusively the language of elites. Within sixty years of its advent in Europe, printing permitted the rapid spread of the seditious ideas that sparked the Reformation, a broad rejection of the legitimacy of the Catholic Church. The Reformation led to the formation of Protestant denominations, dividing Europe along religious lines and launching more than a century of warfare and persecution. It also encouraged a radical expansion in the power of many European states, partly because prolonged religious warfare required governments to extract ever more resources from their population and environments. Paradoxically, the Reformation simultaneously created havens for comparatively free intellectual inquiry by undermining prior sources of authority and elevating the value of individual learning through direct engagement with evidence. In Europe, knowledge gained through observation and experimentation began to supplement, and eventually replace, ideas grounded in theology and ancient philosophy.[14]

By the early seventeenth century, a series of cultural, economic, intellectual, and political upheavals had prepared scholars both to use the telescope to create knowledge, and to broadly disseminate the knowledge they developed. Further developments in astronomy and religion encouraged them to search for evidence of change in solar system environments. At the beginning of the sixteenth century, European philosophers accepted the Earth-centered, or geocentric, model of the solar system. The geocentric picture had been worked out by a succession of classical scholars, and seemingly perfected by the second-century polymath Claudius Ptolemaeus (Ptolemy). Most of these scholars also accepted Aristotle's contention that the Moon, Sun, and stars were made of a fifth element, "aether." Aether was supposed to be perfect and immutable, fundamentally different from what composed Earth. To early sixteenth-century Europeans, Earth was not just at the center of creation—it was the only world, the sole realm with environments capable of change, and so the only place with a history.[15]

The trouble was, in order to predict celestial movements with reasonable accuracy, geocentric cosmology required bizarre complications that did not quite place Earth at the heart of all things. The planets had to spin in little circles, called epicycles, as they revolved around a cosmic center that resided somewhere between Earth and a point called the equant. To simplify matters, a Catholic canon by the name of Nicolaus Copernicus proposed a heliocentric cosmology, in which the planets orbited the Sun in perfect circles. The heliocentric model of the solar system was indeed much simpler than the geocentric alternative, but it was less capable of accurately predicting planetary motion. The astronomer Tycho Brahe suggested a compromise system in which the planets orbited the Sun, and the Sun orbited Earth. Then in 1605 the mathematician Johannes Kepler proposed that the planets' orbits around the Sun were not circles but ellipses. With that revision, the Copernican model finally rivaled the predictive accuracy of its Ptolemaic competitor.[16]

In a heliocentric system, the other planets were not unchanging, aethereal bodies but dynamic worlds, like Earth. For many seventeenth-century astronomers who subscribed to the new Copernican model, finding similarities between Earth and other bodies in the solar system—by identifying changes in their appearance, for example—undermined aspects of received Aristotelian and Ptolemaic wisdom. This quest to find analogues for Earth in the solar system strengthened and was strengthened by the intellectual, cultural, economic, and political conditions that enabled it. It fortified the Protestant Reformation, since the Catholic Church at first condemned heliocentrism and then argued that it could not be proved. It shored up a system of experimentation, observation, and inductive reasoning that coalesced in the Scientific Revolution and inspired efforts to repurpose, refine, and create instruments, such as the telescope, that could enhance human senses. Descriptions and depictions of previously unimagined environments and peoples on Earth, including in regions that Aristotle had assumed were uninhabitable, shaped how astronomers interpreted planetary environments. Observations of those planetary environments also inspired new efforts to calculate longitude aboard the ships that launched Europe's colonial era, and in time led to the creation of new outposts, dedicated to astronomy, along the frontiers of European empires.[17]

In short, by the early seventeenth century European scholars had the motivation to search for environmental changes in outer space, the technology that made the search possible, and the ability to interpret, and then communicate, what they saw. Cosmic environments that had thus far been hidden for the entire history of our species shimmered into view and began to influence affairs on Earth.

New Histories of Outer Space

Histories of human engagement with outer space have long focused on the contributions of influential scientists or engineers, the experiences of astronauts or cosmonauts, and the policies or technologies that enabled their achievements.[18] Today, space historians draw on new forms of history to also consider the cultural, social, and economic contexts of the discoveries and voyages chronicled in earlier accounts.[19] This book contributes to that revisionary effort primarily by emphasizing the role of cosmic environments in the history of space exploration.

In this book I also elevate histories of scientific failure. While searching for evidence of environmental changes beyond Earth, figures such as Percival Lowell, William Pickering, Johann Hieronymus Schröter, and Immanuel Velikovsky practiced what American chemist Irving Langmuir has called *pathological science.* According to Langmuir, when scientists gather data at the very limits of observation or measurability—evidence for a subtle and ephemeral change in a planet's environment, for example—cognitive biases and professional pressures can lead them to conclusions unsupported by their evidence. Once scientists stake their reputations on these conclusions, they may double- and triple-down on them, even in the face of mounting contradictory evidence, employing increasingly tenuous justifications. The concept of pathological science is inherently normative, judging certain scientific practices as incorrect or misguided relative to an imagined ideal. Such judgments suggest that bad science inevitably results in failure—in careers that cannot advance human knowledge. Normative critiques of this kind frequently appear in popular histories of space science. Scholars who either insisted on the existence of cosmic environmental changes that did not really take place, or misinterpreted real environmental changes to propose sensational theories that turned out to be wrong, are accordingly ignored or ridiculed in influential accounts of astronomy's history.[20]

Like other studies in the history of science, this book demonstrates that it is not so easy to distinguish between the practice of good and bad science. Moreover, the book shows that grand failures, including those brought on by pathological tendencies, can more powerfully and productively shape the course of scholarship than modest successes. Real or imagined changes in the heavens repeatedly encouraged sweeping ideas that were false, or at least incomplete and unprovable. But often these ideas anticipated or inspired new, interdisciplinary forms of research. Some mistaken ideas prompted discoveries about the nature of planetary environments and profound insights into the course of human transformations of Earth. Many spurred widespread public interest, and thereby broadly influenced cultural, economic, and political developments. Often scholars who, as we can see in hindsight, were wrong on the specifics turned out to have grasped some fundamental truth that their more careful colleagues had not perceived.

This book is a history of science in that what I identify as proximate causes of historical change are frequently scientists, or else entrepreneurs and government officials moved by science. Yet the causes I identify as ultimate causes are not human—indeed, not even alive—so this book is first and foremost an environmental history. The small but diverse and growing subfield of environmental history that considers outer space can be split between four distinct approaches. First, some environmental historians write about outer space by tracing how efforts to create livable habitats for astronauts in the Space Age—the period beginning with the 1957 launch of *Sputnik*—reflected and provoked new ideas about humanity's relationship with nature.[21] Second, others consider how space travel encouraged new, politically powerful ways of observing and understanding Earth's environments.[22] Third, a growing number investigate how scholars and journalists imagined heavenly bodies as worlds with environments akin to Earth's.[23] Fourth, a few point out that outer space can itself be considered an extreme environment that has both shaped and been shaped by human action.[24]

In this book I draw on many of these approaches to the environmental history of space, but I also offer something new. The scale here is unusually vast, stretching out across the solar system, and the focus is on environmental changes far beyond Earth. We will begin in the solar system's fiery core—the Sun—then move out past Venus, the Moon, and Mars, before ending amid the swarms of asteroids and icy comets that take their marching orders from the giant planet Jupiter. These case studies highlight, to steal a phrase from physics, the many strange actions at a distance that created causal connections between environmental and human changes almost unimaginably far removed in space, and occasionally by time.[25]

PART I
Sun

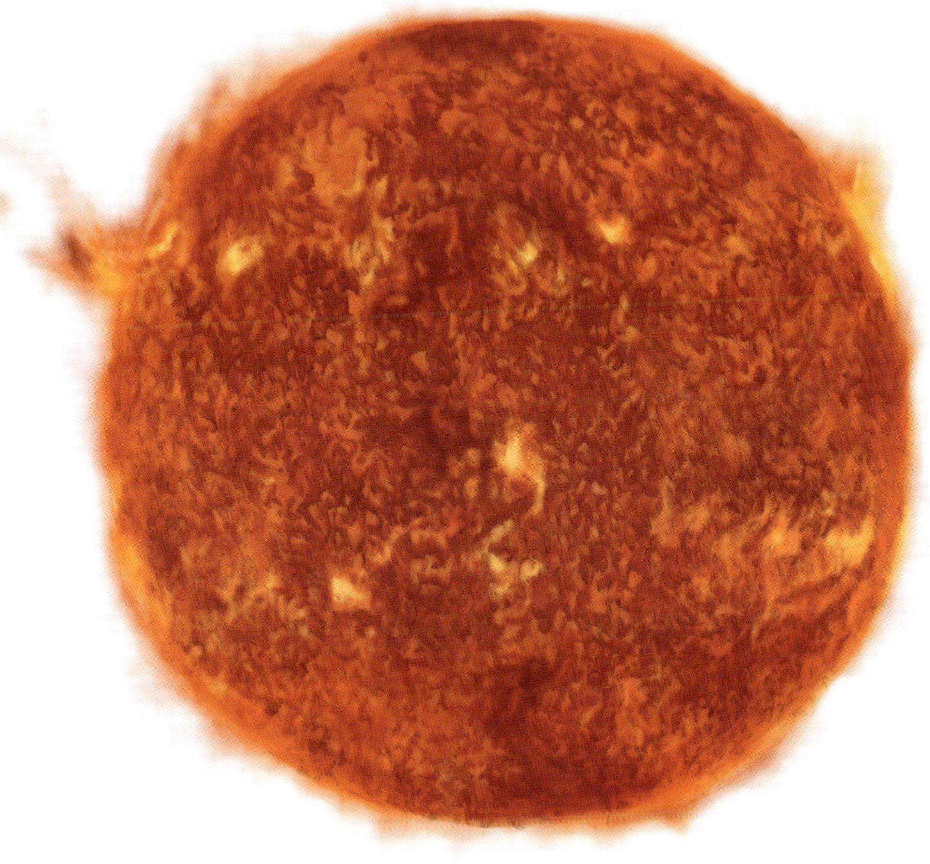

A coronal mass ejection.

“The apparent vulnerability of worlds like our own to relatively small perturbations is an important and sobering discovery of modern planetary research.”
—Carl Sagan, “Planetary Technology” (1985)

“Round about the accredited and orderly facts of every science there ever floats a sort of dust-cloud of exceptional observations, of occurrences minute and irregular and seldom met with, which it always proves more easy to ignore than attend to.”
—William James, *The Will to Believe and Other Essays in Popular Philosophy* (1897)

On most clear nights in Washington, DC, I slip outside with a telescope in one hand, a tripod in the other, and a backpack full of heavy eyepieces. After the Sun sets or before it rises, I sneak into a nearby park, bleary-eyed from lack of sleep. Crouching amid the fireflies, I admire the ring shadows of Saturn, the swirling storms of Jupiter, the ice caps of Mars, the desolate peaks of the Moon, the delicate tendrils of star-forming nebulae, the ghostly glow of unimaginably distant galaxies. Early in the morning, only birdsong breaks the silence. Then the Sun rises, the city wakes, and I stagger home.

In the shadow of a nearby church, below a burned-out streetlight, in a knot of bushes in the darkest corner of a local park, I experience an echo of the wonder shared by the protagonists of this book, as they too peered into the night sky. I have learned how difficult it is to see anything at all, and more importantly how much the heavens can seem to change from night to night. The knowledge is hard-won. There was the lack of sleep, of course, but also the occasional moments of terror. Once, with a shriek and a snarl, a fox chased a rabbit into my lap. Another time, near the Russian Embassy, three glowing cigarettes appeared out of the gloom, accompanied by an increasingly animated conversation in Russian. As the smokers drew closer, I called out to avoid being trampled. They shouted and ran off, leaving me alone—but more than ready to go home.

After one especially tiring night several years ago, I imagined how much more I might sleep if I observed during the day. I asked myself: What if I turned my focus to the star nearest Earth? I bought a solar telescope, an instrument that blocks all light save for that which escapes from a narrow layer of hydrogen just above the Sun's incandescent surface.

One sunny afternoon I climbed onto the roof of my apartment building, solar telescope in hand. When I peered through the eyepiece, I immediately noticed a sunspot: a cooler patch on the blazing solar surface, hollowed out by magnetic fields powerful enough to inhibit convection. Eventually I saw how those same magnetic fields had warped plasma—gas suffused with so much energy that its electrons peel free from its atoms—into wispy filaments around the edge of the solar disc. Yet, as I began to sweat in the baking midday heat, something started to gnaw at me. I felt like I was watching the irregular beating of my own heart, on an unimaginably vast scale. It was one thing to know, but another thing entirely to experience, that the foundation of all human existence was on some level unstable, and unpredictable. Unnerved, I sold that telescope.[1]

Just about everything on Earth depends on the Sun. Nearly all of the solar system's mass is bound up in its star, a conglomeration so immense that it distorts the fabric of space and time. The Sun's core can hold the combined volume of the solar system's planets nearly seven times over. Its temperatures can reach 16 million degrees Celsius, and its pressures approach two hundred billion times that of Earth's atmosphere at sea level. This crushing inferno squeezes hydrogen atoms together, igniting nuclear fusion on such a scale that the Sun's core releases a million times more energy, *every second,* than would be unleashed by every nuclear weapon on Earth detonating at the same time.[2]

Over thousands and even millions of years, the solar inferno changes only gradually. The cause is a feedback loop in which the output of a process influences the input. Negative feedbacks reduce fluctuations in output and thereby promote stability. Positive feedbacks, as we will see, increase those fluctuations and can create runaway changes. The Sun's core provides a classic example of negative feedback. If the core's rate of fusion increases, the core heats up, expands, and becomes less dense, which reduces the rate of fusion. If that rate drops lower still, the core cools down, shrinks, and grows denser, increasing the rate of fusion. In fact, for reasons that remain mysterious, the Sun's brightness seems to vary less than that of otherwise similar stars, and Earth therefore enjoys an unusually dependable supply of warmth and light. Maybe that is a prerequisite for life.[3]

Yet as I saw on my rooftop, the solar environment also fluctuates across shorter timescales: decades, years, even minutes. Even though these changes may be modest by cosmic standards, the Sun so fundamentally shapes life on

Earth that even its slightest fluctuations can have profound consequences for humanity. The subtlest tremor of a sleeping elephant may dislodge the mouse on its trunk.

The chapters in Part I reveal how tremors in the Sun's environment have influenced life on Earth—by altering Earth's climate and our understandings of it, by stimulating scientific discovery and disrupting technological infrastructure, and by exposing existential risks and provoking responses to them. The legacy of prehistoric changes in the Sun's capacity to warm Earth sparked efforts to confirm biblical mythology, and in turn inspired the creation of interdisciplinary tools for identifying climate change. Meanwhile, changes in the environments of Earth and the Sun led to waves of cooling across the Northern Hemisphere, and eventually much of the world, that persisted until the nineteenth century. The complex manifestations of this Little Ice Age (LIA) in diverse environments led some to believe that regional weather patterns, which previously had been thought to be stable, could in fact vary across decades. When the LIA worsened shortages of vital commodities, increasingly capable states encouraged intellectual programs that eventually revealed how climate change could threaten civilization.

The discovery of solar variability in the eighteenth century ultimately suggested a mechanism by which the world's entire climate could change. In the nineteenth century, scholars in ever more specialized and professionalized disciplines sought to identify solar causes for climate anomalies on Earth, and thereby worked out many of the principles of modern climate scholarship. Meanwhile, the spread of industrialization and electrification rendered populations increasingly vulnerable to another form of solar variability. Geomagnetic storms created by bubbles of supersonic solar plasma disrupted telegraph communications and prompted new debates about the link between environmental variability on the Sun and on Earth. Owing to the dramatic expansion of electrical infrastructure and ultimately satellite networks, solar plasma bubbles caused increasing devastation in the twentieth century, nearly provoking nuclear war.

By the end of the twentieth century the effects of solar changes on Earth were increasingly perceived as existential threats. Scientists, and eventually public officials, realized that these grave risks could be profoundly exacerbated, or mitigated, by human action. It has even been found that some technological developments that make human life most vulnerable might also be sources of potential resilience. Some of the world's deadliest weapons, for instance, eventually helped strengthen Earth's magnetic lines of defense. By the dawn of the twenty-first century, humanity's fate increasingly seemed to depend on how skillfully governments and corporations mediate the relationship between Earth and its star. The chapters in Part I show how solar variability has helped shape the course of modern science—and has threatened the technological civilization that science made possible.

1

Ice Ages Great and Small

In the early eighteenth century the Swiss naturalist Johann Jakob Scheuchzer ventured into his country's alpine wilderness. It was an alien world, even for his well-traveled contemporaries. At the time, European aesthetic tastes favored cultivated landscapes, and potential visitors to the Alps were daunted by rumors of demons, dragons, ghosts, and witches. Scheuchzer, a correspondent within the intellectual community known as the Republic of Letters, wrote that reputable locals had indeed glimpsed a dragon or two. Yet he reported that the mountains harbored other wonders that were only superficially less spectacular. Scattered across their slopes and valleys were immense boulders that appeared detached from their surroundings. Scheuchzer mused that the waters of Noah's ancient Flood must have pushed those boulders down from the mountains. In their own way, they were no less frightening than dragons. They were reminders that the world had once been radically different, and could be so again.[1]

We now know that expanding glaciers, not floodwaters, carried Scheuchzer's boulders out of the mountains thousands of years ago when the Sun's capacity to warm Earth temporarily waned. But if Scheuchzer had erred on the details, he was right about the big picture. The world had changed, and it was changing again. Scheuchzer did not know it, but he lived in a time of slumping solar activity, and the signs were all around him. "Horrible" glaciers, he wrote, stretched into mountain valleys like the legs of a sprawling monster.[2]

Fluctuations in solar radiation had spurred the environmental changes, ancient and contemporary, that Scheuchzer observed in the Alps. The cultural impact of these changes, separated by some twenty millennia, informed a growing belief in Scheuchzer's time that Earth's climate had once been very different, that regional climates were unstable, and that climate change could threaten the very survival of societies. These ideas created a foundation for the later development of climate science, even though a natural explanation for global climate change remained stubbornly out of reach.[3]

Memories of the Pleistocene

Earth preserves a durable account of changes in the balance between the amount of solar energy that reaches Earth and the amount that escapes from Earth back into space. Sediments on the ocean floor, for example, reveal that we live (yes, even now) in a remarkably cold period in Earth's history. Such sediments serve as archives of nature because they abound with microscopic shells created by single-celled marine organisms. The oxygen atoms in these shells come in different variants, or isotopes. The heavy isotopes have more neutrons, and the light isotopes have fewer. Light oxygen isotopes evaporate from the oceans more readily than heavy ones, and eventually fall to Earth in rain or snow. When Earth was cold, more of these light isotopes wound up in ice sheets, so that ocean oxygen grew heavier. Because marine organisms build their shells using chemicals in ocean water, the ratio of heavy to light oxygen isotopes in microscopic fossils provides a *proxy*—an indirect indicator—of how the world's climate changed in the distant past.[4]

Ocean sediments show that ice sheets appeared on Earth's surface about thirty-five million years ago. Temperatures continued to decline, owing to the diverse effects of tectonic forces that drained carbon dioxide from the atmosphere, lowered the salinity of ocean water, reshaped atmospheric circulation, and, perhaps most importantly, isolated Antarctica from warm oceanic currents. Finally, by about two and half million years ago, the multimillion-year cooling trend had begun to destabilize Earth's climate by rendering it sensitive to recurring fluctuations in the Sun's capacity to warm the planet.[5]

The shape of the Sun and the gravitational pull of other planets had long created regular variations in the characteristics of Earth's rotation and orbit, which in turn altered how much solar radiation reached different parts of Earth at different times of the year. When Earth was sufficiently cold, these Milankovitch cycles could overlap to trigger self-reinforcing processes, or feedbacks, that were together responsible for periods of extreme glaciation. As the tilt of Earth's rotation on its axis gradually declined, for example, summers cooled and ice sheets began to grow. More incoming solar radiation is reflected by ice than by water or land, and so the expansion of ice unlocked a positive feedback in which more cooling led to more ice, further cooling, and so on. Sea levels slumped as water evaporating from the oceans ended up trapped in ice. The exposed land was gradually colonized by vegetation. Rocks crushed by glaciers nourished plankton by releasing dust into the ocean. Proliferating plants pulled carbon dioxide out of the atmosphere, as cooling waters dissolved the gas with growing efficiency and sea ice covered regions where carbon dioxide had bubbled to the ocean surface. On land and at sea, expanding ice therefore created another positive feedback by pulling greenhouse gases out of the atmosphere.[6]

When Milankovitch cycles converged in just the right way, atmospheric carbon dioxide fell by about a third, and Earth's average temperature collapsed by about 6 degrees Celsius. Ice sheets grew to such an extent that the largest one in North America, the Laurentide Ice Sheet, covered more territory than Antarctica does today. With so much water trapped in ice, sea levels dropped by more than 120 meters. Exposed land expanded and connected the continents. Cold air can hold less water than warmer air, and the ice sheets created new patterns of atmospheric circulation, resulting in sharply reduced precipitation across much of Earth. Our planet was not only far colder than it is now; it was also much drier, and dustier.[7]

From about two and a half million years ago to the present, Earth oscillated no fewer than fifty times between long, frigid, arid glacials (glacial periods) and short, warm, wet interglacials. The most recent glacial—known to climatologists as the Last Glacial Maximum (LGM)—reached its chilliest point around 21,000 years ago. Thereafter, Earth warmed in fits and starts. About 12,900 years ago, torrents of glacial meltwater cascaded into the Atlantic Ocean,

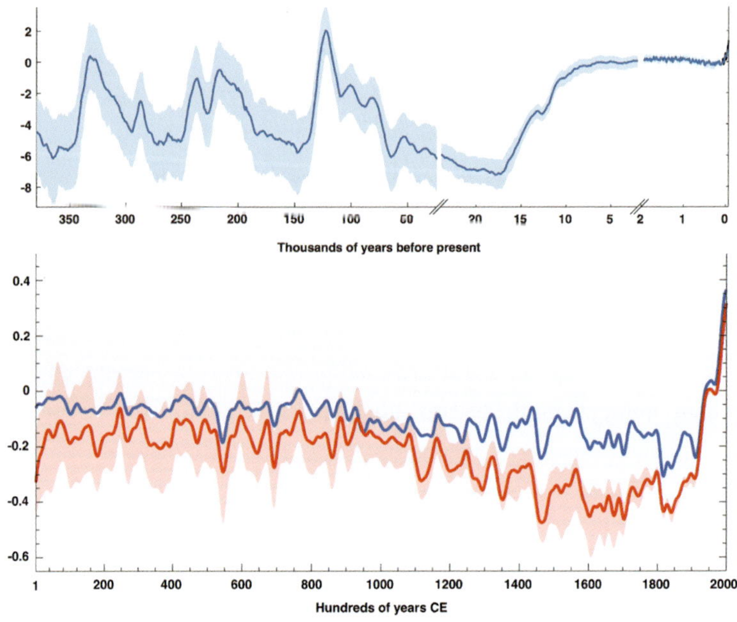

Global mean surface temperature changes over the 300,000-year history of our species (top) and over the past two thousand years (bottom), reconstructed from proxies in distinct natural archives. The first graph is broken into three parts, each covering a different timescale with progressively greater precision. Temperatures are shown relative to the estimated mean of the "preindustrial period" (the late nineteenth century). The top graph reveals that large temperature changes unfolded across thousands of years during the Pleistocene geological epoch, before 11,700 years ago. However, the bottom graph demonstrates that distinct statistical methods for interpreting proxies yield meaningfully different climate reconstructions even over the past two millennia, when increasingly precise proxies can be extracted from natural archives. This graph nevertheless reveals that different statistical methods and proxies tell the same broad story, which is that the Little Ice Age moderately cooled much of the globe between the thirteenth and nineteenth centuries. The top and bottom graphs both indicate that global temperature changes in the Holocene geological epoch were, until recently, far smaller in magnitude than they could be in the Pleistocene.

interrupting the flow of salty water that pushed heat from the equator—and ultimately from the Sun—into the Arctic. In a matter of years, renewed cooling set in across the Northern Hemisphere, and this extreme cold spell endured for more than a millennium. This Younger Dryas, named after a cold-tolerant flower that bloomed at low latitudes, abruptly gave way to renewed warming around 11,700 years ago. In the view of geologists, the epoch distinguished by recurring glaciations, known as the Pleistocene, had at last faded into the warmer, more stable, and in many places wetter Holocene. Sea levels rose by as much as two meters per century, until the oceans reached their present extent.[8]

Scheuchzer and his contemporaries had little inkling that matter is made of atoms, much less that differences between the nuclei in atoms could reveal the deep history of climate change. Yet the emergence of complex, recordkeeping civilizations in the mid-Holocene may have meant that the archives of nature no longer kept the only account of the Pleistocene's aftermath. Long after the onset of the Holocene, diverse populations across much of the world passed along similar legends that traced human origins to a catastrophic flood. Whether by dredging channels, building a giant ship, or migrating to new lands, the protagonists in these ancient origin stories survived and established civilizations that thrived. It is possible that occasional floods of the Nile, Tigris, Euphrates, Indus, and Yellow Rivers provided sufficient inspiration for most of these stories. Yet it is equally plausible that these tales preserved historical memories of a staggering rise in sea levels at the end of the Pleistocene.[9] The archives of societies—the evidence of climate change recorded in human culture—could register a common trauma stemming from the loss of vast, previously inhabited regions to relentless and at times sudden flooding.[10]

Or were legends focused on water actually inspired by drought? Although the Holocene was generally warmer and wetter than the Pleistocene, globally modest but regionally severe variations in temperature and precipitation continued unabated. Across much of Earth, transregional megadroughts posed the gravest risks for ecosystems and the human populations that depended on them. The drying of the Sahara, for example, spurred migration from what was once a vast grassland, and likely played a role in congregating populations around the Nile Delta. Egyptian civilization would subsequently develop a water-centered mythology that included descriptions of both apocalyptic flooding and rebirth. The most influential of all ancient flood stories was arguably the Sumerian *Epic of Gilgamesh.* This epic took shape around the time of the so-called 4.2K BP event, a period of profound and perhaps centuries-long drought that stretched across swaths of North America, Africa, Asia, and the Middle East. The art historian Sugata Ray has argued that historical droughts inspired cultural expressions that focused on increasingly precious water; perhaps the same held true in the first states.[11]

The Hebrew account of Noah's Flood copied almost every detail originally found in the Gilgamesh epic. As a result, the spread and durability of Abrahamic religions preserved one of the earliest accounts of ancient, perhaps prehistoric, environmental change. This example from the archives of societies did not inspire a reappraisal of the archives of nature, however, until the sixteenth century. Lydia Barnett, a historian of science, has traced how that era's natural historians sought physical evidence for Noah's Flood as a way unifying the increasingly frayed culture and politics of contemporary Europe. Barnett and other scholars show that the bewildering diversity of the world revealed by journeys of exploration, along with the upheavals spurred by the Reformation and the Scientific Revolution, led to a widespread sense of unraveling among Europe's literate elites. One answer was to search for the origins of contemporary disunity, and that search naturally led scholars to the biblical story in which humanity was reduced to a single family.[12]

By the eighteenth century, Barnett argues, naturalists in Britain and Switzerland sought to forge a world-straddling network of fossil hunters whose evidence might confirm that the biblical deluge had truly inundated the whole Earth. Erratic boulders like those sighted by Scheuchzer were reimagined as signs that Noah's Flood had profoundly altered Earth and its climate. Fossilized fish found on dry land and apparent elephant tusks exhumed in Britain, France, and Siberia were interpreted in the same way. Barnett identifies a sinister element in the enterprise. By confirming that the Flood covered the entire world, scholars also hoped to establish that all peoples were descended from Noah. This would mean that Christ's offer of salvation applied to everyone, providing ideological justification for military and missionary expeditions that subjected a growing portion of the world's population to colonial, Christian governments.[13]

As Scheuchzer rambled through the Alps, however, the Flood confirmation effort was losing momentum. At the dawn of the Catholic Enlightenment, correspondence in the Alpine Republic of Letters—the network of scholars centered in northern Italy and Switzerland—fostered a growing rejection of what was now perceived as a British Protestant effort to read the Deluge into the archives of nature. Patriotic Catholic scholars increasingly emphasized that local inundations, rather than a universal Flood, had long ago given special qualities to their nations.[14]

Even though the Flood lost its unifying power in European scholarship, the effort to find its traces in nature had transformed contemporary science. By integrating local evidence into a model of global change, and interpreting the past to forecast the future, the effort pioneered the kind of multi-scalar thinking that would later reveal the existence of global warming. The Flood confirmation enterprise began to demonstrate the potential of natural archives to tell the story of the deep history of climate change, and it hinted at the power of combining these archives with those made by societies. Not all of its concepts proved durable, though. By the turn of the nineteenth century, for example, the emerging scientific discipline of geology coalesced around a reversion to the uniformitarian assumptions of Aristotelian and Stoic philosophy, according to which the forces shaping today's Earth are identical to those that operated in the distant past. That ruled out a world-altering Deluge. Still, proponents of the Flood narrative had worked out a set of tools that would later reveal the spectacular history, and frightening future, of climate change—and in turn, of the Sun's variable influence on Earth.[15]

Experiencing the Little Ice Age

In the sixteenth to eighteenth centuries, the archives of nature and societies were not the only evidence of climate change. Owing to the LIA, ordinary people could feel a change in climate through their everyday experience of weather. Yet if the term *Little Ice Age* suggests something akin to a Pleistocene period of glaciation, it is a misnomer. Global temperatures varied only modestly during the Holocene, and the forces responsible for the LIA paled in comparison to those that triggered the LGM. Still, they also involved changes in the quantity of radiation that reached Earth's surface from the Sun.[16]

Fluctuations in solar environments were likely responsible for some of the cooling associated with the LIA. For reasons still not fully understood, the Sun's radiant output waxes and wanes ever so slightly in a regular, approximately eleven-year cycle. Occasionally the Sun's output slumps for decades at a time, during grand solar minima in which the telltale signs of an active Sun—sunspots and solar flares—largely disappear. During these minima, lows in total solar irradiance likely had a modest cooling effect across much of the globe. Reductions

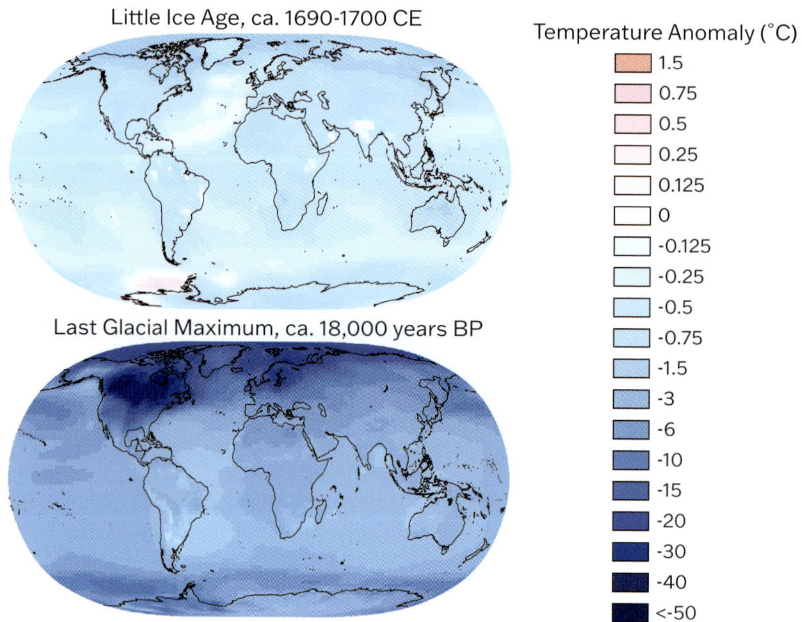

Ice Ages great and small. Top left: Temperature anomalies for the years 1690–1700, an especially cold decade of the Little Ice Age. Bottom left: Temperature anomalies 18,000 years ago, during the Last Glacial Maximum. The temperature baseline used for the top left map (the 1960–1990 global mean temperature) is similar to the baseline used for the bottom left map (the 1800–2000 global mean temperature). The maps were created using different proxies with different levels of precision, and the proxies were interpreted using distinct statistical methods. The maps therefore cannot provide a perfect comparison between different climatic periods, but they do reveal that cooling in the Little Ice Age was far less severe than it was in the Last Glacial Maximum.

in the Sun's ultraviolet radiation, in particular, seem to have lowered temperatures in the stratosphere, especially in the tropics. Systems of atmospheric, then oceanic, circulation changed in response. It now seems that the impact of these circulation changes on the extent of Arctic sea ice partly accounts for the onset of the LIA.[17]

The coldest stretches of the LIA also coincided with clusters of explosive volcanic eruptions that may have had an even more substantial influence on Earth's climate. Violent eruptions hurl sulfuric gases into the stratosphere, the atmospheric layer some fifteen to fifty kilometers above Earth's surface. There the gases mix with water vapor, creating droplets of sulfuric acid that scatter incoming solar radiation and cool Earth's surface. In the LIA, stratospheric winds repeatedly blew these aerosols across much of Earth. Yet because each eruption released distinct plumes of sulfur dioxide at different latitudes, and because stratospheric winds did not evenly distribute volcanic sulfuric "veils" across the whole Earth, the cooling effect differed from region to region. Moreover, volcanic cooling quickly faded as aerosols fell from the sky, unless a truly massive eruption or a cluster of smaller eruptions set off positive feedbacks in Earth's climate system. The expansion of sea ice in the wake of these eruptions, for example, may have helped trigger the LIA by reflecting more solar radiation back into space than darker water would have.[18]

These and other forces were at work in the thirteenth and fourteenth centuries, when temperatures across the Arctic and eventually much of the Northern Hemisphere started to decline. Waves of cooling subsequently spread across much of the Northern Hemisphere, and eventually most of Earth, in the early 1400s, the late 1500s to the early 1600s, the late 1600s to the early 1700s, and the late 1700s to the early 1800s. Global temperatures during these waves changed only modestly, no more than several tenths of a degree Celsius, relative to average temperatures midway through the twentieth century. Nevertheless, explosive eruptions repeatedly lowered regional and seasonal temperatures to such an extent that Europeans described "years without summer."[19]

Due in part to their influence on hemispheric temperatures, modest fluctuations in the quantity of solar radiation that reached Earth's surface altered atmospheric and oceanic circulation during the LIA and thereby disrupted regional patterns of precipitation. Extreme rain battered much of early thirteenth-century Europe and eighteenth-century Mexico. Severe or prolonged droughts afflicted fourteenth-century Cambodia, fifteenth-century Central America, nineteenth-century India, and other places. Moreover, changes in patterns of atmospheric circulation played a role in reducing the predictability of weather in Northern Europe, eastern North America, and other areas, in part by increasing the frequency of severe storms.[20]

Around the world, the telltale weather of the LIA was hard to ignore. A substantial share of the world's roughly five hundred million people lived within the contested borders of agrarian empires. There, millions depended on a few common grains, but agricultural yields were incomparably poorer and less dependable, and supply chains were unimaginably less efficient and more fragile, than they are today. Governments were far weaker and less stable, information was much harder to come by, and social safety nets were inadequate and largely beyond the purview of the state. Few enjoyed access to safe drinking water or diets with adequate calories and nutrients. The absence of effective medication and sterilization, let alone vaccination, transformed into ruthless killers what today are only common illnesses and injuries. Owing to overcrowding, open sewage, air pollution, inadequate housing, nonexistent or deficient emergency services, and widespread infestation by insects and vermin, more people died in cities than were born in them. Even in the wealthiest and healthiest corners of Earth, about half of all children did not survive to adulthood.[21]

Empires were continuously at war, and agricultural laborers were forced to join armies that, when accompanied by merchants and prostitutes, outnumbered all but the largest cities. Owing to the inadequacies of military logistics, these roving communities despoiled and depopulated vast territories by living off the land. Soldiers or bandits also formed gangs that killed, tortured, or robbed in areas where the authority of the state was weak or contested. The common wartime practice of destroying environmental infrastructure, such as tilled fields and irrigation canals, erased lives and lifeways on a grand scale, and settlers destroyed infrastructure with particular sadism on the periphery of expanding empires. Dehumanizing servitude was widespread, and nowhere crueler or more destructive than in the plantations that devastated Caribbean ecosystems, exploited African slaves, and introduced mosquito-borne diseases into the Americas.[22]

Above all, most people who experienced the LIA lived in a state of scarcity—of food, information, mobility, and security. Even in the richest empires, scarcity created compounding, synergistic vulnerabilities to even modest climatic fluctuations. For example, the sixteenth-century Ottoman Empire ruled some twenty-five million subjects across five million square kilometers, yet its stability was precariously dependent on harvests of barley and wheat from marginal farmland. Reports by Ottoman administrators, and a stretch of thin growth rings in trees that revealed climatic conditions unsuitable for plant growth, both suggest that frost damaged these crops in the late sixteenth century. Worse, changes in atmospheric circulation soon brought ruinous drought to Ottoman lands. As harvests failed, food prices soared. Instead of taking action to ameliorate suffering, the Sultanate pursued a ruinous war with the Austrian Habsburgs that demanded transfers of grain from the Ottoman countryside to the army. When famine and starvation left many vulnerable to outbreaks of epidemic disease, desperate farmers gathered in bandit gangs that metastasized into rebel armies. By the early seventeenth century, millions had died.[23]

A similar story seems to have unfolded across China in the Ming Dynasty. A stretch of narrow tree rings indicates that droughts and cold snaps intensified across much of northern and central China in the early seventeenth century. By then the dynasty had become too dysfunctional and corrupt to muster an effective response. Once again, food prices soared and starvation may have encouraged sedition, which soon broadened into civil war. Especially severe droughts in the grasslands that bordered northern China spurred local Jurchen chieftains to seize Chinese grain supplies. As drought and civil war intensified in China, however, abundant rain began to fall in Jurchen territories across Manchuria. With plenty of grass to feed their horses, the Manchu—as the Jurchens had come to be known—exploited the chaos in China to overthrow the Ming Dynasty. Millions had died during decades of starvation, civil war, and foreign invasion.[24]

Contemporary observers did not guess that the Sun's capacity to warm Earth had waned, or that local disasters were part of a global trend. Still, some did connect famines, epidemics, and political disasters to a regional shift in the frequency or severity of extreme weather. Nowhere did such connections seem clearer than in China. While monarchs around the world claimed that they possessed a divine right to rule, in China and Korea that right, ostensibly granted by heaven, was widely seen as provisional. According to an idea first articulated by the Confucian philosopher Dong Zhongshu in the second century BCE, heaven would rescind the right to rule if the emperor governed unjustly. In these circumstances, ordinary people could legitimately rise up against the emperor. Because destructive droughts, cold spells, and floods were considered signs of divine displeasure, climate change posed a mortal threat to Chinese dynasties. That was especially true in decades when preexisting political or military crises increased the vulnerability of agriculture and the likelihood that people would seek signs of divine anger. "Now we are endowed by heaven with a special opportunity," the leader of one rebellion proclaimed as Ming rule unraveled, "for our masters are all weak and feeble." With rebels closing in, the Chongzhen Emperor, the last ruler of the Ming Dynasty, accepted responsibility for the natural disasters that had undermined his regime and committed suicide.[25]

Meanwhile, across Europe, widespread perceptions that climate was deteriorating as a result of divine displeasure or demonic malevolence may have joined complex legal, cultural, and political changes—not to mention personal grudges—to fuel explosions of violence against marginalized groups. In the central European countryside, sixteenth-century peasants murdered so-called witches for the "sudden and violent raising" of "storms and tempests," which had indeed grown more common. Townspeople ostracized Jews for allegedly profiteering off soaring food prices. Across Eurasia the fact that groups blamed for extreme weather were scapegoated suggests that ordinary people were aware of regional changes in climate, even if they did not imagine that these were caused by the Sun's waning capacity to warm Earth.[26]

In the seventeenth and eighteenth centuries, literate observers described popular beliefs that seem to have reflected widespread awareness of climatic trends in Europe. Around 1793, for example, the Reverend John Grant of Karmichael reported an apparently "universal" concern across the Scottish Highlands that the climate had grown chillier and rainier in recent decades. Some locals not only remembered that the region had been warmer earlier in the century; they also recalled that their parents had described frigid, wet conditions toward the end of the previous century. They had in effect reconstructed the LIA across two periods of waning solar activity and two waves of subsequent cooling across the Highlands.[27]

Even where written sources shed little light on popular perceptions of climate change, art and architecture provide telling clues. In northern India, the positioning of sixteenth-century buildings, and the riparian views available from those buildings, suggest that architects shared a preoccupation with water. Artists, meanwhile, crafted vivid waterscapes, some of which spilled beyond the borders of their paintings. These developments in both art and architecture date to a period of drought and famine that appears connected to the LIA's influence on atmospheric circulation. Water is always fodder for the imagination, yet the strengthening of a "hydroaesthetic" culture does hint at broad awareness of climatic patterns in sixteenth-century India.[28]

Other adaptations to the LIA provided more tangible benefits for populations that faced unfamiliar and often threatening weather. In North America, sixteenth- and seventeenth-century Iroquoian communities, for instance, either migrated south and relied on trade to share dwindling agricultural surpluses, or else abandoned agriculture altogether in favor of hunting. Mojave settlements responded to climatic variability by developing new storage technologies that permitted trade for beans, maize, and squash grown by nearby Kwatsáan farmers. Wabanaki hunters used snowshoes, an ancient technology perfected across Indigenous North America, to pursue game in long LIA winters and to wage effective warfare against English settlers and soldiers who initially lacked snowshoes and in deep snow could not stray far from agricultural settlements. Such ingenious responses to LIA weather within the First Nations of North America seem to reveal a broad capacity to identify and confront regional trends in climate.[29]

To an extent that may now be difficult to imagine, the vast majority of those who experienced the LIA were vulnerable to the vagaries of weather. At sea or in the field, on the battlefield or in the kitchen, extreme weather was a matter of life and death. In that context, it should not surprise us that ordinary people perceived trends in the frequency of severe winters or dry summers, and that they understood the threats and opportunities that such weather could bring. Yet because they did not quantify weather trends, they could not easily sense changes in average, day-to-day weather. Nor could they imagine that the Sun's waning capacity to warm Earth bound the local changes they perceived to broader patterns that unfolded across much of the globe.[30]

Controlling Climate, Building Civilization

During some of the chilliest decades of the LIA, European intellectuals sought to determine whether, as the Parisian correspondent Marie de Rabutin-Chantal put it, "the behavior of the sun and of the seasons has completely changed."[31] They interviewed the elderly, studied commemorative landmarks, inspected signs of cultivation in abandoned landscapes, and scoured written records. In the seventeenth century, artisan-scholars invented meteorological instruments in order to illuminate through precise measurement the inanimate forces they suspected were responsible for weather. Some proposed that observatories equipped with these instruments be linked in order to form monitoring networks that could shed further light on the mechanisms powering weather. To Francis Bacon, the English pioneer of empirical, inductive reasoning, ways would soon "be found to predict the risings, fallings, and times of winds," which among other things would be useful for planning naval battles in the age of sail. Attempts at long-term weather prediction and *reconstruction* (identifying historical weather patterns) suggested that regional climates cycled through periods marked by meteorological extremes, an idea that roughly fits modern understandings of the LIA.[32]

For much of the LIA, however, widespread belief in regional climate change clashed with the Aristotelian idea that latitude determined climate. Many European scholars adhered to this view even in the seventeenth century, when it was increasingly obvious that climates in Europe and North America were profoundly different at the same latitude. Some scholars adopted the neo-Hippocratic idea that climate reflected not only latitude but also such factors as vegetation and human occupation. Such was the way of contemporary scholarship. The Scientific Revolution had elevated observation and experimentation as pathways to knowledge, yet most scholars used experience not to refute classical wisdom but instead to refine or choose between combinations of ancient world systems.[33]

Neo-Hippocratic climatology suggested that the character of human settlement could shape the climate of a region, and the implications were profound. In Europe, increasingly powerful states had sought to expand the area of cultivated land, partly to avoid harvest failures during extreme weather. By the eighteenth century, elite "improvers" used burgeoning networks of scholarly exchange to promote an empirical approach to agriculture that discarded traditional best practices. Now it seemed that the very act of cultivating land could improve the prospects for cultivation—a positive feedback that worked through the mechanism of climate change. The historian of science Frederik Albritton Johnsson has shown that newly motivated improvers recommended a raft of diverse and contradictory strategies to bring the Scottish Highlands under cultivation.[34]

Prominent improvers decided that the spread of civilized settlement had already altered climate across much of Europe. Therein, it seemed, lay the explanation for the otherwise mysteriously cold climate of the Western Hemisphere, relative to that of Europe at the same latitude. Settlers argued that because the Indigenous occupants of the Americas had not brought the hemisphere under cultivation, they had not tamed its climate. Colonists in New France, in North America, now embarked on a vast logging campaign to bring about a regional warming that would aid European agriculture. Doctors suggested an even more comprehensive deforestation program in tropical colonies to "improve ventilation and dispel harmful miasmas." By the early eighteenth century, some settlers believed their efforts were bearing fruit. The climate of New France appeared to be warming, and indeed it was. The cause, though, was not deforestation but instead the ebbing of volcanic eruptions and the resumption of normal levels of solar activity. Nevertheless, the warming trend provided justification for the brutal conquest of Indigenous land.[35]

Colonialism might well have altered Earth's climate, but in a manner opposite to that imagined by eighteenth-century settlers. Beginning in the sixteenth century, some sixty million people succumbed to a genocide unleashed by Spanish conquest, forced migration, slave labor, and pathogens that caused epidemics of bubonic plague, measles, mumps, smallpox, and other diseases. Death on such a scale led to the abandonment of vast tracts of previously cultivated farmland and woodland. This permitted the expansion of rainforests, which absorbed more carbon dioxide than the carefully managed landscapes they replaced. Bubbles of captured atmosphere, trapped in cores exhumed from ice sheets, suggest that the concentration of carbon dioxide in Earth's atmosphere may have dipped ever so slightly as a result, reducing the quantity of solar heat that Earth's atmosphere could retain. Reforestation across South and Central America may have contributed to cooling during an especially cold stretch of the LIA, which means that, in turn, reforestation brought harvest failures and catastrophe to Spanish territories in Europe.[36]

In an era of incessant warfare that may have been partly sparked by climate change, European governments developed new systems of forestry administration to secure stable supplies of wood for naval construction. By the eighteenth century, "scientific" foresters joined improvers in using empirical methods, notably quantification, to more efficiently manage landscapes that supplied the state with vital resources. According to the environmental historian Paul Warde, improvers and foresters realized that resources and energy circulated through both societies and environments. By finding links between these circulations, scholars began to clarify the interconnectedness of environment and society. Meanwhile, imperial expansion introduced Europeans to the ruins of collapsed civilizations in deserts across North Africa and the Middle East. By the early nineteenth century, scholars began to suggest that these civilizations had degraded a once-temperate climate by interfering with the circulatory systems that sustained soils and forests. Climate change, it seemed, could bring about the collapse of civilizations that were foolish enough to cause it.[37]

By the turn of the nineteenth century, in short, three different but connected strands of evidence, each inspired partly by changes in the Sun's capacity to warm Earth, had together created a foundation for the later development of climate science. European scholars had developed and combined different academic disciplines to identify evidence for climate change in the archives of nature and the archives of societies. Many now agreed that the climate had once changed, and that ordinary people were right about trends in precipitation or temperature that can today be associated with the LIA. Diverse thinkers suspected that these changes could imperil civilizations by transforming the environments on which states depended. Yet these intellectual currents could not coalesce into a complete understanding of climate change without an accurate sense of the forces, or what climatologists call the forcings, by which the world's climate could change. Also needed was a method for quantifying both the mechanisms responsible for climatic trends, and the impact of those trends on society. The study of the Sun and other stars would soon begin to supply the answers.

2

Changing Stars, Changing Climates

In 1757 Wilhelm Herschel, a nineteen-year-old oboist in the Hanoverian Guards Band, immigrated to England after the defeat of his regiment in the Seven Years' War. Now going by the name William, he worked as a musician and composer. Six years later he welcomed his younger sister, Caroline, when she too immigrated from Hanover. Performing at his concerts, she soon acquired a reputation as a talented vocalist. Her singing career stalled, however, when William's passion for astronomy eclipsed his interest in music. This passion came to define both of their lives.[1]

Intellectually curious and wishing to engage with the learned gentlemen of his day, William Herschel read widely in natural philosophy. The similarities he perceived between sound and light quickly attracted him to astronomy, and in 1773 he decided to try stargazing. Dissatisfied with the quality of contemporary lenses, he began building his own telescopes, assisted by his loyal sister, who fed him, read to him, and spent hours polishing mirrors. Together they manufactured instruments of groundbreaking quality. Using one of his cutting-edge telescopes in March 1781, Herschel made a shocking discovery: a new planet. The planet, eventually named Uranus, earned him a royal stipend, an observatory at Windsor Castle, and, several decades later, a knighthood.[2]

Nevertheless, in some respects Herschel remained an outsider in the discipline of astronomy. Since the publication of Isaac Newton's theory of gravity in 1687, astronomy had been dominated by increasingly sophisticated efforts to precisely calculate the movements of heavenly bodies. Scientists hoped thereby to measure the scale of the universe (Chapter 5), find new ways of calculating longitude (Chapter 9), and most importantly, confirm and further develop the Newtonian physics of motion.[3] Indeed, as soon as Herschel discovered Uranus, mathematicians and astronomers rushed to calculate the planet's movement. As an amateur with a background in music, not mathematics, Herschel could not join their efforts.[4]

However, his discovery of a new planet revealed that there was still much to learn about the physical reality of the solar system. Herschel saw the cosmos as a network of shifting environments: "a vast dynamic region in which stars evolved from clouds of nebulous material," rather than a "static backdrop against which to measure planetary positions." His thinking owed debts to Prussian philosopher Immanuel Kant, who had argued in 1755 that stars and planets are formed from rotating gaseous clouds that slowly collapse and flatten, and to the natural historians with whom Herschel conversed at the Bath Philosophical Society. These natural historians were using the new system of taxonomy, developed in 1735 by Carl Linnaeus, to apply order to the vast diversity of life that the age of discovery had revealed. Increasingly, Herschel imagined the whole universe taxonomically. He classified celestial "species," considered how they related to one another, and studied how they evolved and changed.[5]

In some respects, curiosity about the nature of cosmic bodies marked a return to the preoccupations of an earlier age, before Newtonian physics came to dominate professional astronomy. Yet the Herschels had pioneered what contemporaries came to see as a "new astronomy." It involved the iterative refinement of novel astronomical instruments, and the application of emerging theories in other sciences to interpret what those instruments seemed to reveal about the cosmos. The new astronomy would create unprecedented links between shifting environments in the solar system and human affairs on Earth. That was perhaps especially true of the Sun, for William Herschel's studies seemed to reveal connections between solar activity, terrestrial weather, and the welfare of contemporary states. In pursuing these studies, Herschel synthesized some of the intellectual currents that had established a foundation for the discovery of climate change. But he went beyond them by identifying how exactly climate could change, and quantifying both the nature of that change and its influence on humanity. Subsequently, the study of solar variability would play a central—and until now, largely overlooked—role in the emergence of climate science.

Solar Changes and Climate Crises

William Herschel married the widow Mary Baldwin Pitt in May 1788. Four years later they welcomed the arrival of a son, John. By that time William had started observing the Sun in the hope of detecting another new planet moving across its surface. At first he tracked sunspots, in order to avoid mistaking them for a planet. Soon he noticed something that fascinated him. Most seventeenth- and eighteenth-century astronomers had believed that sunspots were elevated above the solar surface—the Italian scholar Galileo Galilei had proposed that they were clouds—but Herschel's superior telescopes revealed the opposite. Sunspots were, in fact, beneath the brilliant solar surface that surrounded them. To Herschel, they looked like holes in the solar atmosphere that permitted glimpses of the Sun's dark surface. Instead of being clouds, sunspots revealed the absence of cloud. It seemed to Herschel that the Sun's clouds might be similar to Earth's glowing aurora borealis, the Northern Lights. Because the aurora did not seem to heat up Earth, Herschel reasoned that the Sun's surface could be cool enough for life.[6]

The key to Herschel's thinking was the method of analogy, in which perceived similarities between things known and unknown allowed the characteristics of the known thing to reveal the nature of the unknown.[7] As John Locke had remarked in 1689, analogical reasoning seemed like the only way for natural philosophers to deduce anything about those parts of reality that, "either for their smallness in themselves or remoteness from us, our senses cannot take notice of," such as "whether there be any plants, animals, and intelligent inhabitants in the planets."[8] Herschel believed he had demonstrated that the Sun was similar to Earth: it had clouds and a surface visible through sunspots. Reasoning by analogy suggested that the Sun must also be inhabited, like Earth.

Such thinking was commonplace among seventeenth- and eighteenth-century astronomers. European explorers had found life wherever they went, including in "torrid zones" around the equator that Aristotle had believed were too hot for life. Scholars therefore reasoned that God had not created environments that were lifeless; for them, lifeless was synonymous with useless. Some speculated that life must therefore exist on other planets, and even that organisms traveled through space between planets. This notion of a "plurality of worlds" was popularized in 1686 in a best-selling and widely translated book by the French writer Bernard le Bovier de Fontenelle. At the time, the Sun had seemed sufficiently distinct from the planets to discourage analogies between terrestrial and solar environments. Now Herschel argued that the Sun, too, was a living world in the cosmic ecosystem.[9]

Ironically, it was Herschel who would soon prove that the Sun had little resemblance to Earth. While attempting to build an eyepiece that more effectively filtered the Sun's light, he realized that sunlight shining through red-tinted glass was unusually hot. Intrigued, he used a prism to separate sunlight into its constituent colors, and then a series of thermometers to measure the temperature of each color. When he slid one thermometer just to the side of the red color, he was astonished when the thermometer recorded a much hotter temperature than the others. The greatest heat seemed to come from invisible light that was redder than red; we now call it infrared. Not only had Herschel provided the first indications of the true breadth of the electromagnetic spectrum, but he had also conclusively determined that Earth's heat came from the Sun. The solar surface, he admitted, would be too hot for life, after all.[10]

Herschel's discovery led him to consider the volatility of stellar light and heat. In his nighttime observations, he joined other astronomers in studying "variable" stars whose brightness seemed to change. He speculated that large numbers of spots like those he saw on the Sun might be dimming the luminosity of these stars. Then he wondered what would happen to Earth if a massive outbreak of sunspots diminished the Sun's radiance. He suspected that the consequences for human civilization would be disastrous, because life on Earth depended on the light and warmth of the Sun.[11]

It was the right time for gloomy thoughts. The French Revolution had sparked fears of social disorder in England, and now war with France had led to a blockade of grain shipments from the continent that coincided with weak harvests amid a spate of bitterly cold winters and wet springs. The price of food was soaring, and hunger was spreading. Moved in part by the crisis, and by reports of the environmental constraints on Indigenous and settler societies in the New World, the English cleric Thomas Malthus formulated his famous theory that the production of food could never keep up with increases in population. Whereas Utopian theorists had believed that society could be perfected, Malthus argued that demographic collapse was inevitable. Mass starvation could only be delayed by taking steps like abolishing laws that empowered parishes to aid the poor, which Malthus saw as promoting unsustainable population growth. Herschel feared an even worse fate. If the Sun were to dim as other stars did, the resulting decline in agricultural output would make even a Malthusian collapse look trifling by comparison.[12]

Herschel argued that a sustainable society would not only regulate its own resources but also attempt to forecast and prepare for any natural disruptions in the supply of those resources. It was "presumptuous," he wrote, to assume "the stability of the present order of things," in part because it was not certain that the Sun's brightness remained constant. He now believed it was crucial for the future of civilization to study the Sun and determine whether its light and heat were variable. Its changes needed to be understood, anticipated, and prepared for.[13]

Sunspot Studies and Climate Science

In the closing weeks of 1799, Herschel began to record the weather whenever he observed the Sun with his telescope, which was now almost daily. Because he believed sunspots were responsible for diminishing the Sun's light and heat, he expected to find evidence that the weather cooled when sunspots were more plentiful. As the months passed, Herschel thought he could indeed discern a correlation between sunspots and weather, but it was the opposite of what he had expected. In early 1801, Herschel wrote in his log that "openings," as he was now referring to sunspots, and other signs of activity on the Sun "may assist us possibly to expect a copious emission of heat, and therefore mild seasons," whereas their disappearance "will on the contrary denote a spare emission of heat and may induce us to expect severe seasons." On February 14, 1801, when his land steward asked "if we were likely to have sharp weather," Herschel felt confident enough to reply "that we should not have anything severe," apparently

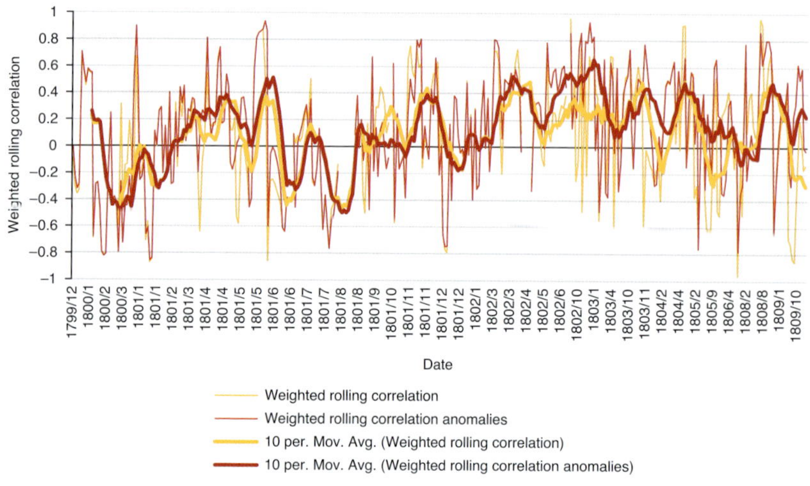

Correlations between sunspots and temperatures during Herschel's solar observation program. This graph shows the correlation (the y-axis) between English temperatures and sunspots recorded by Herschel (over time, the x-axis). The correlation ranges between −1 (a perfect negative correlation, meaning two variables move in precisely opposite directions, with one increasing when the other decreases), 0 (no correlation), and 1 (a perfect correlation, meaning two variables move in sync). The graph depicts the correlation not only by tracking English temperatures during each day that Herschel recorded sunspots (yellow), but also by tracking temperature anomalies (meaning deviations from average temperatures). To make this graph, Herschel's sunspot counts (384 entries in all) were first arranged on a simple ordinal scale: 0 for no sunspots, 1 for some sunspots, and 2 for thirty or more sunspots. The Central England Temperature (CET) series was then used to identify temperatures on days when Herschel recorded sunspots (yellow). Temperature anomalies for those days (red) were found by subtracting daily CET temperature averages from the mean monthly values over England in the period 1991–2021. A rolling correlation analysis, weighted based on the proximity of dates and with a ten-entry window, was performed using OpenAI Data Analyst GPT in order to account for the discontinuous nature of Herschel's sunspot observations. Trend lines depicting the ten-entry moving average were added to highlight changes in the correlation during 1801 especially.

due to the appearance of abundant sunspots.[14] However, by the end of 1801 Herschel had started to doubt that sunspots could impact daily weather. In December he acknowledged that the "influence is not so immediate," as temperatures had dipped below freezing despite plentiful sunspots.[15]

The oldest continuous weather record compiled with thermometers is the Central England Temperature (CET) series. It provides temperature data with far higher precision than proxy sources derived from natural archives. Moreover, by comparing the CET's daily information with long-term or modern weather measurements, it is possible to identify when temperatures were *anom-*

alous, meaning unusually warm or cold for the time of year. The CET reveals that when Herschel began his careful observations of the Sun, there was a modest negative correlation between sunspots and temperatures, meaning that sunspots tended to be plentiful when temperatures were low. Yet in early 1801 the correlation flipped. Precisely when Herschel began to suspect that sunspots could cause warming rather than cooling, those sunspots were suddenly more common in warm weather. Because daily changes in sunspots and weather actually have no connection to each other, however, the correlation was unstable. By the end of 1801 it flipped again, just as Herschel lost faith in the link between sunspots and daily weather. Although he lacked the statistical tools needed to identify correlations between sunspots and weather, Herschel's findings seem to have reflected real changes in the environments of the Sun and Earth.[16]

Herschel eventually abandoned his idea that the number of sunspots had an immediate effect on daily temperatures. But he found a connection on a broader scale between annual trends in sunspots and weather. Combing back through years of previous solar observations in his logbook, he noticed that there were very few sunspots between 1795 and 1800. Intrigued, Herschel pored through scholarly journals, seeking published sunspot observations that could help him piece together a longer record of solar activity. Despite the scattershot nature of these observations, Herschel identified previous periods of low solar output in 1650 to 1670, 1679 to 1684, 1686 to 1688, 1695 to 1700, and 1700 to 1713. He had uncovered what would be the most compelling evidence yet assembled for solar minima (periods of low solar activity) and the influence of solar variability on terrestrial climates.[17]

Caroline Herschel had become one of many weather observers across Europe employing thermometers to record daily temperatures. Some institutions were beginning to compile such measurements. Yet William Herschel had no access to long-term, standardized weather data, and so he lacked a way to compare the sunspot trends he had identified to a climate reconstruction. He landed on the idea of using grain prices, compiled in Adam Smith's *Wealth of Nations,* as a proxy for weather. Grain harvests should be higher, and prices lower, Herschel reasoned, when solar activity produced warm weather on Earth. Sure enough, he found an association between low grain prices and periods in which astronomers recorded many sunspots. It seemed plain to Herschel that the Sun should now be carefully monitored and, he wrote, "provision made for those years when sunlight and heat are deficient and the harvest bad."[18]

Herschel's study of the Sun was a turning point in the history of climate science. He had identified a genuine, natural force that can change Earth's climate. He had discovered resemblances between the Sun and other stars. He had found that the Sun's output varies across annual and decadal timescales, and that its output declines when there are fewer sunspots, then rises when sunspots multiply. He had refined the use of proxy sources and identified a social response to weather—alterations in commodity prices—that is still used by scholars to support (or substitute for) climate reconstructions compiled with natural archives.[19] The evidence he uncovered allowed him to, for the first time, quantify the relationship between climate change and human affairs, which in turn allowed him to propose—accurately—that there can be lag between forcings (causes of climate change) and weather. He had even called for a prioritization of redundancy over efficiency in food production, a form of climate adaptation still proposed by scholars and policy advisors. Changes in the appearance of the Sun had, in a matter of months, inspired Herschel to develop many of the key themes and methods of climate scholarship and policymaking.

Herschel's innovations were widely ridiculed. An article in the *Edinburgh Review* expressed a sentiment common among Herschel's learned contemporaries, likening his "hasty and erroneous theory concerning the influence of the solar spots on the price of grain" to the ideas of Jonathan Swift's Laputans, who in *Gulliver's Travels* obsessed pointlessly over the consequences of changes in celestial bodies. Stung by such criticism, Herschel turned to new projects.[20]

It would be decades before Herschel's ideas were rehabilitated by the intersection of scientific discovery and environmental change. The discovery of the eleven-year solar cycle suggested that the flux of solar radiation changed, as Herschel had argued. Growing evidence for Pleistocene glaciation indicated that it had changed much more dramatically in the past, with profound consequences for Earth. According to the historian of science Deborah Coen, Austro-Hungarian scientists established a network of weather stations that enabled them to reconstruct the mechanics of atmospheric circulation in their empire, and to speculate about the solar forcings that might cause it to change, all in an effort to justify the union of diverse peoples under Austro-Hungarian rule.[21]

Meanwhile, Herschel's son, John, ascended the ranks of professional astronomy. Beginning in the 1830s he played a central role in the construction of an even larger network of weather stations across the British Empire. By enabling an emerging class of professional meteorologists to track the movement of weather systems, the network promised to aid imperial navigation, commerce, and conquest. John Herschel hoped it would also clarify the relationship between Earth and the Sun. In 1872 weather stations in the network recorded the absence of the usual Indian monsoon across territories that were then some of the most important of the empire. Drought spread first across Bengal and Bihar, and then, more catastrophically, over southern India. Millions died in famines exacerbated by colonial mismanagement (Chapter 14).[22]

With British rule in India seemingly imperiled by the disaster, the empire's scientists launched new investigations into how weather might be predicted. Building on the foundations established by decades of solar and climatological science, astronomers finally revived William Herschel's idea that changes in solar output influenced weather and food production on human timescales. If this were true, then observations of the Sun could be used to stabilize the empire. At the Royal Observatory in Greenwich, mathematically gifted children were set to work scouring standardized datasets of sunspot counts and weather measurements that William Herschel could have only dreamed of. These children soon uncovered suggestive correlations between the solar cycle and air pressure over India; storms in the Indian Ocean; temperatures in Scotland and South America; and rainfall around the Great Lakes. The economist William Stanley Jevons, meanwhile, expanded Herschel's work, which he called "prophetic," to identify patterns in historical registers of commodity prices, economic performance, and deaths due to plague that seemed to match the solar cycle.[23]

After they claimed to find correlations, scholars like Jevons proposed mechanisms to explain them. The monsoon failed during periods of low solar activity, Jevons argued, triggering famines that reduced agricultural exports from India, drove up commodity prices, reversed economic growth, and created the conditions for epidemics that eventually worked their way to Britain. Jevons inspired a wave of scholarship that identified similar correlations between solar and social cycles, proposed new causal explanations for them, and used them to forecast the future. In 1889, predictions of imminent drought, based on the solar cycle, sparked widespread concern in Australia. However, skeptics pointed out that the periods of the agricultural or economic cycles proposed in the new studies did not always precisely match the periodicity of the solar cycle, that correlations between cycles dissolved when datasets were interpreted using different statistical methods, and that causal explanations for the correlations were based on speculation more than science.[24]

By the early twentieth century, few scholars accepted that solar changes could be clearly tied to social changes, but most agreed that the Sun's variability could affect Earth's weather and climate. To understand the influence on Earth of trends in solar output, the astronomer Andrew Ellicott Douglass soon developed the first long-term climate reconstruction using tree rings (Chapter 16). Scientists studying ocean sediments and ice sheets subsequently worked out how shifts in the Sun's capacity to warm Earth had unlocked the self-reinforcing feedbacks responsible for the glacial periods of the Pleistocene. By the 1970s this research was beginning to reveal how human changes to atmospheric carbon dioxide concentrations could profoundly alter Earth's climate (Chapter 8). Spurred in part by this revelation, scholars resurrected the idea that climatic changes had influenced human history and could imperil humanity's future.[25]

Changes in solar environments therefore played a crucial role in synthesizing approaches to climate change in eighteenth- and nineteenth-century Europe. By sparking new ideas among scholars, they helped create a transdisciplinary study of climate change that incorporated the full complexity of past, present, and future connections between climatic, environmental, and social variability. By the late nineteenth century, though, it was clear to some astronomers that solar changes influenced more than terrestrial weather and climate. There was another, previously hidden link between shifting solar and terrestrial environments, and it appeared to threaten the industrial technology that was just beginning to alter Earth's atmosphere.

3

Hidden Connections and Solar Science

When thirty-three-year-old Richard Carrington settled down to sketch sunspots on the morning of September 1, 1859, he could not have known that he was about to witness the most important change ever glimpsed in a cosmic environment. His career, once so promising, seemed to be slipping away. The son of a successful brewer, at the young age of twenty-five, he had gained entry to the Royal Astronomical Society, and shortly thereafter he had built his own observatory. There, over the next three years, he created a catalog of precise stellar measurements that was so immediately useful to navigators that the Lords of the Admiralty directed it be printed at public expense. The discovery of the solar cycle now moved Carrington to undertake a rigorous study of sunspots that soon earned him a reputation as Britain's premier solar observer.[1]

Before dawn on July 18, 1858, Carrington was suddenly called home. "In arriving," he wrote to a colleague, "I . . . found my father no more." It fell to him to manage the family brewery. The work gave him the money he needed to sustain his career, but it also consumed the time he needed to stay at the forefront of solar astronomy. On the first morning of September 1859, Carrington must have felt that solar observation was a luxury he could not count on for much longer.[2]

Then, at exactly 11:18 a.m., something changed on the solar surface. "There suddenly broke out," Carrington wrote, ". . . a kind of conflagration," which for two full minutes outshone the rest of the Sun. He quickly suspected that he had witnessed a flare of previously inconceivable power. Seeking independent confirmation, he visited a colleague at Kew Observatory, near his brewery. He learned that the observatory's magnetic instruments had recorded a minor disturbance in Earth's magnetic field at exactly the moment that he had seen the flare. About eighteen hours later, brilliant, blood-red auroras surged toward Earth's equator. Electric telegraph lines across Europe and the Americas shorted out. Other lines worked even when switched off.[3]

The "Carrington Event," as it came to be known, could not rescue Carrington's career, but it did secure his immortality in the annals of science. It suggested the existence of a profound, previously hidden connection between sudden environmental changes on the Sun and on Earth. The connection was perceptible in 1859 because previous discoveries of the Sun's variable influence on climate had fostered the development of solar science. It seemed important because it appeared capable of disrupting new electrified communication and transportation networks that were beginning to give industrializing countries unprecedented control over terrestrial environments. For the rest of the nineteenth century and into the twentieth century, magnetic storms, triggered by colossal solar explosions, repeatedly disrupted those networks. Each storm sparked a wave of scientific inquiry, until, in the early twentieth century, scientists finally understood why electrifying societies had grown precariously vulnerable to environmental upheavals on the Sun.

Coronal Mass Ejections and Magnetic Storms

The storms that disrupted nineteenth-century infrastructure on Earth had their origins deep within the Sun. Such storms occur when a vast interior region of the Sun that is too hot for convection collides with the cooler region nearer the Sun's surface, creating a magnetic field of unimaginable power. The magnetic field arcs above the solar surface into the billowing solar corona, where temperatures reach an extraordinary 20 million degrees Celsius. In this frenzied heat, about one million tons of charged particles careen into space every second. This solar wind thins as it races ever farther from the Sun, but even around Earth up to ten trillion particles pass through every square meter of space every second. Because these particles are charged, the solar wind extends

the Sun's magnetic field into interplanetary space. When the Sun travels through a relatively empty expanse between stars, as it has for some two million years, then it is only in the outer solar system that the pressure of the solar wind drops so low that it can no longer hold back the pressure of interstellar dust. Everything within the solar wind is inside the Sun's extended atmosphere, known as the *heliosphere,* which means that, in a sense, Earth orbits through the Sun itself.[4]

The energetic particles of the solar wind would wreak havoc on terrestrial life were it not for the constant sloshing of molten iron in Earth's outer core. There, interactions between streams of liquid iron generate a magnetic field one quintillion times more powerful than that of an ordinary lab magnet (but thousands of times weaker than the Sun's). Earth's magnetic field radiates from our planet's poles, arcs deep into space, and creates a vast, magnetic bubble, known as the magnetosphere, that protects us from the solar wind.[5]

But Earth's inner dynamo does not provide a perfect defense. The solar wind smashes into the leading edge of Earth's magnetosphere with supersonic speed, creating a shockwave that can come closer than one hundred thousand kilometers to the planet's illuminated side. The wind then sweeps past Earth along a magnetotail, an extension of Earth's magnetic field away from the Sun, that can be millions of kilometers long. When the magnetic fields of Earth and the solar wind point in opposite directions, they can link at points of contact on either the shockwave or the trailing edge of the magnetotail, opening a portal directly to the Sun's corona. Charged particles originating from the corona surge into the gap but are soon snared by two great magnetic loops—the Van Allen Radiation Belts—that usually keep them from Earth's surface.[6]

Even these belts do not offer complete protection. A new magnetic flux continually arises within the Sun and surges into the corona. Convection within the Sun and the differential rotation of the Sun's surface warp the coronal magnetic field, as do links to the interplanetary field created by the solar wind. The coronal magnetic field is therefore constantly under immense strain, and when that strain becomes too great, its tangled magnetic field lines snap open. The result can be an eruptive solar flare: a colossal release of electromagnetic energy. The most powerful flares usually detonate from large sunspot groups, where the Sun's magnetic field can be a million times stronger than Earth's. Gigantic

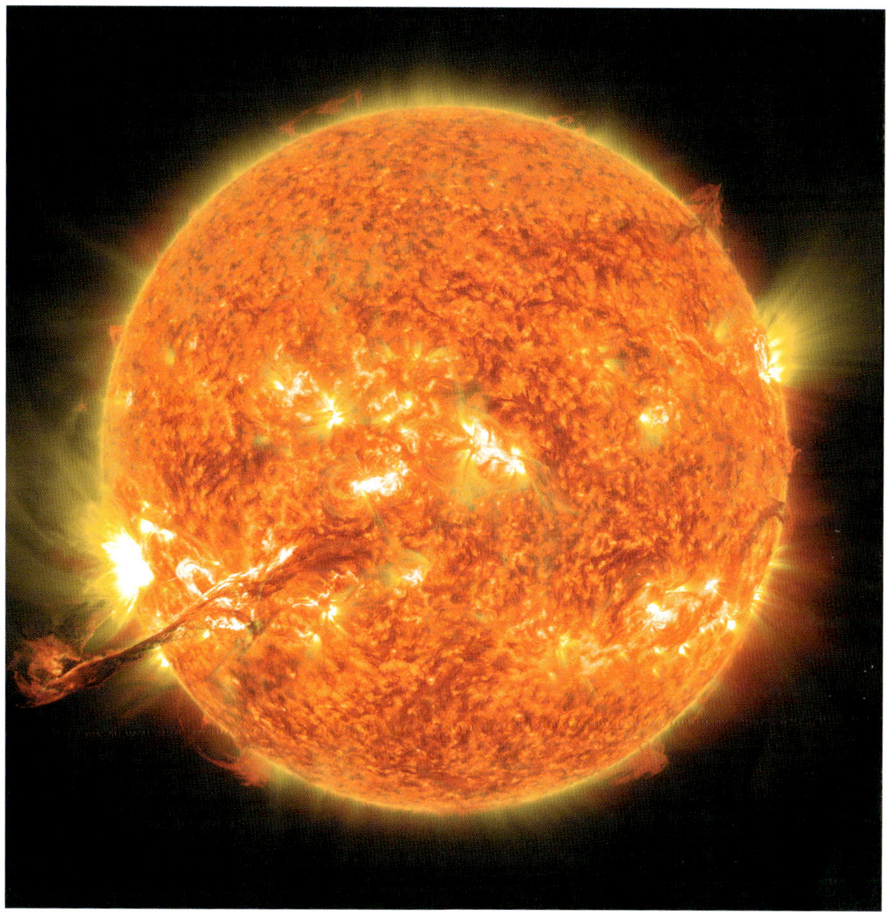

In this close-up image of the Sun, taken by NASA's Solar Dynamics Observatory, a massive CME launches into space (bottom left) in August 2012.

magnetic bubbles also break free from the Sun as the solar magnetic field reconfigures, launching up to fifty billion tons of plasma into interplanetary space. Powerful flares may accompany the departure of these coronal mass ejections (CMEs); both are examples of solar storms.[7]

Electromagnetic emissions from flares travel at light speed and therefore reach Earth in around eight minutes. Radio signals then rush through Earth's atmosphere. X-ray and extreme ultraviolet emissions enhance the ionization of the ionosphere, the layer of the upper atmosphere charged by solar radiation.

Several minutes later, electrons and protons may arrive in enormous quantities, dramatically increasing radiation in Earth's magnetosphere.[8]

CMEs, meanwhile, hurtle through space at up to three thousand kilometers per second. Most take several days to travel Earth's distance from the Sun, and the overwhelming majority sail harmlessly into the outer solar system. Some, however, collide with Earth's magnetic field. When the lines of a CME's magnetic field and those of the Earth's magnetic field are pointed in the opposite directions, the CME can unite with the Earth's magnetosphere and fill it with vast quantities of solar energy. The result is a geomagnetic storm that amplifies the ionospheric and radio disruptions caused by flares and saturates the radiation belts surrounding Earth to such an extent that new belts can suddenly form. The density of the upper atmosphere also changes at midlatitudes, while magnetic substorms explosively release stored magnetic energy. Geomagnetically induced currents—or GICs—surge across Earth's surface. The auroras intensify, expand toward the equator, and often turn a vivid blood red. Because Earth's magnetic field is twice as intense around the poles as it is at the equator, the magnetic consequences of a solar storm are far greater at higher latitudes.[9]

Arctic Climates, Solar Changes, and Big Data

For some three hundred thousand years, auroras were the only signs of geomagnetic storms that human populations noticed. Then, in the thirteenth century, refinements to an ancient technology, the compass, led to its widespread adoption by sailors. No doubt some sailors noticed that compass needles no longer pointed toward the magnetic north pole during auroras. This was the first way in which geomagnetic storms disrupted transportation technology.[10]

Early in the eighteenth century, scholars first described in writing the correlation between auroras and erratic compass movements, but they had no way of knowing that environmental changes on the Sun were the cause. Later that century the rising importance of waterborne commerce for Europe's maritime powers helped motivate efforts to map Earth's magnetic field. If they had precise knowledge of the continually shifting difference between the magnetic and geographic north poles at every point of their journey, sailors would be able to use their compasses more effectively for navigation. The Prussian baron

Alexander von Humboldt carried out the best-known of these mapping efforts in Central and South America. When he returned home, he resolved to study how magnetic fields changed across space and time. He was measuring shifts in magnetic declination—the angle between geographic and magnetic north poles—in a shed outside Berlin when, on December 21, 1806, his compass needles started moving erratically just as auroras danced across the sky. Humboldt coined the term "magnetic storm" to describe the magnetic disturbances that auroras seemed to cause.[11]

It was a change in the influence of the Sun on climate that would begin to clarify the connection between solar changes and magnetic storms. Sulfuric aerosols trapped in ice cores suggest that a truly massive volcanic eruption took place in 1809, followed by a succession of smaller eruptions that culminated in the staggering explosion of Mount Tambora in 1815. All of these eruptions in a period of low solar activity sent average global temperatures plummeting to lows they had never reached before during the LIA. Records from that period indicate the climatic effects. Arctic explorers, such as the British whaler William Scoresby, struggled to navigate expanding sea ice. In 1816 the stratosphere was saturated with Tambora's dust, and sea ice nearly wrecked Scoresby's ship off the eastern coast of Greenland. Yet in the following year Scoresby reported "a remarkable diminution of the polar ice" in the same location. Climatologists now know that volcanic dust was beginning to fall out of the atmosphere at that point, but marine sediments, tree rings, and other evidence extracted from archives of nature suggest that Arctic temperatures had not recovered. Instead, the continued influence of Tambora may have affected atmospheric pressure in the northern Atlantic Ocean, altering wind patterns to push sea ice from the Greenlandic coast. If so, Scoresby unwittingly described the effects of a shift in the quantity of solar radiation that reached Earth's surface.[12]

According to the historian of science Vidar Enebakk, Scoresby's report intrigued proponents of scientific discovery and Arctic exploration. Soon they had ample motivation to join forces in exploiting the apparent thawing of the far north. The Northwest Passage Act, which offered a reward to any Briton who could chart a shipping route through Arctic Canada to Asia, had been repealed in 1818. Its replacement, the Longitude Act, which incentivized both polar exploration and improvements to navigation, had the effect of aligning the goals of scientific and military personnel. As a side effect, it elevated officers in the Royal Artillery, the Royal Engineers, and the Admiralty Hydrographic Office, whose expertise in precise measurement suddenly seemed essential in voyages of polar exploration. One participant was a young Irish officer named Edward Sabine.[13]

In 1819 Sabine joined a crew aboard the *Hecla,* a refitted "bomb vessel"—a ship armed with mortars—that, like others in its class, had been named after a volcano. Off the coast of Baffin Island, the crew approached the magnetic north pole, where Sabine took measurements of the intensity and inclination of the local magnetic field. It was the beginning of a lifelong obsession. Over the next two decades, his magnetic measurements would take Sabine from Brazil to Svalbard, a Norwegian archipelago on the edge of the Arctic Ocean. Eventually he assumed a leading role within the "magnetic lobby," a group of influential scientists and naval officers who recognized the practical and scientific potential of measuring Earth's fluctuating magnetic field. Owing in large part to Sabine, the British government financed the construction of magnetic observatories around the empire. Because John Herschel believed that the correlation between magnetic storms and auroras revealed both to be meteorological phenomena, the new observatories would be equipped with weather instruments that, in time, would permit the first global measurements of climate change.[14]

Sabine would soon mastermind the largest scientific enterprise of his era. From 1839 to 1857, his observatories, equipped with magnetometers and other measuring devices, operated around the clock, six days a week (they closed on Sundays). An unprecedented torrent of environmental data flowed into Sabine's "magnetic department" in Woolwich. There, Sabine led a team of mathematicians—women and men known as "computers"—who tabulated and analyzed measurements on a scale that would have been difficult to imagine in an earlier age.[15]

Sabine's wife, Elizabeth Leeves, was a scientist in her own right. In the 1850s she translated Humboldt's *Cosmos,* a sweeping compendium that popularized the discovery of the sunspot cycle (Chapter 2). When Leeves introduced the cycle to Sabine, he realized to his astonishment that it precisely matched a mysterious pattern in the magnetic disturbances he had charted with his computers. The correlation seemed to suggest that space was alive with invisible forces that bound together the variable environments of the Sun and Earth.

"We stand on the verge of a vast cosmical discovery," John Herschel wrote after hearing the news, "such as nothing hitherto imagined can compare with."[16]

New Vulnerabilities, New Discoveries

Scientists like Sabine, Leeves, and Herschel also stood on the threshold of modernity. Perhaps above all, the new era involved the rapid expansion and growth in density of networks that enabled unprecedented movement of commodities, people, information (including the measurements recorded by Sabine's observatories), and invasive organisms. Those networks, powered in the late eighteenth century by wind (for most waterborne transportation), muscle (for most land transportation), and, increasingly, light (for emerging telegraph networks), were revolutionized in the nineteenth century by the widespread exploitation of fossil fuels and, before long, electricity. Electric telegraph lines and railroads encouraged and were encouraged by a new age of imperial expansion, commodity extraction, industrialization, urban growth, global migration, rising population, and scientific development, among other things. The world was coming together, economically, culturally, and biologically, but it was doing so through the self-serving machinations of imperial governments in a handful of wealthy cities.[17]

In deeply unequal ways, the new era of networked industrialization and electrification began to alleviate the conditions of scarcity that had rendered many communities so vulnerable to the climatic disruptions of the LIA. The Carrington Event in 1859, though, suggested that the networks responsible for increasing the resilience of populations in the face of changes in the Sun's influence on climate also made those populations vulnerable to mysterious solar fluctuations that had previously merely illuminated the night sky. Meanwhile, as we have seen, monsoon failures suggested to some British scientists that the imperial periphery remained at the mercy of the Sun's influence on weather, with potentially dire consequences for the empire as a whole.[18]

While undertaking a magnetic survey of the Indian subcontinent, Colonel Alexander Strange of the Royal Artillery had experienced firsthand the human toll of drought. In 1872, as drought ravaged India, he proposed that the British government finance the construction of a unique, nationally funded laboratory. Among other things, it would maintain continuous surveillance of the Sun, using a spectroscope, an instrument that disperses light into its constituent colors to reveal the chemical properties of its source. Strange hoped that spectroscopic surveillance could predict the Sun's influence on weather, which he thought could aid efforts to avoid another horrific famine in India.[19]

Strange wanted a facility separate from the Royal Observatory at Greenwich because he felt that the observatory's director, George Airy, no longer understood how astronomy could serve the state. It was a common refrain among British scientists of the nineteenth century, many of whom sought to create new institutions for specialized disciplines or else reform old ones to be more inclusive and responsive to advances in scholarship. By the 1870s many astronomers had flocked to the exciting frontier that the techniques of spectral analysis (called the "physics of astronomy," or astrophysics) seemed to open in the new astronomy. The aging Airy had devoted his observatory to stellar measurements that now produced trifling improvements to navigation. By aiding mariners, Airy believed, such measurements provided a tangible service to the state that the newfangled spectroscopy might not be able to match. Still, he was savvy enough to forestall Strange's challenge to his authority. He undermined support for a new observatory by introducing a program of solar photography and spectroscopy at the Royal Observatory.[20]

In 1873 twenty-two-year-old Edward Walter Maunder was hired as the spectroscopic assistant for this program. The son of a Wesleyan minister, Maunder had attended King's College London, a working-class alternative to Oxford and Cambridge. As a teenager he had once glimpsed the Sun "low down in the west, shining red through the mist," he later recalled, and noticed "a round black spot, just as if a nail had been driven into him up to the head." The next time he saw the Sun just above the horizon, that sunspot had moved, and on the following occasion it was gone. Maunder, enthralled, would devote the rest of his life to the study of the Sun.[21]

On a foggy November morning in 1882, Maunder noticed another, even bigger spot on the crimson Sun. It was so obvious that soldiers marching nearby pointed it out to one another. Using the spectroscope, Maunder detected a hydrogen flare arcing out from the sunspot. That night the telegraph network near London collapsed while an aurora lit up the sky. When he discovered the next morning that a magnetic disturbance had been measured at Greenwich, Maunder resolved to find the statistical relationship between sunspots and magnetic storms that had long eluded solar astronomers.[22]

Maunder soon found that even big sunspots did not always coincide with magnetic storms. He wondered whether sunspots could launch their magnetism into space through flares. Earth, a tiny target in the vastness of space, would often be spared from being hit by these magnetic projectiles. The science author Stuart Clark notes that the concept was nonsensical to contemporary scientists, who assumed that magnetism could not break free from a magnet and that the Sun's energy radiated equally across its surface.[23]

After seventeen years at the Royal Observatory, Maunder was finally allowed an assistant. Influenced by the progressive Methodist faith in which he was raised, Maunder was one of his profession's most outspoken advocates for women's participation in science. He hired the first woman ever employed by the Royal Observatory: a brilliant computer named Annie Russell, who soon proved his equal in solar astronomy. United by their shared fascination with the Sun's environmental changes, Annie and Walter wed in 1895.[24]

Annie was then forced to resign from her job, because married women were not allowed to hold positions in the civil service. Nevertheless, she continued her work. Among other things, she refined methods for photographing the solar corona during eclipses. After studying her pictures, Walter realized that the corona was not homogeneous. Instead it was full of the magnetized structures he had proposed earlier. Then in 1903 a mammoth sunspot failed to cause a large storm, but a smaller spot coincided with one so massive that, as one solar astronomer put it, "practically the world's whole telegraph system was upset." In a foreshadowing of graver threats to come, this time GICs also disrupted the electrical signals that powered the rail network in London and surged through telephone lines around Chicago.[25]

In response, the Maunders pored through reams of sunspot and magnetic observations until they found the correlation that established the causation solar astronomers had long sought. Magnetic storms, they realized, followed a twenty-seven-day cycle, matching the Sun's rotation. During every storm, a sunspot had revolved into position near the apparent center of the solar surface, like the turret of a battleship lining up its target. Drawing on an idea proposed by the Swedish physicist and chemist Svante Arrhenius, Walter suggested that rays of charged particles might shoot out of sunspots and smash into Earth. This theory gradually won widespread acceptance—although it would be another seventy years before instruments in space revealed the magnetic bubbles, not rays, that we today call coronal mass ejections.[26]

A unique combination of solar environmental changes, climatic disruptions, magnetic storms, technological advancements, and astronomers being in the right place at the right time had revealed a previously hidden connection between the Sun and Earth. This connection would become increasingly important over the twentieth century, as the rapid development of technological infrastructure created ever-expanding and increasingly dangerous sensitivities to geomagnetic storms.

4

Solar Storms and Existential Risk

In late May 1967 a series of violent solar flares accompanied the collision of a colossal CME with Earth's magnetosphere. According to the atmospheric scientist Delores Knipp, the timing could not have been worse. Soviet and US warships had recently collided in the Sea of Japan; US and South Vietnamese forces had just crossed over the Demilitarized Zone that separated North from South Vietnam; and tensions were soaring across the Middle East on the eve of the Six-Day War. For the first time, the US and Soviet nuclear arsenals included more than ten thousand warheads each, far more than any country possesses at the time of this writing.[1]

On May 23 the largest radio burst in the history of the twentieth century overwhelmed the high-frequency radar stations in Alaska, Greenland, and England that constituted the American Ballistic Missile Early Warning System. Officers at the North American Air Defense Command (NORAD), which operated the system, had received an update on the solar storm from the Air Weather Service (AWS) Solar Observing and Forecasting Network. AWS personnel had expanded their purview to forecasting the Sun's environmental changes, because it was now well known that solar storms could disrupt radar. Yet the unprecedented magnitude of the radio burst that scrambled US radar stations in May 1967, and the tense atmosphere that prevailed then, both conspired to convince Air Force officers that Soviet jamming had taken down their early warning system. The ability to launch a retaliatory nuclear strike depended on early detection, so Cold War commanders interpreted radar disruptions on this scale as an act of war. For all they knew, hundreds of Soviet missiles were inbound across the shortest-possible route from Russia to the United States: over the Arctic.[2]

US strategic bombers readied for takeoff. Their crews had trained to carry out a nuclear attack on preassigned targets unless they were called off. According to Knipp, though, the solar storm had disrupted high-frequency communications to such an extent that it might have been impossible to communicate with the bombing crews. Even if every crew had been successfully called back, a full-scale launch of US strategic bombers could have invited a commensurate Soviet response. For a few tense hours, the survival of humanity may have depended on accurately interpreting how solar environmental changes were affecting Earth. Luckily, in the nick of time AWS forecasters managed to convince NORAD officers that a solar storm, not Soviet jamming, likely accounted for the radar disruptions. The bombers never took off, and civilization survived another day.[3]

The 1967 near-apocalypse belongs to the hidden history of the twentieth and early twenty-first centuries. Scholars have shown how, in this period, a "Great Acceleration" in population growth, energy use, industrial production, resource extraction, commodity consumption, technological development, urbanization, and interconnectedness transformed Earth, especially its atmosphere, in ways that would have been hard to imagine even in the nineteenth century.[4] But they have not considered how the technologies that permitted, and were permitted by, this unprecedented era of expansion inadvertently connected humanity to the heliosphere: the invisible outer atmosphere of the Sun that envelops the magnetic Earth. New technological infrastructure was vulnerable to fluctuations in the solar atmosphere, caused by solar storms, so this *space weather* could increasingly imperil human life. Powerful CMEs occurred in just about every decade of the twentieth century, and the pattern continues thus far in the twenty-first century. Military technologies also triggered vast and often unintentional changes to the solar atmosphere near Earth, replicating the effects of space weather but also providing some protection against it. After a particularly severe geomagnetic storm in 1989, government officials and utility company engineers came to suspect that solar storms could jeopardize the continued survival of technological civilization, and they belatedly struggled to respond. Modern civilization, it turned out, might exist on the whim of the Sun, and therefore it might not exist much longer.

Radio, Empire, and the Volatile Near-Earth Environment

It was the invention of radio that most clearly connected human infrastructure to the fluctuating environment of the Sun's outer atmosphere. Theoretical advances in the study of electromagnetic radiation led, in 1895, to the first wireless transmission using radio. At first the applications seemed limited. Radio waves traveled in straight lines, so unlike a telegraph wire it seemed they could not follow the curvature of Earth. That ruled out communicating over the horizon, across meaningful distances.[5]

By 1901 a team managed to send a radio signal across the Atlantic Ocean. The link between auroras and geomagnetic disruptions had led some geophysicists to theorize that the upper atmosphere included a conducting layer. Now it seemed like that layer really did exist, and that it had reflected the radio message back to Earth. Scientists would soon confirm the existence of the layer, which they called the ionosphere, and determine that it extended far into space. Short-wave (high-frequency) radio communication that exploited the reflective properties of the ionosphere in near-Earth space eventually eclipsed the telegraph, partly because radio more effectively unified the far-flung empires that dominated the early twentieth century. Yet it would soon grow apparent that because solar activity so profoundly influences the ionosphere, new technologies that strengthened empires also rendered them far more vulnerable to the once-invisible instability of the Sun.[6]

The Second World War provided the first hints of just how dangerous the new vulnerability could be. In September 1941, for example, an intense geomagnetic storm disrupted radio transmissions around the world, while a blood-red aurora brightly illuminated a Canadian convoy for a squadron of lurking U-boats. Yet for the Allies, a far more concerning development would surface in the following February. By then engineers had developed radar systems that could detect ships and aircraft by bouncing radio waves off of them.

After the outbreak of the Second World War, radar provided a crucial technological edge to Britain in its long-shot defense against German invasion. Yet for several hours in February, radio bursts from the Sun temporarily brought down the entire British radar defense network. Fortunately, the Luftwaffe did not know about Britain's dependence on radar and did not take advantage of the disruption.[7]

Military radar matured rapidly as the Second World War gave way to the Cold War. After both the United States and the Soviet Union acquired the hydrogen bomb, radar provided the ultimate insurance against a nuclear attack. Peace depended on maintaining the belief, among a critical mass of Cold War leaders, that a nuclear first strike would be detected using radar and then avenged with enough nuclear weapons to destroy the attacker. To uphold the new paradigm of "mutually assured destruction" (MAD), militaries used radio waves to surveil the atmosphere for incoming bombers; later they surveilled outer space for inbound missiles. They also used radio to maintain contact with the Air Force bases and, eventually, the submarines that would launch any nuclear counterstrike.[8]

Reliance on radio made the precarious peace established by MAD spectacularly vulnerable to solar storms. Both civilian and military observers monitored the state of the ionosphere and warned of any expected radio propagation disturbances. Observatories across the United States, Greece, and the Philippines, for example, hosted US Air Force officers who provided real-time data to the Air Weather Service.[9]

Militaries exploited outer space, and the tenuous outer atmosphere of the Sun itself, not only to communicate or to detect enemy forces, but also to deliver their deadliest weapons. After the 1957 launch of *Sputnik* demonstrated the Soviet Union's capacity to target the United States using powerful missiles, the Greek physicist Nicholas Christofilos at Lawrence Livermore National Laboratory suggested that detonating a nuclear bomb in outer space could replicate the effects of a solar storm, and thereby create enough energetic electrons to disable inbound warheads. In 1958 his tireless advocacy for the idea, and the fear provoked among military officials by the launch of *Sputnik*, led officials in the Defense Atomic Support Agency (DASA), which managed the US military's nuclear weapons program, to detonate three atomic bombs above the atmosphere of the Pacific Ocean. The tests, collectively called Operation Argus, confirmed that a nuclear weapon could create an artificial belt of high-energy electrons in Earth's magnetic field. The radiation was not strong enough, however, to quickly destroy Soviet warheads.[10]

DASA officials fast-tracked Operation Argus because Cold War leaders had come to agree that nuclear tests posed unacceptable risks to public health, and more importantly that an international ban on nuclear testing would slow the worldwide proliferation of nuclear weapons. Mere weeks after the final Argus explosion, the US and Soviet governments agreed to a moratorium on nuclear testing. Yet the moratorium lasted only three years. Cold War tensions surged following the destruction of an American U2 surveillance plane over the Soviet Union. Officials at the Atomic Energy Commission (AEC), a newly established federal agency that oversaw the development and regulation of atomic power, joined DASA representatives in planning for a spectacular escalation of nuclear testing in outer space. They hoped to determine the extent to which hydrogen bombs, which are far more powerful than atomic weapons, could alter the outer space environment.[11]

Months before the Cuban Missile Crisis, in 1962, their efforts began with two top-secret launches. Both failed, but on July 9 a nuclear test code-named Starfish Prime successfully exploded some 400 kilometers above Earth's surface, 1,450 kilometers off Hawaii, with a power nearly one hundred times greater than that of the atomic bomb that leveled Hiroshima. An electromagnetic pulse (EMP) far more powerful than expected shut down streetlights in Hawaii, and brilliant auroras suddenly appeared over the detonation site and halfway around the world. To a much greater extent than the Argus explosions, the test reproduced the effects of a solar storm by creating a new belt of high-energy electrons in Earth's magnetic field. AEC officials were astonished when the belt crippled seven satellites, one-third of the spacecraft then in orbit. The test was the beginning of a devastating new tactic in nuclear war: detonating nuclear bombs in outer space to create electromagnetic pulses that collapse electric grids before other warheads arrive at their targets on Earth. The temporary transformation of the space environment near Earth also revealed just how dangerous nuclear testing had become. It helped persuade the governments of the United States, Soviet Union, and United Kingdom to agree to the

1963 Partial Test Ban Treaty, which prohibited atomic testing in the atmosphere, under water, and in outer space.[12]

Cold War militaries used more than nuclear explosions to transform the near-Earth portion of the heliosphere. Soon after Operation Argus, Christofilos pointed out that by disrupting the ionosphere, solar storms or high-altitude nuclear explosions would scramble military communications, giving the Soviets a window to attack. In response, scientists at the Massachusetts Institute of Technology (MIT) devised a plan, which they called "Project Needles," to create an artificial ionosphere by dispersing copper filaments, each about two centimeters long, some thirty-five hundred kilometers above Earth. In the early 1960s they renamed their plan Project West Ford, after a town where MIT operated a radio antenna, and released nearly five hundred million filaments in Earth's orbit.[13]

Government officials assured scientists that the project was only an experiment, and that the filaments would soon fall back to Earth, but astronomers around the world protested. Many feared that if the experiment was successful, the Air Force would make it permanent by releasing billions of filaments into orbit, which would interfere with future optical telescopes in outer space and especially with radio telescopes on Earth. Soviet officials accused the United States of contaminating outer space, rendering it unsafe for crewed spacecraft. Ambassadors at the United Nations demanded an explanation for the United States having acted unilaterally, without prior consultation, to alter the space environment. Environmentalists decried the potential impact of the re-entry into Earth's atmosphere of the copper filaments. It was obvious that Cold War militaries had obtained the power to alter the cosmic environment and transform the relationship between Earth and the Sun. Protests around the world condemned the experiments as destabilizing the balance of power on Earth. Diplomats in the UN Committee on the Peaceful Uses of Outer Space advanced a new vision of space as a global commons and enshrined it in the 1967 Outer Space Treaty, which created legal penalties for altering the cosmic environment.[14]

As the last electrons of the Starfish Prime test finally left space, the 1967 solar storm and nuclear close call encouraged the Air Force to dramatically expand its efforts to monitor the Sun. Given the now-obvious threat posed by solar storms to spacecraft and astronauts, officials funded by NASA and the newly established Environmental Science Services Administration—the precursor of today's National Oceanic and Atmospheric Administration (NOAA)—also kept the Sun under constant surveillance. Meanwhile, the US Navy created a solar research program to study the effect of solar radiation on radio communications. Satellites launched in the program helped reveal how solar flares disrupt the ionosphere by releasing X-rays.[15]

No amount of improved monitoring, though, could prevent a solar storm from again disrupting US military operations. In early August 1972, brilliant flares accompanied the departure of two modest CMEs from the solar surface. Bubbles of magnetic plasma can occasionally break free from the Sun in quick succession, and when they do, the first CME hollows out a pathway through the solar wind that others can follow. With minimal resistance, second and third CMEs can smash into Earth's magnetosphere at extremely high speeds, giving them tremendous kinetic energy. The magnetosphere buckles and retreats in the face of such a superstorm, and when it does, vast numbers of charged solar particles surge through Earth's atmosphere and across the planet's surface. In 1972 that is exactly what happened when a third CME crashed into Earth's magnetosphere at record speed.[16]

Several months earlier, the People's Army of Vietnam had undertaken the Nguyen Hue Offensive, a massive invasion of South Vietnam. To strangle the flow of supplies to the offensive, in May the US Navy began to deploy some eleven thousand magnetic-influence mines (mines that explode when a magnetic field, like that of a ship, comes near them) around North Vietnamese ports. As brilliant auroras rippled toward the equator on August 4, American pilots off Vietnam noticed a wave of explosions across the sea surface, all within a matter of seconds, and sailors aboard nearby destroyers felt the shock of the detonations. Later assessments determined that the severe magnetic storm created by the third CME had simultaneously detonated some four thousand US mines. The weakened US blockade still managed to hinder the offensive, but the US Navy scrambled to replace its mine designs with versions less vulnerable to geomagnetic storms. Meanwhile, the Air Force introduced a more capable space weather warning system.[17]

Transformers and Transformed Risks

The 1972 magnetic storm was treated as a curiosity by the US press, but it provided some of the first hints of a new vulnerability to space weather, one that struck at the heart of modern society. Studies undertaken by power companies had already suggested that geomagnetic storms could interfere with power systems, especially at high latitudes and in areas where igneous bedrock provided high electrical resistance. Now, as GICs swept across North America, power companies in the northern United States and in Canada reported numerous disruptions. None seriously interfered with power generation or transmission, and most engineers in the power industry agreed that geomagnetic storms were incapable of causing blackouts or seriously damaging equipment.[18]

In early March 1989, however, no fewer than eleven powerful solar flares heralded the arrival of back-to-back CMEs. The result was the most intense geomagnetic storm of the twentieth century. Just as the second CME collided with Earth's magnetosphere, a substorm—a sudden injection of energy from the tail of the magnetosphere to the ionosphere—created a sudden change in the magnetic field over much of Canada, and produced intense GICs in the igneous rock that covers Quebec. The sprawling Hydro-Quebec power system was one of the largest hydroelectric complexes in the world, supplying power to millions of customers across Canada and the northeastern United States. On the evening of March 12, power system operators noticed steadily increasing magnetic disturbances, but managed to retain control of voltage over the network. Then just before 3:00 a.m., currents suddenly surged through the system's La Grande network. Before operators could respond, the currents tripped seven of the static compensators that regulated the grid's voltage. The frequency and voltage of the network suddenly changed, tripping its transmission lines. In a matter of moments, the La Grande network and the 9,400 megawatts of power it produced were entirely isolated from the Hydro-Quebec network. The rest of the system could not compensate for the loss. Most of its transmission lines tripped within seconds, and the entire network collapsed. Some six million people were now without power.[19]

Because the collapse had crippled critical equipment that had to be repaired or replaced at generating stations and substations, restoring most of the system took nine hours. Due in part to chilly weather in a climate colder than it is today, when parts of the system were restored, they repeatedly overloaded because of the demand for heating, causing further damage. Most importantly, overvoltage in the initial seconds of the outage at one of the La Grande stations had damaged two transformers: hulking devices that are essential to electrical grids because they transfer electricity between circuits. Nor were the effects of the storm confined to Quebec. Some two hundred disruptions or alerts cascaded across the North American power grid, damaging or shutting down equipment that included another large transformer at Salem Nuclear Power Plant in New Jersey.[20]

The 1989 superstorm shocked engineers in utility companies, standard-setting organizations, and regulatory agencies across North America. Engineers in utility companies had relied on past estimates of geomagnetic storm strength that provided measurements of gradual changes in Earth's magnetic field. They had failed to appreciate the greater importance of minute-by-minute fluctuations in the field, which can generate the most powerful GICs. It was the speed of these fluctuations during the 1989 storm that overwhelmed the Hydro-Quebec power system.[21]

Engineers had also failed to understand that the electric power network had grown more vulnerable to geomagnetic storms since the minor disruptions that accompanied the 1972 CME. In response to a spectacular blackout in 1965 that originated in another Canadian hydroelectric generating station, near Niagara Falls, new institutions, technologies, and monitoring systems had together created a vast high-voltage transmission system across much of eastern North America. Power generation near the region's largest cities had not kept up with rising demand, so increasingly long transmission lines carried more and more electrical energy across immense distances between power stations and consumers. Partly because they crossed more of Earth's fluctuating magnetic field, long transmission lines were unusually sensitive to GICs. Worse, the new system could not easily provide the occasional boost to the *reactive power* it used to push electricity through power lines, a potentially fatal vulnerability during a geomagnetic storm. Owing in part to their latitude, proximity to igneous rock, and the type of transformer they used, Canadian hydroelectric power stations that had become essential to the grid were uniquely vulnerable to geomagnetic storms.[22]

As early as the mid-nineteenth century, solar storms had threatened communications networks, then transportation networks, and then military systems that controlled nuclear arsenals. Now, at last, they seemed to imperil the foundations of modern civilization: the networks that channel energy from where it is produced to where it is consumed. The sudden realization of the risk spurred a wave of activity. In the United States, agencies like the Defense Nuclear Agency and the Department of Energy sponsored conferences and studies to reconstruct how exactly the 1989 superstorm had disrupted power systems across North America. Industry organizations and utility companies soon established systems to monitor GICs and measure their impact on power grids. These systems informed the development of new procedures that power system operators were instructed to follow during geomagnetic storms. In the United States, space forecasting centers in the Air Force and NOAA were enlarged, reorganized, and enhanced by the launch of spacecraft that monitored solar activity in real time.[23]

Solar Storms, Satellites, and the Evaluation of Existential Risk

The expansion of electrical infrastructure on Earth paralleled the growth of satellite networks in near-Earth space. By the turn of the century, hundreds of satellites transmitted television, telephone, and internet signals; surveilled governments and infrastructure; monitored environmental changes, including weather; and, through the recently completed Global Positioning System (GPS), provided essential navigation data for a growing number of militaries and industries. The technological networks that sustained modern civilization increasingly extended into the heliosphere itself, creating unprecedented vulnerabilities to space weather.

In solar storms, electrons can cause destructive electrostatic discharges on or within spacecraft and degrade microelectronics by charging insulation. Protons and ions, meanwhile, can reduce the efficiency of microelectronic materials by undermining their crystalline structure, or break them directly in single event effects caused by ionization and nuclear interactions. Earth's atmosphere also expands during solar storms. This expansion creates drag that can alter satellite orbits, interfere with satellite tracking, and even create torque that destroys attitude control systems, ending satellite missions prematurely. Most satellites have defenses against these threats. Still, differences in their design, function, and orbit render some far more vulnerable than others.[24]

By the 1991 Gulf War, no military relied more on satellite networks than the US military. These networks produced a battlefield advantage but also a vulnerability that adversaries did not share. In the 1990s, US military officials worried about telecommunications satellite failures that coincided with geomagnetic storms. In 2001 a commission that had been chaired by incoming defense secretary Donald Rumsfeld warned that in the event of a crisis, the "relative dependence of the U.S. on space" could tempt an enemy to launch a "Space Pearl Harbor." Several months later Rumsfeld would oversee the beginnings of the US military's response to a radically different surprise attack, by militants who could not credibly threaten US satellites. Still, the subsequent "war on terror" increased the US military's dependence on satellite networks. The recommendations of the Rumsfeld Commission led to the unified US Space Command, and eventually to the creation of the US Space Force.[25]

When the House Committee on Science convened in October 2003 to discuss the utility of what were now called space weather forecasts, it was the Sun itself that revealed the magnitude of the risk posed by solar storms. As happened in 1989, eleven intense solar flares, including the most powerful ever observed, accompanied gigantic CMEs that barreled from the Sun toward Earth at progressively faster speeds. Anomalies affected at least forty-seven satellites, destroying a $640 million climate-monitoring satellite operated by the National Space Development Agency of Japan. NASA even instructed astronauts aboard the International Space Station (ISS) to seek shelter in the shielded *Zvezda* Service Module.[26]

The 2003 solar storm led national academies and militaries to commission reports and convene conferences to quantify the risks posed by space weather. Such work took on greater urgency in July 2012, when two enormous CMEs broke free from the Sun, collided with one another, and narrowly missed Earth. Had they smashed into the Earth's magnetosphere, the resulting storm might have rivaled the Carrington Event.[27]

Owing in part to the efforts of John Kappenman, an American power industry analyst, risk assessments came to focus on the largest storms the

The aurora australis, photographed in 2014 by the Expedition 40 crew aboard the International Space Station.

Sun could produce, rather than smaller fluctuations in geomagnetic activity that caused routine damage to satellites, power grids, and such. Kappenman had assessed the risk posed by geomagnetic storms since 1977. As a junior engineer at Minnesota Power, he had analyzed how geomagnetic storms could interfere with a new network of transmission lines that carried electricity from hydroelectric plants in Manitoba, Canada. The company had assigned its least experienced engineer to investigate the issue because at that time geomagnetic storms were widely regarded as a nuisance rather than a serious risk. Kappenman, however, was fascinated. And after the 1989 superstorm he was alarmed. He came to view severe geomagnetic storms as perhaps the worst natural disasters that could befall the United States. "We've expended an enormous amount of geopolitical capital to preserve the security of oil supplies to the country," he explained to one alarmed interviewer, but electricity provided more than twice as much energy to Americans as oil. "We've been very fortunate that our electric supply has not been threatened," he concluded, adding that "the Sun could change that equation in an instant."[28]

But how much could the Sun's behavior change in ways that would affect Earth? In 2012, mere months before Earth narrowly missed another Carrington Event, studies led by two Japanese scientists suggested that the answer could be: a frightening amount. In May 2012 the first study revealed that the luminosity of apparently Sunlike stars could abruptly surge during superflares, which detonate with the power of a trillion hydrogen bombs—far more powerful than anything that had been observed on the Sun. In June 2012 the second study identified a surprising increase of the isotope carbon-14 in tree rings dated 774 to 775 CE. In solar storms, energetic particles from the Sun create carbon-14 when they strike Earth's atmosphere, but the increase in the eighth-century tree rings was so extreme that the cause must have been far more powerful than the Carrington Event. Other scholars have now identified a series of such carbon-14 surges, named Miyake Events after the Japanese scientist who discovered them, in the tree ring record of the past ten thousand years. Their cause is still debated, but they suggest that our Sun is capable of unleashing superflares that have no precedent in modern history. Recently a survey of nearly sixty thousand Sunlike stars determined that such stars unleash superflares roughly once every century.[29]

While solar physicists and astronomers attempted to determine how powerful solar storms could get, engineers debated how vulnerable modern civilization had become. Kappenman argued that transformers were the critical weak points in modern infrastructure. Electrical grids could not function without them, but their size and complexity made them exceptionally difficult to replace. The damage caused by a geomagnetic storm therefore could be disproportionate to the severity of the storm. The damage inflicted by the 1989 storm could be repaired within hours. But a storm comparable to the Carrington Event could overload and thereby destroy hundreds of transformers. A disaster of that magnitude would lead to the collapse of electrical grids in any country. Across much of the United States, transformers and electricity would not be restored for years. This would lead to a total breakdown of overlapping national systems that provided food, running water, medication, and services for emergencies, transportation, communication, and banking. Kappenman and other analysts stressed that a resulting global economic meltdown, caused in part by the disruption of supply chains, could slow or even thwart any recovery by preventing the construction of new transformers.[30]

Such doomsday forecasts depended on the assumption that most transformers were equally vulnerable to geomagnetic storms. To verify that theory, engineers at utility companies and regulatory authorities disassembled transformers damaged by GICs, passed intense currents through transformers, and modeled how transformers would respond to high internal heat. Their studies found that some transformers were indeed sensitive to GICs, but most would likely survive even a Carrington Event.[31]

Today it is difficult to pin down the nature of the risk posed by geomagnetic storms. Over the past century, a series of geomagnetic storms have spurred new regulatory measures, electrical grid infrastructure updates, emergency protocols, and the development of space weather monitoring systems that have collectively reduced the risk posed by at least moderate CMEs. It is now possible to monitor the departure of a CME from the Sun in near-real time. Precautionary responses, such as adjusting or temporarily shutting down power systems, might mitigate potential catastrophic impacts on Earth. It is difficult, though, to predict the exact trajectory or severity of a CME. This uncertainty could result in extreme preventative measures not being implemented because those measures themselves would disrupt societies. Staggering increases in the scale and complexity of technological systems have also created new vulnerabilities to such storms. If a single storm really is capable of causing a Miyake Event, then any week the Sun could end modern civilization as we know it.[32]

Recently, however, satellite measurements have revealed a surprising source of resilience: an invisible transformation of the space environment near Earth, unwittingly carried out by the world's largest militaries. During the First World War, the US Navy began to use very low frequency (VLF) radio transmissions to communicate with sailors around the world. Because these transmissions could penetrate seawater, the British Navy soon adopted them to contact submariners.[33]

In 1958 Nicholas Christofilos proposed that even lower-frequency radio waves could better penetrate water and reach submarines when they were far from the surface and so much harder to find. Because they seemed capable of surviving a first strike, submarines carrying nuclear missiles were emerging as crucial components of Cold War nuclear arsenals. To increase their ability to escape detection, Cold War navies planned to use extremely low frequency (ELF) transmissions to communicate with them. Low frequencies travel on long wavelengths, so they require long antennae for effective transmission. To send ELF radio waves, the US Navy proposed to build an antenna nearly ten thousand kilometers long, which would be buried to survive a nuclear attack. But due to popular concern over the environmental consequences, and the development of Soviet missiles capable of destroying underground infrastructure, Navy officials instead opted to build two more modest, if still enormous, transmitting facilities.[34]

ELF signals not only reached underwater, but also radiated thousands of kilometers into space. Scientists now know that they interacted with particles in the magnetic Van Allen belts that surround Earth. Those interactions seem to have gradually pushed the radiation in the belts farther from Earth, providing greater protection to satellites and astronauts in low Earth orbit. Ironically, the physicist whose ideas had inspired an effort to replicate the effects of solar storms, and to end the threat of nuclear annihilation, had also informed the creation of a system that defends against those storms but could also someday transmit the order to unleash nuclear war.[35]

CONCLUSION

Harnessing Our Star

Borrowing a concept from the British science fiction author Olaf Stapledon, in 1960 the physicist Freeman Dyson proposed that a technological civilization would grow until it confronted planetary constraints in the supply of energy and matter. Dyson speculated that a truly advanced civilization could respond by building around its parent star an artificial biosphere: a swarm of megastructures capable of harvesting stellar energy with radical efficiency, and providing far more living space than any planet, or collection of planets, ever could. The construction of such a Dyson Sphere could produce otherwise inexplicable infrared radiation. It could also dim the light of the star—and, in time, possibly hide it altogether from the rest of the universe.[1]

In late 2015 a team led by Tabetha Boyajian, an astrophysicist at Louisiana State University, announced that citizen-scientists, combing through telescope data, had found mysterious fluctuations in the brightness of an ordinary star nearly fifteen hundred light years from Earth. The cause? Boyajian and her colleagues suggested a swarm of comets. Others proposed that a planet or moon had been destroyed. Still others speculated that an alien civilization more advanced than our own had started to envelop Tabby's Star with a Dyson Sphere. That explanation now seems improbable. But a survey of five million stars within a thousand light years of Earth has since revealed anomalous infrared radiation leaking from seven stars. Could it be that Dyson Spheres surround these stars, providing a vision of our distant future? Will we, or perhaps our artificially intelligent progeny, someday enfold the Sun fully within our civilization—and separate it from everything else?[2]

Maybe. Until then we remain on our little world, at the mercy of a variable star. Changes in the Sun's influence on Earth have influenced weather and inspired ideas about climate change that long predate global warming. Explosions on the Sun's surface have threatened modern technology and motivated efforts to mitigate existential risk. Yet technological development—spurred in part by the demands of world-threatening war—has also helped protect our planet. For now, our distance from the solar surface provides our greatest defense against the Sun's occasional extremes. As we will see, the same cannot be said for Venus, the planet long regarded as Earth's twin.

PART II

Venus

A drawing of Venus in ultraviolet light.

"Every great matter has, among its circumstances, some that are ludicrous or pitifully banal, which does not mean that they do not play an integral role."

—Stanislaw Lem, *His Master's Voice* (1968)

"Truth emerges sooner from error than from confusion."

—Francis Bacon, *Novum Organum* (1610)

In just about every course I teach about climate change, and after nearly every public lecture I give, someone inevitably asks me: Are we doomed? It seems like a simple question, but of course that word—*doomed*—means different things to different people. Some wonder whether global warming will worsen and, if so, whether it will cause widespread devastation. There the answer is, sadly, yes.

Imagine the weight of everything built by humans, added to the weight of every organism on Earth. The total, some two trillion tons, roughly equals the weight of the greenhouse gases added to the atmosphere by human action since the nineteenth century. All of this pollution has increased the atmospheric concentration of carbon dioxide by about a third: roughly as much as it decreased during Pleistocene glacial periods (Chapter 1). As of this writing, emissions of carbon dioxide and other greenhouse gases continue to increase. Computer simulations forecast that current trends in greenhouse gas emissions will likely warm the Earth by an average of nearly 3 degrees Celsius by the end of this century, relative to where global temperatures stood in the late nineteenth century, just after the Little Ice Age. Climate change on that scale would profoundly reshape our planet. The disintegration of polar ice sheets would accelerate; sea levels would soar; oceanic circulation could break down; tropical cyclones would intensify more rapidly; wildfires would burn longer and with greater ferocity; heat waves, droughts, and torrential downpours would be commonplace. For thousands, perhaps millions of years, Earth could be profoundly less habitable than it is today.[1]

Many wonder whether climate change will literally drive our species to extinction. That seems unlikely, though much depends on how quickly governments and corporations can reduce greenhouse gas emissions, and how thoroughly societies can adapt to a transformed planet. Global carbon dioxide emissions may be plateauing, but emissions of other greenhouse gases are soaring, the rate of warming could be increasing, the weather extremes enabled by warming are remarkably severe, and the consequences of warming for many local environments already appear irreversible. Because we have no precedent for our modern technological society or for the extreme heating it has caused, the past cannot definitively tell us whether our civilization will cope with global warming. So time cannot give us clear answers. Can space?

In 1950, the story goes, four prominent nuclear physicists sat down for lunch at Los Alamos National Laboratory. Enrico Fermi and Edward Teller were leading the effort to design the hydrogen bomb. Emil Konopinski had, with Teller, proved that such a weapon would not destroy all of Earth. Herbert York, the fourth man present, would later become an outspoken advocate for arms control. When the conversation turned to a recent wave of UFO sightings, Fermi asked something like, “Where is everybody?” UFO reports did not seem credible, but it was logical to suppose that intelligent life had evolved on many worlds. Alien technology should be visible throughout the cosmos—but it did not seem to be.[2]

Not long after that lunchtime conversation a young astronomer, Frank Drake, calculated that new military radar systems could enable communication over interstellar distances. In 1958 Drake accepted a job at the National Radar Observatory at Green Bank, West Virginia, where in 1960 he would use the facility’s largest radio dish to listen for signals from two nearby stars. He heard nothing, but the project spurred widespread interest in extraterrestrial life. Eventually Drake scribbled a famous equation on the blackboard at Green Bank:

$$N = R^{*} fp\, ne\, fl\, fi\, fc\, L$$

where the number of technological civilizations capable of contacting Earth (N) equals the annual rate of star formation (R^{*}) multiplied by the fraction of those stars with planets (fp), the average number of habitable planets per star system (ne), the fraction of habitable planets actually inhabited by life (fl), the fraction of inhabited planets with intelligent life (fi), the fraction of civilizations that develop technology capable of sending signals across interstellar distances (fc), and the time span in which that technology can be detected (L).[3]

The Drake equation provided a clear heuristic, perhaps even a roadmap, for answering Fermi’s question. The more of its variables that scientists could replace with numbers, the closer they would come to determining whether we are alone in our galaxy. In subsequent decades scholars refined and critiqued the formula, but it still underpins today's increasingly sophisticated search for extraterrestrial intelligence (SETI).[4]

In 1961 astronomers knew only the value of the first variable—the rate of star formation. In 1992 radio astronomers discovered a planet orbiting another star, and since then scientists have found so many exoplanets that they now believe the value of the second variable—the fraction of stars with planets—to be nearly all. The next variables to come into focus will likely be the third and fourth: the average number of habitable planets in a star system and the fraction of those planets that are actually inhabited. Astronomers can begin to find these values by exploring our own solar system. The discovery of other worlds that have microbial life—there are many candidates—would start to fill in the value of the fourth variable. The value of the third variable, however, can perhaps best be found by studying one planet, the one that approaches nearest our own: Venus.[5]

In some ways Venus is Earth's twin. Its radius is almost the same as Earth's; its mass and surface gravity are just a little lower. It seems to have a similar internal structure, and volcanoes have engulfed its vast plains with basalt, the same material that constitutes much of Earth's oceanic crust. The chemical composition of its atmosphere may suggest that it was always too hot for life. Yet atmospheric chemistry could also indicate that Venus was once a water world, like Earth.[6]

If it was, then it is Venus's proximity to the Sun that explains why it differs from Earth. Although the Sun seems to be an unusually stable star, it has slowly grown brighter in the 4.6 billion years of its existence. Earth is far enough from the Sun that its temperature could fluctuate and even decline in spite of this gradual increase in solar output, owing primarily to changes in the concentration of carbon dioxide in Earth's atmosphere. For Venus, there was no escape. The planet's oceans began to evaporate as the Sun's heat increased. Water vapor is a potent greenhouse gas, and a positive feedback of frightening power established itself. The more the oceans boiled, the hotter Venus became. As the oceans dried up entirely, ultraviolet radiation broke the bonds between the hydrogen and oxygen atoms in the suspended water vapor. The lightweight

Two artist's depictions of the surface of Venus, painted for NASA just before *Pioneer Venus* departed for the planet.

hydrogen atoms careened into space, and before long Venus was bone-dry. Its atmosphere was still thick, and it trapped so much solar radiation that the planet's average surface temperature soared to about 462 degrees Celsius: not only hotter than an oven, but hot enough to melt an oven.[7]

The deep history of Venus hints that Earthlike worlds are fragile, and that the value of the third variable in the Drake equation, *ne,* may be low. But could the history of Venus also shed light on the final variable, *L*? What if life emerges quickly and evolves readily on worlds awash with solar energy, as Venus is and was? Could a technological civilization have established itself on Venus, launched an industrial revolution, and, billions of years ago, found itself in the same predicament we face now? It seems vanishingly unlikely, but that is partly because, at present, we simply cannot quantify all the variables in the Drake equation.

These questions are not entirely new. Beginning in the early seventeenth century, scientists and eventually governments saw in Venus a means of illuminating exactly where humanity stood in the universe. First, the movements of Venus seemed to provide a tool for measuring distance in the cosmos, though the variable appearance of the planet foiled efforts to use that tool with precision. Second, changes on Venus appeared to suggest that inhabited worlds were common but diverse, owing perhaps to the different ages of planets that all passed through a similar evolutionary process.

Then, from the nineteenth century on, scholars, environmental activists, entrepreneurs, and ultimately policymakers reimagined Venus as an example, a warning, or even a cause of environmental changes so vast that they transformed entire plants. Some even speculated that Venus had changed profoundly in antiquity, and, by passing close to Earth, had altered our planet as well, nearly driving humanity to extinction. This catastrophism flew in the face of the uniformitarian consensus among contemporary geologists, but it appeared to be supported by the discovery that Venus's atmosphere was extremely hot and toxic. In the countercultural movement that emerged with the dawn of the Space Age, Venus-centered catastrophism provided an alternative to the theories of mainstream science. On university campuses, it offered a way to confront the hegemony of science in US culture—and, indirectly, the unjust political system that science seemed to uphold.[8]

Scientists contended, accurately, that Venus and Earth had not dramatically changed each other in humanity's distant past. Yet scientists who discovered the real cause for the scorching heat of Venus—its extreme greenhouse effect—also imagined how changes to the planet's environment could impact humanity. They proposed seeding the Venusian clouds with genetically engineered life that could kick-start a runaway process of climatic cooling. This reversal of Venus's ancient warming would be managed from cities that would be created to float in the Venusian atmosphere. Once Venus was cold enough, a second Earth would be available for settlement.

By the 1970s, though, close study of the Venusian atmosphere suggested both that it could not be easily cooled and that it had once been much cooler. The discovery of ancient climate change and ongoing ozone depletion on Venus alarmed scientists and led them to investigate whether similar processes were afoot on Earth. Studying Venus helped reveal the danger posed by anthropogenic climate and ozone degradation, and then contributed to political efforts to find a solution. In the 1980s, Venus became a symbol for the worst possible outcome of environmental crises on Earth, even as it made some of those crises far less likely to occur.

5

Measuring the Universe

In August 1768 a crew under the command of Captain James Cook left the Thames aboard HMS *Endeavour,* a coal-hauler retrofit for scientific exploration. After a long, stormy, and deadly journey, Cook and his crew anchored off Tahiti on April 13, 1769. Although he had impressed upon his crew that the resident Tahitians were to be treated with "all imaginable humanity," Cook ordered the immediate construction of a fortress, within range of the *Endeavour*'s cannons. The threat of violence would leave Cook's men free to pursue their purpose in Tahiti: to observe Venus as it passed in front of the Sun. It appeared to do that at a slightly different time in Tahiti than at distant observation sites in Norwich, England, for example, or at the Cape of Good Hope. The difference in timing would enable astronomers to precisely triangulate the distance between Earth and the Sun, and thereby gain a sense of the scale of the universe.[1]

Although some of his sailors had already died for that dream, Cook soon found that the British Admiralty had used the Venus project to disguise its real ambitions. When the expedition's astronomer, Charles Green, finished his observations, Cook opened a letter from the Admiralty. Only then did he learn that his mission was to search for Terra Australis, the southern continent that scholars believed must exist to balance the otherwise strangely unequal distribution of land in the Northern and Southern Hemispheres. In an imperial age, the Admiralty's interest was obvious.[2]

Cook and his crew followed their new orders by sailing west until they encountered what is today New Zealand. After a bloody skirmish with a group of Māori, they mapped the islands of New Zealand and claimed them for Britain. It was clear, though, that they had not found a landmass big enough to be the fabled southern continent. They sailed west again and became the first Europeans to chart the eastern coast of a truly continent-sized island (today's Australia). Setting a course for home, they ran aground in the Great Barrier Reef, then limped to Batavia (present-day Jakarta) with a broken ship. After enduring a deadly bout of dysentery, they struggled back to London with a skeleton crew, three years after setting forth. Charles Green did not survive.[3]

Cook's expedition was only one among many. In the 1760s the era's superpowers sponsored rival voyages to measure the passage, or "transit," of Venus across the face of the Sun. Because it promised to illuminate the nature of humanity's place within the cosmos, the effort appealed to the learned members of scientific institutions that now viewed scholarly discoveries as a reflection of imperial power. As with the Space Race to the Moon three centuries later, scientific goals were easily subverted by imperial ambitions. And like twentieth-century astronauts, eighteenth-century astronomers risked their lives to bring about a "giant leap for mankind." Every time success seemed within grasp, however, the changing environments of Venus and Earth pulled it just out of reach. The interplay of two dynamic atmospheres ensured that every superpower lost the space race of the eighteenth century.

Venus, Key to the Cosmos

It was Venus's unchanging appearance that most vexed sixteenth- and early seventeenth-century astronomers. The planet had phases, like the Moon, whether it spun in an epicycle located between Earth and the Sun (as it did in Ptolemaic cosmology), or around the Sun, which then orbited Earth (as in the Tychonic alternative), or around the Sun, like Earth (as in the Copernican system). Phases should have altered the planet's brightness, yet its light remained constant. Kepler hypothesized that Venus glowed with its own light, but that explanation did not satisfy because it undermined the Copernican reimagining of all planets as being similar worlds orbiting the Sun.[4]

Galileo understood that a close-up of Venus could help solve the mystery. Better yet, it would reveal whether Venus could ever be exactly halfway illuminated. Such a dichotomous phase would be observable only when Venus was to one side of the Sun as viewed from Earth, which of course would be impossible in Ptolemaic cosmology. Armed with his telescope, Galileo saw that phase in December 1610. His observations did not convince everyone, for the telescope was newfangled technology. "Who are you," one young Florentine objected, "who presumes to descry, / Such distant things . . . Glancing from your bench with a feeble eye?" Yet as the telescope spread throughout Europe, Jesuit astronomers confirmed Galileo's observations, and for a while both Tychonic and Copernican systems gained legitimacy.[5]

Even before Galileo glimpsed the phases of Venus, Kepler realized that the accurate predictions enabled by his version of the Copernican model, in which planets orbited in ellipses rather than circles, could allow him to calculate when Mercury and Venus would transit the Sun. A successful prediction would reveal the superiority of his revised Copernican model, and a transit itself would provide strong evidence that the planets orbited the Sun rather than Earth. Because Mercury is so tiny and distant, its transits were hard to follow with available telescopes. Venus was an easier mark, but its orbit is inclined by more than 3 degrees relative to the ecliptic, the imaginary plane around the Sun created by Earth's orbit. This tilt means that, from our perspective, Venus only transits the Sun in one decade every century—but it does so twice in that decade.[6]

It was nearly thirty years after Galileo sighted the Venusian phases that a young astronomer named Jeremiah Horrocks refined Kepler's tables of planetary motion and used them to determine the exact date of the next transit of Venus. By extraordinary luck, he found that the transit would occur just one month in his future. On a cloudy afternoon on November 24, 1638, he waited with his friend, the astronomer and merchant William Crabtree, in the little town of Much Hoole. Horrocks and Crabtree had prepared a helioscope, an instrument that projected the Sun's image onto a screen so it could be safely observed. Suddenly, Horrocks later wrote, "the clouds, as if by divine interposition, were entirely dispersed," and the Sun appeared on the screen with a black dot on its surface. Horrocks and Crabtree had become the first observers to see what they knew was a planet moving in front of the Sun.[7]

While he was preparing a manuscript on the transit of Venus, Horrocks died of unknown causes, likely illness, at just twenty-two years of age. Before his death, however, he had recognized the significance of Kepler's greatest insight: that the period of a planet's orbit is related to its distance from the Sun. This harmonic law strengthened the popular assumption that God had created an orderly, rational universe, and it also provided a method to create a scale map of the entire solar system. All that was missing was the scale. If the distance between Earth and the Sun could be worked out, the dimensions of the entire solar system would snap into focus, and, with them, accurate values for the mass and mutual attraction of all its worlds. By observing the transit of Venus, Horrocks was able to estimate Venus's diameter and the ratio of its orbit to Earth's. Now he was able to guess the distance of the Sun from Earth: just over ninety-five million kilometers, a figure about one-third lower than the real value.[8]

To do more than guess required a method that could yield a precise measurement. In 1663 a young astronomer in Scotland, James Gregory, proposed that Venus would appear to transit at slightly different times, along a slightly different path, if observed from two widely separated points on Earth. By carefully timing the transit in far-flung locations, astronomers could measure the angular difference between the paths. They could then use trigonometry to calculate the distance to the transiting planet, and then the harmonic law to determine the distance between the Sun and Earth. After working out the distance between Earth and the Sun, the same basic method could be used to calculate differences in the position of the stars at different points in Earth's orbit, which in turn would give the distance to those stars. The first rungs of the cosmic distance ladder—the multistep method of calculating the scale of the entire universe—would be cleared, and the true size of the cosmos would finally come into focus.[9]

Like Horrocks, Gregory died at a young age, suffering what may have been a stroke while showing his students how to use a telescope. His ideas were subsequently forgotten until 1716, when Edmond Halley, Britain's second Astronomer Royal, suggested a suspiciously similar method for determining the distance between the Sun and Earth. By now Isaac Newton had proposed his theory of gravity, explaining how the universe stayed in motion, and scholars had turned their attention to determining how big the moving universe really was. Gregory's ideas, when repackaged by Halley, were suddenly in vogue. Observers dispatched to distant "Halleyan Poles," Halley wrote, would need only a pendulum clock and simple telescope to precisely time when Venus encountered the western edge of the Sun, moved entirely in front of it, touched the opposite limb of the Sun, and finally left the Sun behind. The next transit would occur in 1761, after Halley's death. He urged "the curious investigators of the stars to whom, when our lives are over, these observations are entrusted" to follow his method, so that the "immensities of the celestial spheres, compelled to more precise boundaries, may at last yield to their glory and eternal fame."[10]

The Space Race of the Eighteenth Century

Measuring the upcoming transit became one of the great scientific preoccupations of the eighteenth century. Across the burgeoning colonial world, it was an era of quantification. In Europe the roots of that era stretched back at least into the high medieval centuries, when strengthening state bureaucracies sought ways to measure, and so better manage, their resources. The movement toward quantification gained pace with, among other things, the growth of scholarly networks fostered by the printing press; the expansion of capitalistic economic structures in Italy and then Northwestern Europe; refinements in navigation that permitted waterborne networks of extraction and commerce; and an emphasis on observation and experimentation that fostered the Scientific Revolution. By the 1760s many astronomers believed that the increasingly precise measurement of cosmic phenomena would gradually reveal the nature of the universe, and there was no opportunity for measurement quite like the transit of Venus.[11]

European astronomers had planned to coordinate their transit measurements, but as 1761 approached, the Seven Years' War broke out between Britain and France. Now, rival teams sponsored by hostile monarchies voyaged to the remotest corners of their empires in a competition to track the transit of Venus with greatest accuracy. The teams that measured the movement of Venus most precisely would claim glory for their respective empires by unveiling the cosmos.[12]

Yet it was hard to reach the Halleyan Poles. Travel by land was dangerous and painfully slow, for states were still too weak to enforce the law across their whole territories, and most lacked the ability to construct a dense and durable road network. Travel by sea, in the rat-ridden hulls of cramped and unhygienic ships, posed its own risks—it was to be expected that a share of the passengers and crew would drown or succumb to disease during any long voyage at sea, and it was always possible that all aboard would die in a pirate attack or storm. Indeed, Earth's turbulent atmosphere at first seemed like the gravest obstacle to the success of the transit enterprise. Weather prevented some expeditions from reaching their destinations, then kept others from seeing the Sun on the day of the transit.[13]

A bigger disappointment was still to come. After the survivors of each successful expedition returned home with their measurements, it grew obvious that their efforts would not yield a precise figure for the distance from Earth to the Sun. Two effects were to blame: first, the appearance of a "luminous ring" around Venus as the planet approached the Sun, and second, the distortion of otherwise circular Venus into a "black drop" as the planet began and ended its march across the solar disk. Venus's dense atmosphere accounted for the first effect; smearing caused by optical imperfections and Earth's atmospheric turbulence was to blame for the second. Both unexpected changes in Venus's appearance at critical moments made it nearly impossible for observers to agree on when exactly the planet moved through the phases of its transit.[14]

Nevertheless, the expeditions narrowed the likely range of the distance between Earth and Sun, confirming the basic validity of Halley's method. When the next transit approached in 1769, astronomers prepared to establish seventy-seven observing stations across the furthest reaches of Europe's empires. Some died while attempting to reach them; one took so long to return home that he was declared dead. And again their sacrifices had been in vain. The variable atmospheres of Earth and Venus thwarted every effort by Enlightenment astronomers to acquire the discrete data points they sought. Nevertheless, the transits inspired decades of analysis until, in 1824, Johann Franz Encke carefully evaluated every measurement to calculate, with a much narrower range of uncertainty, a new value for the distance from Earth to the Sun. This figure—153,240,000 kilometers, give or take 660,000 kilometers—seemed authoritative until other methods for calculating the distance, using the movement of other celestial bodies, clustered around a slightly higher figure. The issue was again thrown into question when the next pair of transits arrived, in 1874 and 1882.[15]

Preparations for the first transit of the nineteenth century unfolded on an unprecedented scale, in imperial capitals from St. Petersburg to Mexico City. The many scientists involved were well versed in the optical distortions caused by the unstable atmospheres of Earth and Venus, and were ready to deploy new technologies to overcome them. For example, the French astronomer Pierre Jules César Janssen, who once used a balloon to both observe an eclipse and escape a German army besieging Paris, introduced a device resembling a handgun that took second-by-second snapshots of the Sun. In time, this *revolver photographique* would be developed into the cinematographic camera. Armed with such contraptions, astronomers, photographers, cooks, carpenters, mechanics, and even hunters fanned out across the colonial world in expeditions that soon constructed a raft of fortified transit stations. A wave of popular interest accompanied both transits, fanned in part by sensational coverage in newspapers that increasingly targeted the general public.[16]

Widespread fascination with the transit expeditions reflected the extent to which science had come to shape the culture of all social classes in colonial nations, especially the growing bourgeoisie. Scientists had forged a new, specialized, and highly educated professional class. Their laboratories and observatories resembled factories, enabling a systematic approach to scientific inquiry. Widely communicated through mass literature, public lectures, and museum exhibitions, interpretations of scientific findings and theories informed public health reforms, labor laws, environmental conservation efforts, and eugenics movements.[17] The transit expeditions of the nineteenth century contributed to growing public interest in the sciences. The dynamic atmospheres of Earth and Venus, however, kept the expeditions from identifying a precise figure for the distance from Earth to the Sun. With no transit scheduled until 2004, it would be the nature of Venus, rather than its movement, that would most interest scientists at the turn of the twentieth century, especially when that nature seemed to change.

6

Venus as a Changing Earth

One painfully humid morning in August 2020, I hauled a telescope almost as large as myself to the darkest corner of a dewy field in Washington, DC. After I doused myself with bug spray, I noticed Venus, blazing like an oncoming aircraft above the predawn horizon. Its brilliance surprised me, though I knew the cause: its proximity to us and its exceptionally reflective sulfuric clouds. When I turned my telescope to the planet, its wavering light shone through my eyepiece like a beacon in a dense, roiling fog. Then, as my eye adjusted to the glare, I noticed a spark just to its left, a mote that moved with the planet and looked for all the world like a companion, a satellite. For a moment, I thought that our Earth and our Moon might look similar when viewed from Venus.

My little spark was an illusion, caused perhaps by the morning atmosphere or the imperfections of my telescope. Venus has no moon. Yet for centuries, observations of the planet with telescopes inferior to my own led many astronomers to glimpse the spark I saw, continually shifting its position relative to Earth. For most of the eighteenth century, astronomers convinced themselves that the planet really did have a moon. Apparent fluctuations in the appearance of Venus itself, meanwhile, led many to believe that Venus was a world like Earth, an idea that supported and was supported by the theory of the plurality of worlds. Then at the turn of the twentieth century, apparent changes in environments on Earth and Venus inspired a broad reimagining of the planet as an alien place, perhaps inhospitable to terrestrial life.

Scholars have long assumed that, before the Space Age, Venus's thick and apparently unchanging clouds allowed scientists and popular authors to imagine the planet's surface however they wished. In fact, it was the shifting appearance of the planet that encouraged people to see on Venus the environmental transformations many deemed most important on Earth.[1]

Discovering Earth's Twin

The notion that Venus had a moon stemmed from Kepler's belief that satellites orbiting other planets revealed those planets to be worlds similar to Earth, which in turn implied that Earth could not be the unique center of creation. Moon sightings—and eventually observations of change on Venus itself—were, in part, illusions that both strengthened and were encouraged by astronomers' support for a Sun-centered model of the solar system. Later they reflected widespread belief in the plurality of worlds, the idea that every planet in the solar system resembles Earth in harboring life (Chapter 2). So it was that efforts by astronomers to telescopically observe Venus were from the start warped by the application of confirmation bias to analogical reasoning. In 1645 an optician in Naples named Francesco Fontana announced that he had identified not one but two satellites orbiting Venus. By the end of the eighteenth century, no fewer than twelve astronomers had reported that they had seen at least one moon near Venus.[2]

Still, most observers understood that Venus is ferociously difficult to observe. Because it is never far from the horizon after the Sun has set, it shines through more of Earth's obscuring atmosphere than all the planets save Mercury. When our atmosphere is even a little unstable, Venus is a quivering blur through any Earth-bound telescope, and even in the best of times the planet's extraordinary brilliance magnifies every optical imperfection. Small wonder, then, that most astronomers needed firmer proof than the occasional sighting to accept that a moon really did revolve around Venus.[3]

The eighteenth-century transits of Venus promised to provide that proof, because the brilliant Sun would clearly frame any Venusian moon. When those transits offered no evidence for such a moon, astronomers struggled to explain earlier sightings. Were they caused by optical illusions, or undiscovered planets between Venus and the Sun, or even wayward wisps of solar atmosphere? The answer threatened to call into question the methods, instruments, and especially the observers responsible for advancing and professionalizing astronomy, striking at the foundation of the entire discipline. Well into the nineteenth century, some astronomers kept searching for a moon. Finally, in 1887, Paul Henri Stroobant, a Belgian astronomer, proposed that earlier observers had simply misidentified background stars around Venus. Here was an explanation that did not make a complete mockery of the profession, and most professional astronomers coalesced around it. Amateurs, however, continued to speculate about a moon until the first spacecraft reached Venus in 1962.[4]

In 1645 Fontana also claimed to see a dark feature in the center of Venus. At that time the appearance of clouds, land, and sea on Earth, especially from mountain prospects, led some natural philosophers to believe that the Moon's dark plains were seas and that its bright highlands were continents (Chapter 9). Many would apply the same analogical reasoning to the planets. In 1726, for example, Francesco Bianchini, a clergyman and astronomer, grew convinced that he could discern markings on Venus that resembled those on the Moon. Under the pellucid sky of twilit Rome, he wrestled with a refractor sixty-six feet long to draft the first comprehensive map of Venus, complete with oceans, straits, and promontories he named after kings, explorers, and colonizers. In his view, he was pioneering a new science, "celidography," from the Greek word for spots. To promote his vision of Venus, he crafted a globe of the Earthlike planet he believed he could see through his eyepiece. It was the first globe to depict a world other than Earth.[5]

Most importantly, the spots of Venus seemed to be changing. Not only did they appear more visible on some nights than others—they seemed especially clear when viewed through a red aurora—but their shape also fluctuated. In the *Encyclopédie,* the great compendium of French Enlightenment thought, the mathematician Jean le Rond d'Alembert wrote that these changes revealed Venus to be a dynamic world, like Earth. He believed that the cause of change on Venus was clear: an atmosphere thick enough to carry clouds.[6]

Not all astronomers agreed, but in 1761 the transits of Venus once again provided a means of verification. In St. Petersburg the Russian scholar Mikhail

Lomonosov concluded that the luminous ring that appeared around Venus as it approached the Sun could only be caused by light refracting through an atmosphere "equal to, if not greater than, that which envelops our earthly sphere." The idea received little attention until an astronomer in the Holy Roman Empire, Johann Hieronymus Schröter, observed Venus using one of William Herschel's telescopes. Schröter noticed that light blurred between the illuminated and unilluminated sides of Venus, and that the "horns" of Venus in its crescent phase could wrap around the planet's unilluminated hemisphere. Only a thick atmosphere could cause those effects, Schröter concluded. Herschel agreed—not least because of the "faint, changeable spots" he had seen on Venus.[7]

In the context of widespread belief in the plurality of worlds, the existence of a cloud-bearing atmosphere seemed to confirm that Venus was an inhabited planet like ours. Further resemblances to Earth were apparently revealed when Fontana sighted irregularities in the Venusian terminator, the line between the illuminated and unilluminated halves of the planet. These irregularities, he argued, could only be caused by mountains. Others, including Bianchini, saw them too, yet Schröter's telescope showed them with unique clarity. By studying Venus during the day, when it rose high above the obscuring atmosphere near Earth's horizon, Schröter found moreover that the cusps of crescent Venus changed shape, and that specks of light were visible in the darkness beyond the terminator. He deduced that mountains must be catching or blocking the light of the Sun, just as they do on Earth. Yet he reasoned that the mountains on Venus must be much taller than those of Earth to be so clearly visible.[8]

Then in 1813 the German astronomer Franz von Paula Gruithuisen noticed unusually brilliant patches on crescent Venus. These, he believed, could only be a southern ice cap. Explorers had not yet found Antarctica, but here, it seemed, was evidence from Venus that a glaciated continent likely existed on Earth, too. Photographs of the transit of 1882 subsequently seemed to reveal a gigantic bulge on Venus. In the wake of the recent discovery of glacial periods in the deep history of Earth (Chapter 1), the bulge appeared to be a range of icy mountains. Yet how could so much ice exist near the Sun? Some scientists wondered whether changes in the shape of Venus were caused by the same forces responsible for its supposed dark patches. Agnes Clerke, an influential historian of astronomy, speculated, for example, that apparent peaks and valleys on Venus revealed the

A nineteenth-century drawing of an imaginary Venus landscape with towering peaks. Camille Flammarion speculated that "perhaps we would not feel so out of place if we arrived before a landscape on Venus."

"uprush in some places of vast masses of superheated air or gases," and in others "atmospheric compression under influence of cold currents." Nevertheless, most scientists agreed that soaring mountains existed on Venus.[9]

In reality, mountains are never visible above the planet's suffocating atmosphere, and the Venusian atmosphere is too hot and dry for glaciers. Yet reasoning by analogy did not always lead astronomers astray. In the hundred years from 1783 to 1883, some of the most powerful volcanic eruptions in the history of the Holocene profoundly disrupted Earth's atmosphere. Colonial expansion and the closely linked development of science meant that there were plenty of observers to track how the explosions altered the atmosphere. It may be that something similar was happening on Venus. Spacecraft have since observed prolonged brightening of Venusian clouds, owing to plumes of sulfuric acid lofted into the planet's upper atmosphere by explosive volcanic eruptions. The mountains and glaciers ostensibly seen by scientists may have been volcanic gases similar to those wreaking havoc on Earth.[10]

What about the dark regions so clearly identified by the likes of Bianchini and Schröter? For most observers, including me, such features are too subtle to make out. Yet when viewed in ultraviolet light, Venus has striking dark patches that change shape and occasionally persist for months, owing to a mysterious ultraviolet-absorbing substance in the clouds. The planetary scientist David Grinspoon has suggested that those who detected dark regions on Venus were among the minority of humans who can see into the ultraviolet part of the electromagnetic spectrum.[11]

Volcanic eruptions and shifting, ultraviolet-absorbing clouds may have together fooled eighteenth- and nineteenth-century scientists into believing that Venus closely resembled Earth. Herschel, for example, reasoned that changes in Venus's appearance revealed the existence of a thick atmosphere that would reflect most incoming sunlight, and thereby cool the surface to permit a temperate, appropriately European environment for life. Like Herschel, Schröter was a pluralist (someone who believed in the plurality of worlds), but he disagreed. The atmosphere of Venus had to be transparent, because "Providence" would ensure that the Venusians would share in the joy "of discovering, like a Herschel, still more and more distant regions of the universe."[12]

In 1862 the famed French astronomer Camille Flammarion summed up the pluralist vision of Venus. He wrote that the increasing "absurdity" of the idea that the Sun "should be employed exclusively to illuminate and heat" one little planet—Earth—became "even more striking" for astronomers and philosophers "when Venus was found to be a planet of the same dimensions as Earth, with mountains and plains, seasons and years, and days and nights analogous to our own." Because the two planets were "alike in their physical characters," Flammarion concluded, "they must be alike also in their role in the universe: if Venus were without population, then the Earth must be similarly lacking."[13]

But if the characteristics of Venus, including its apparent changes, seemed to establish that the planet was inhabited, what were the Venusians like? Due to the more "vigorous and active influence" of the Sun on Venus, Fontenelle reasoned, the planet's inhabitants would resemble the Moors of Grenada, who were ostensibly "Witty, full of Fire, very Amorous," and "much inclin'd to Musick and Poetry." By the eighteenth century, however, many agreed that the supposed inferiority of colonized peoples stemmed from hot climates that discouraged the development of civilization. It was an assumption that seemed to legitimize European imperialism, and it had implications for the inhabitants of Venus. In 1755 Immanuel Kant decided that to cope with the planet's great heat, the inhabitants of Venus would be of "grosser build and sluggishness" than Earthlings.[14]

Pluralist visions of a mountain-covered but Earthlike Venus had, by now, encouraged science fiction that imagined visits to other worlds. In 1758 the Swedish natural philosopher and mystic Emanuel Swedenborg, for example, combined genres by publishing an account of his own travels to Venus, which he claimed had been permitted by Jesus. Dialogues with the dead and reflections on the plurality of worlds allowed eighteenth-century writers to craft entertaining critiques of society. These narratives often drew on the colonial trope of idealizing distant lands, such as Tahiti, as places where people had figured out how to live perfectly, away from European corruption and complexity. Despite its popularity, Swedenborg's narrative was roundly criticized because Swedenborg claimed that the experiences he described were real. He wrote that he had seen Venusians who were indeed hulking brutes, but that others led a life of virtue and simplicity that he thought Earthlings would be wise to emulate.[15]

Of course, quantity mattered as much as quality. Swedenborg speculated that there were about as many Venusians as Earthlings. The English cleric Thomas Dick, by contrast, extrapolated that if the planet were as densely populated as England, its population might approach fifty-four billion. Both Swedenborg's mysticism and Dick's calculations were controversial at best. The apparent discovery of high mountains on Venus led some to question whether the planet had the flat, fertile soils that they assumed were necessary for life to thrive. In 1869 the inventor and poet Charles Cros hypothesized that the bright patches seen by astronomers like Schröter might have been attempts by Venusians to contact Earth. He suggested a variety of plans for sending a reply. "The eternal isolation of the spheres," he concluded, would soon be "vanished," and the inhabitants of the solar system could at last discuss their shared interests. It was one of the first proposals for sending a message to another world (Chapter 15).[16]

Reimagining an Alien World

By the late nineteenth century it seemed increasingly clear that Venusians would have to contend with more than exceedingly tall mountains. After Bianchini mapped Venus, Jacques Cassini, the son of the astronomer Giovanni, used the map and his father's observations to work out the length of a Venusian day. By happy coincidence, at 23 hours 21 minutes it closely resembled Earth's. Schröter gave a similar figure, and most astronomers accepted it until the pluralist assumption that every celestial world resembled Earth began to unravel in the late nineteenth century.[17]

The English scientist and Anglican priest William Whewell argued that pluralism depended on analogies that could not be proved through observations of other worlds, and that there was little theological foundation to the idea that God created only habitable worlds. Many scientists rejected Whewell's arguments, but others used them to explain why new astronomical instruments provided data suggesting that the planets and their moons were diverse, and their moons differed from Earth. In 1877 Giovanni Schiaparelli, a respected observer, concluded that markings on Venus did not move at all, suggesting that the planet was very different from Earth. Venus, he contended, must have a day as long as its year, which meant that the same hemisphere would always face Earth. That hemisphere, forever shaded from the Sun, would have to be bitterly cold, while the hemisphere that faced the Sun would be scorching hot, and a "permanent hurricane system" must swirl between them. Could life have emerged on such a bifurcated world?[18]

In 1877, the year he seemingly established the rotation period of Venus, Schiaparelli unwittingly laid the groundwork for an answer by charting a network of lines crisscrossing Mars (Chapter 14). As we will see, Percival Lowell, heir to one of the great US fortunes, soared to fame by mapping these "canals" and identifying their origin in an alien attempt to transform a planet. In 1896 he announced that he had discerned similar features on Venus. These lines were thicker and less straight, however, so Lowell judged them to be "perfectly natural" rather than the work of intelligent builders. The trouble was that, for the most part, only he and his staff could see them. When astronomers started to doubt that Lowell had seen canals anywhere, including on Mars, he suffered a nervous breakdown and temporarily retracted his observations.[19]

In 1915 a hydraulic engineer named Charles Edward Housden published a theory that seemed to explain the variable markings sighted on Venus by observers from Fontana to Lowell. Housden assumed that Venus was tidally locked to the Sun—meaning one of its hemispheres forever pointed sunward—and that its atmospheric composition resembled Earth's. Drawing on theories of atmospheric circulation in the fast-developing field of climatology, he speculated that atmospheric convection patterns that, on Earth, move heat and moisture latitudinally from the equator to the pole, would work longitudinally on Venus. Moisture thereby moved from the sunlit side of Venus to the dark hemisphere, dividing the planet between a scorching, bright desert and frigid, starlit icefield. Between these extremes, Housden theorized, a thin temperate strip would permit the emergence of life. The wavering appearance of this "collar" from Earth, Housden argued, explained the irregularities that astronomers had previously noticed along the Venusian terminator.[20]

Both Lowell and Housden had been influenced by Herbert Spencer, an English polymath who popularized the idea that the principles of Darwinian evolution also operated in society. Social Darwinism seemed to justify the depredations of capitalism and colonialism alike, for the "survival of the fittest" meant that society inevitably had winners and losers. On both Venus and Mars, Housden believed, the relentless struggle for survival in challenging environments had driven civilizations to build canal systems. The canals glimpsed by Lowell on Venus, Housden argued, funneled water from the planet's dark, glaciated half to strips of irrigated, temperate land that bordered the scorching desert of the sunlit hemisphere. Brilliant markings on Venus were reflective ice; shifting, dark markings were farmland visible through gaps in clouds. Housden claimed that his experience as a hydraulic engineer helped him see things that astronomers had missed. A recent surge in canal building on Earth had revealed that conveyance of water on a massive scale was "not only possible but financially profitable"; aliens, of course, would not let such an opportunity slip by.[21]

Even though academic disciplines were becoming professionalized, this had not yet completely closed them to outside interventions. Housden's claims seemed to have integrated the intellectual currents of his time so skillfully that astronomers responded with polite criticism rather than outright dismissal. Yet by 1915 celestial canals had largely lost their luster, owing in part to the debut of telescopes that resolved straight lines into discontinuous and clearly natural features. Moreover, scientific understandings of Venus had changed. In the nebular hypothesis first elaborated by Pierre-Simon Laplace, the planets formed one after another from rings of gas swirling around the nascent Sun. Many astronomers therefore believed that Mars was at a more advanced stage of development than Earth—a conviction that, as we will see, encouraged widespread belief in its canals—and that Venus was in a younger, thus hotter, period in its history. Proof seemed to lie in its comparative cloudiness. Ancient Mars had cooled and dried, so clouds were rare in its atmosphere; Venus was wet and hot, so its clouds were excessively abundant. Venus was an "an orb doubly swathed in a steaming ocean," as Clerke put it, "and in the steaming vapors exhaled by it." Earth, of course, occupied the happy middle.[22]

Despite widespread acceptance of Schiaparelli's rotation period, most scientists, and in turn most science fiction authors, began to imagine the planet as one giant Carboniferous swamp, with atmospheric currents that equalized temperatures between the equator and poles. A hot, stable planet seemed to preclude the emergence of animal life, because Darwinian evolution depended on environmental change. When interpreted through the lens of new evolutionary theories, the atmosphere of Venus complicated the old vision of a plurality of worlds populated by humanlike beings. Still, science fiction authors, such as the Americans Edgar Rice Burroughs, H. P. Lovecraft, Clark Ashton Smith, and Lester del Rey, seized on the idea that Venus was an early Earth to thrill readers with the ultimate example of a colonial fantasy: a jungle world full of exotic cultures, beautiful women, and prehistoric animals.[23]

When, in the 1930s, rigorous spectroscopic studies of Venus revealed abundant carbon dioxide in its atmosphere, some scientists thought the planet might be too wet even for a swamp. Land, it seemed to them, would have pulled carbon dioxide out of the atmosphere through erosion, so the surface of Venus must be one world-straddling ocean. Others reached the opposite conclusion. By the 1930s the relentless expansion of mechanized, market-driven farming across vast stretches of the American Southwest had disturbed the sod—the upper layer of earth, fortified with roots—that once protected the region from dust storms. Scorching temperatures and drought enabled winds to carry away soil and ruin farmland on such a scale that some 2.5 million people migrated out of the Southwest. Some astronomers now figured that windblown dust accounted for both changes in Venus's appearance and the yellow tint of the planet's atmosphere. Venus in this view was one big Dust Bowl, and it seemed unlikely that life could survive there.[24]

Impressions of Venus had come a long way in nearly three hundred years. In the seventeenth and eighteenth centuries, astronomers in expanding empires saw there the exotic environments and peoples of a new world, a planet like Earth. Eventually they believed they could detect signs of the volcanic eruptions that cooled Europe in the late eighteenth and early nineteenth centuries. Industrialization encouraged new attempts to discern on Venus a vision of humanity's technological future, while the advent of evolutionary theory spurred efforts to reinterpret the planet as a glimpse of our distant past. Then, during the Dust Bowl, one of the great self-inflicted ecological disasters in human history, Venus seemed increasingly uninhabitable, just like the southern Great Plains.

In the 1950s the British astrophysicist Fred Hoyle proposed a model of Venus that truly fit the era. The development of the internal combustion engine, the growing exploitation of oil fields in the Middle East, and the economic boom that followed the Second World War all ensured that by the 1950s, oil shaped the fortunes of empires and fueled modern life. Carbon dioxide in the atmosphere of Venus, Hoyle suggested, revealed a planetwide ocean not of water, but of petroleum, "beyond the dreams of the richest Texas oil-king." This "Hoyle oil" would soon serve as evidence for one of the most bizarre, and influential, scientific heresies of the twentieth century: one that proposed the most spectacular planetary changes imaginable.[25]

7

Worlds in Collision

From the seventeenth century on, there was one mysterious change on Venus that, in hindsight, clearly revealed the planet's true nature. In 1643 Italian priest Giovanni Riccioli noticed that the hemisphere of Venus not illuminated by the Sun seemed to glow ever so faintly. Was it the inner light hypothesized by Kepler? Other observers sporadically noticed the glow. Gruithuisen, writing just after the Battle of Waterloo in 1815 ended the Napoleonic Wars, suggested a cause: "general festivals of fire," owing perhaps to the ascension of some Napoleon to a Venusian throne. Such festivals would be "easily arranged" on Venus, Gruithuisen speculated, because the warm climate of Venus must nourish "tree growth . . . far more luxuriant than in the virgin forest of Brazil." Later he proposed that the Venusians had instead illuminated their planet by burning enormous tracts of jungle to make way for farmland, just as settlers had done in South and Central America.[1]

Given the consensus that life existed on other worlds, these ideas were not as absurd as they seem now—nor were they the most fanciful that Gruithuisen would propose (Chapter 9). Still, they hardly satisfied his colleagues. Some suggested instead that Earth could be illuminating Venus, but most correctly suspected that Earth's light would be too dim at that distance. Others proposed the glittering phosphorescence of planet-straddling oceans, or the dull gleaming of a vast ice sheet. Some blamed the glow on widespread lightning or spectacular auroras. Many concluded that was merely an illusion.[2]

But in 1983 astronomers in Australia discovered that the otherwise dark hemisphere of Venus shimmered with a patchwork of infrared light. The thick cloud layers of Venus, astronomers realized, could be translucent for both visible and infrared light, permitting glimpses of the surface that were swamped by sunlight in the illuminated hemisphere of the planet. Here, at long last, was an explanation that made sense. Gaps in the clouds exposed the baking basalt of the Venusian surface.[3]

The shifting appearance of Venus had given observers a hint of the planet's true identity and, as we will see, a warning for Earth's future. Yet no astronomer had imagined that the planet could be hot enough to glow. Only a popular author widely derided as a crackpot had predicted it. By guessing right, he would influence a widespread rejection in Europe and the United States of a new form of a science, one that was so allied with Cold War governments and corporations that it seemed to imperil the needs, aspirations, and values of ordinary people.

Venusian Heresies in the Early Cold War

In 1939 Immanuel Velikovsky arrived in New York City. He had been born in Vitebsk, in what was then the Russian Empire, and then traveled widely. In 1921 he had joined a foundation his father had established in Berlin to translate scientific publications into Hebrew. The work helped modernize Hebrew as a scholarly language, and it introduced Velikovsky to Elisheva Kramer, a violinist who worked as the foundation's secretary. The two wed, and in 1924 they joined thousands of Zionists migrating to Palestine. There, Velikovsky practiced medicine and grew increasingly fascinated with the subconscious. After striking up correspondence with Sigmund Freud, he found Freud's book *Moses and Monotheism* in a Tel Aviv bookstore next to Hitler's *Mein Kampf.* He bought Freud's book and realized that Freud had suggested that Moses was an Egyptian follower of the god Akhenaten, rather than a Hebrew prophet. Velikovsky was dismayed. To him, the Jewish claim to Palestine, and in turn the Zionist movement, depended partly on God's gift of the land to Moses. Now Velikovsky traveled to New York, where he planned to use archived collections of Egyptian papyri to substantiate the biblical account of the Exodus. Soon, though, he stumbled across what seemed to him an even more important idea. Ancient texts, he believed, documented real catastrophes with natural causes. As the Second World War erupted in Europe, he resolved to write what he called "a book of wars in the celestial sphere," chronicling long-ago events that he believed had nearly destroyed humanity.[4]

Velikovsky knew that he was following in the footsteps of seventeenth-century "catastrophists" who had scoured Earth in search of evidence for Noah's Flood (Chapter 1). His approach to the textual record of antiquity was also far from unique. Even twentieth-century astronomers had consulted mythology to argue, for example, that stars changed color over time. Yet Velikovsky resolved to take a more scientific approach. To him, a single story in the mythology of one people could not establish that an event had really occurred. Even a similar story told by many peoples might only register the transfer of an idea from one culture to another. If diverse accounts of an event could be found in different myths from the same era, however, they likely represented common efforts to grapple with something real.[5]

Velikovsky decided that evidence from the archives of nature could confirm his reading of the archives of societies. He knew that eighteenth-century scientists had uncovered fossils of cold-adapted animals at temperate latitudes, and he consulted nineteenth-century evidence of staggering advances and retreats of glaciers across ancient Europe and North America. He learned that scientists had mapped the extent of these changes, and that some had proposed a connection to cycles in Earth's orbit and rotation. Yet he understood that most scientists did not accept this explanation, partly because the cycles could not by themselves account for the onset of glacial and interglacial periods.[6]

Velikovsky realized that although there was a scientific consensus that the world had repeatedly undergone tremendous upheavals, there was no universally accepted explanation for why they had occurred. The records of ancient societies, he believed, provided telling clues. Obelisks, sundials, and other artifacts appeared to reveal that the length of the day and year, and even the direction and axial tilt of Earth's rotation, had changed since antiquity. Most revealing of all, ancient accounts of the heavens seemed to omit Venus.[7]

Velikovsky theorized that thousands of years ago, an enormous electromagnetic discharge had sent a portion of Jupiter careening into the inner solar system. It coalesced into a planet-sized comet—Venus—which in time nearly collided with Earth. Meteorites from the encounter rained petroleum upon Earth (the origin of today's oil); thereafter Earth's rotation stalled as the two planets attracted one another. Earth's oceans, caught by the gravity of Venus, towered miles into the sky before an electric discharge between planets sent them crashing down. At last Venus wheeled away, but, six weeks later, it returned to exact more punishment. Now Earth's rotation spun in a new direction and the planet followed an orbit slightly farther from the Sun. Earth's poles shifted, and ice sheets expanded.[8]

Velikovsky contended that fifty years later, Venus had yet another go at Earth. This time it tilted Earth's axis but preserved its new rotation. Naturally, survivors across the ruined Earth kept fearful watch, and recorded the devastation. Then, some 750 years later, in the eighth century BCE, Venus, having grazed the Sun, encountered Mars. Unlike Earth, little Mars was no match for Venus; a near-collision devastated whatever life might have existed on the planet, creating scars that Lowell and others had mistaken for canals. Mars, its atmosphere warped by the encounter into all manner of exotic shapes and sizes, saved Earth by stabilizing Venus's orbit. Yet Venus's gravity sent Mars barreling toward Earth for a series of near misses that brought still more misery to ancient peoples and dazzled them with celestial pyrotechnics. Eventually Mars settled into its present orbit, and after 687 BCE the solar system quieted down for several millennia.[9]

Velikovsky believed he had discovered that cosmic collisions had transformed Earth on human timescales. This would mean that Newton's view of the solar system as stable, predictable, and harmonious should be replaced with older, catastrophist theories of sudden change. It also meant that the world's religions had a common origin in cosmic upheavals. Velikovsky argued moreover that his theories permitted a series of bold predictions that, if proved correct, would strengthen his "entire chain of deductions." If he was right, Jupiter would have a powerful magnetic field, Mars would be lifeless, Earth's climate would have changed abruptly, and comet or asteroid impacts would have occurred frequently. All of these ideas seemed implausible in the 1940s. His most important prediction seemed especially unlikely: recently formed Venus would be much hotter than scientists believed possible, and hydrocarbons on its surface would create petroleum fires that seeded a thick, carbon dioxide atmosphere.[10]

When Velikovsky completed his book, *Worlds in Collision,* he sought out leading scholars to test these predictions. In 1946 he approached the American astronomer Harlow Shapley, director of the Harvard College Observatory, to request spectroscopic studies that he thought could reveal, for example, whether Venus was exceptionally hot. Shapley said he was too busy to read Velikovsky's book, but promised to do so if another colleague would vouch for it. One of those colleagues turned out to be fellow American Horace Kallen, dean of graduate faculty at the New School for Social Research in New York City. Kallen read the book and judged that Velikovsky had "built up a serious theory" that could revolutionize "not only astronomy but history and a good many of the anthropological and social sciences." He urged Shapley to perform the studies. Shapley responded that although he had not read Velikovsky's book, it was "pretty obviously based on incompetent data," for "if Dr. Velikovsky is right, the rest of us are crazy."[11]

Eight publishers had rejected *Worlds in Collision,* apparently because of its long footnotes, before the editor James Putnam at the textbook publisher Macmillan agreed to take it on. Gordon Atwater, the curator of the Hayden Planetarium in New York, provided a favorable review in the popular magazine *This Week.* While he doubted the planetary wanderings and near-collisions that Velikovsky described, he took seriously Velikovsky's interpretations of ancient myths and decided to create a show at the planetarium to promote the book's conclusions. When Putnam informed *Harper's* about the upcoming book, the magazine's editor, Eric Larrabee, published two sensational articles promoting its findings.[12]

In response, Shapley went to war. He informed the editorial department at Macmillan that when he met Velikovsky, he had been so "astonished" at his "crackpot" claims that he "looked around to see if he had a keeper with him." Because the book was "intellectually fraudulent," he warned that its publication would provoke outrage from scientists who taught with and contributed to Macmillan textbooks. Given Shapley's lofty stature, his letters were referred to George Brett, the president of Macmillan, who agreed to have *Worlds in Collision* reviewed by new readers. In response, Shapley mentioned a letter he had received from Velikovsky in 1947 that he had ostensibly filed with "the writings of the Flat Earth Society." In that letter Velikovsky confided to Shapley, "What I am afraid of is not to be disputed but to be dismissed without being read." Yet now Shapley solicited and distributed withering reviews of *Worlds in Collision* by colleagues who had, indeed, not yet read the still-unpublished book.[13]

A majority of Macmillan's reviewers urged the press to publish *Worlds in Collision*. When it made its scheduled debut, Shapley convinced his colleagues to refuse to meet with Macmillan representatives, and to write critical letters and reviews instead. One review, by influential biologist John Haldane, even saw in Velikovsky's book a plot by the US military to prepare public opinion for nuclear war. After all, if the ancients had survived a near-collision with Venus, then the modern world could cope with a few thousand hydrogen bombs. Velikovsky complained that the editors of scientific journals prevented him from replying, forcing him to direct his responses to less reputable newspapers.[14]

Shapley claimed to regret that his colleagues "took the matter so seriously." Yet privately he complained that Velikovsky's theories symbolized an "obvious decadence" in US society that would soon lead to a "black future." This was a long-standing fixation of Shapley's. In 1935 he had confided to the American geographer Ellsworth Huntington, the president of the American Eugenics Society, that "conditions in the slums" should not be alleviated because "the individuals who have gravitated" to them had "demonstrated the quality of their stock." At universities, he wrote, "eugenical education" should be carried out because most students constituted "the classes you would propose to encourage" having children. Shapley's zeal for advancing white "civilization" in the United States may have given special animus to his campaign against Velikovsky, a Russian Jew.[15]

In the end Shapley's efforts compelled both Atwater and Putnam to resign. Eight weeks after publishing *Worlds in Collision*, Macmillan convinced Velikovsky to transfer the book to Doubleday, a competitor with no textbook division. Macmillan vice president Harold Lantham—who later regretted his company's capitulation and Putnam's dismissal—informed Shapley of the cancellation. According to *Los Angeles Times* book editor Robert Kirsch, Shapley had organized a response, "which for intensity and hostility was unequaled in twentieth century scientific history."[16]

Ironically, the attempt to cancel its publication created a furor that made *Worlds in Collision* a bestseller in the United States. Meanwhile the reputation of science suffered. In a review of Velikovsky's book—which she had not read—the American astronomer Cecilia Payne-Gaposchkin insisted that, for the scientist, "the discovery of discordant facts is cause for rejoicing, not consternation." The problem with Velikovsky was that he had produced no "real evidence." This was a common refrain among scientists, who argued that the scientific method gave them an objective, self-correcting tool that justified their ascendant position in Cold War society. The political scientist Alfred de Grazia pointed out, though, that in order to appraise evidence, the scholar was obliged to honestly read, review, and in many cases test a colleague's claims. The response to Velikovsky's theories suggested that scientists were not as dispassionate as they claimed to be. They had "acted more like priests of a dogmatic religion," one journalist concluded, "than champions of free inquiry."[17]

Why? Because "a work like this," Payne-Gaposchkin argued, "can scarcely be dignified by the name of science." Many scientists objected to Velikovsky's catastrophism; others considered his prioritization of evidence from human, rather than natural sources, "preposterous." Some scientists suspected that Velikovsky's theories represented little more than a crude attempt to prove the historical veracity of the Old Testament, and the profession had long since turned against such efforts. The American sociologist Norman Storer concluded that McCarthyism, in its sweeping attack on "all intellectual enterprise," had made scientists fearful of another ideological attack on their profession. Once Shapley organized a reaction, Storer argued, "the forces of group loyalty took precedence over the disinterested quest for truth—if such a quest has ever actually existed."[18]

By the 1960s many sociologists and humanists perceived in the scientific reaction to Velikovsky an attempt to suppress ideas that could upend the primacy of the physical science establishment in the United States. After all, if Velikovsky was right, then the evidence and methods of the natural sciences did not provide the only, or even necessarily the best, guide to the history of Earth and the nature of the universe. The chemist and author Isaac Asimov complained that humanists "knew just enough science to cast Velikovsky in the role of a suffering Galileo." As the philosopher David Stove put it, they were only too happy that scientists, "so many having made knaves of themselves . . . made fools of themselves, too."[19]

Some humanists did gain genuine insights into the epistemology of science through the Velikovsky controversy. Criticisms of the scientific response anticipated and informed some of the arguments made by the American philosopher

Thomas Kuhn, who proposed that science always depended on arbitrary assumptions that smoothed over nagging differences between theory and data. Theories, Kuhn realized, underpinned not just professional identities but also personal worldviews, as the Velikovsky affair seemed to show. When theories changed—in what Kuhn called "paradigm shifts"—they gave way to new assumptions that applied fundamentally different, and not always superior, meanings to accepted reality.[20]

Venus and the Counterculture

If science was not quite the dispassionate search for truth that Payne-Gaposchkin made it out to be, there was no denying that the scientific method was a uniquely powerful tool for Cold War governments and corporations. For all its faults when used by scientists, its emphasis on empirical evidence, reproducibility, and falsifiability created a means for increasing knowledge and, when applied to technological development, power. A distinguishing feature of twentieth-century science was therefore its integration within corporations and governments. Even scientists at universities were subservient to the research and development priorities of companies and states that provided funding. By the 1960s, scholars and activists decried the emergence of what the American sociologist Lewis Mumford called the "megamachine": a social system, created by the union of bureaucratic, scientific, and technological power, that prioritized growth at the expense of human values and the environment. Across much of the developed world, young people in particular responded by broadly rejecting mainstream culture, in diverse ways that included everything from New Age spirituality to the Black Power movement.[21]

For many within the new counterculture, Velikovsky's ideas provided a model for an inclusive, human-centered science that did not uphold corporate or state power. His apocalyptic reimagining of ancient history suited the needs of the peace and environmental movements, which called attention to the existential risks posed by nuclear confrontation and pollution. For Velikovsky himself, it was a mixed blessing. He embraced his growing popularity among disaffected youth, giving lectures at some sixty college campuses between 1964 and 1969 alone, all the while launching student clubs devoted to his ideas. Yet he also craved acceptance by the very establishment science his devotees opposed. Like a scientist, he hoped to win converts on the merits of his ideas, rather than their utility for young activists.[22]

The strangest thing about *Worlds in Collision,* and a key reason it was relevant among scholars for decades after its publication, was that its predictions had an undeniable tendency to come true. Some of the world's most prominent scholars took notice. Albert Einstein, for example, lauded Velikovsky's "dynamic talent" but rejected the idea that the planets were charged bodies with magnetic fields that swept through the solar system. Velikovsky responded that their debate could be settled by listening for radio signals from Jupiter. In 1955, when these signals were indeed detected, Einstein proclaimed that he would arrange the experimental tests Velikovsky had long sought. Einstein died nine days later, though, *Worlds in Collision* open on his desk.[23]

The radio telescope, an instrument that receives radio waves from celestial sources and converts them into electrical signals, provided the most compelling evidence for Velikovsky's theories. In 1956, astronomers at the Naval Research Laboratory in Washington, DC, used one of the first large radio telescopes to discover that Venus emitted vast quantities of microwaves, the shortest radio waves in the electromagnetic spectrum. Because hot objects can release large amounts of microwave radiation, and microwaves pass easily through atmospheres, the microwave emissions from Venus seemed to indicate a blistering surface temperature near 300 degrees Celsius. With few exceptions, planetary scientists found that figure implausible. Some suggested that the microwaves came from elsewhere: from a Venusian ionosphere, for example, many kilometers above a temperate surface.[24]

In 1962 the first spacecraft successfully dispatched to another planet—*Mariner 2*—flew within thirty-five thousand kilometers of Venus. Many scientists interpreted its radiometer readings as showing the Venusian surface to be even hotter than terrestrial radio telescopes had indicated. Only Velikovsky had imagined that Venus could be that hot. Meanwhile, news reports and eventually NASA's popular account of the mission noted that *Mariner 2* had uncovered Venusian clouds composed of "condensed hydrocarbons held in oily suspension." Were these the hydrocarbons that Velikovsky had predicted? As Velikovsky's profile rose among some scholars, catastrophism also began to reenter mainstream scientific thought, owing in part to the discovery of Earth-approaching asteroids and the reinterpretation of craters on the Moon and Earth as scars left by cosmic impacts (Chapter 18). Prominent physicists and geologists now argued that it was time to seriously consider Velikovsky's ideas in scientific conferences and journals.[25]

While most scientists continued to reject Velikovsky's ideas, by the 1970s Velikovsky's books were recommended reading in some university courses—in 1973, twenty-seven in North America alone—and he enjoyed standing ovations on one campus after another.[26] Documentaries covered his rise, conferences at leading institutions debated his theories, and celebrities were outspoken advocates. In universities, many humanists and social scientists embraced Velikovsky, while in private industry even some natural scientists supported his challenge to the scientific establishment. The academic journal *Pensée,* founded by an undergraduate in 1964, rebranded itself as a forum for discussing Velikovsky's ideas. Its first issue on Velikovsky had a run of about seventy-five thousand copies. "I survived, as you see," Velikovsky crowed in 1972 to a crowd of a thousand or so Harvard students.[27]

The Decline and Fall of Immanuel Velikovsky

According to the winter 1973 editorial in *Pensée,* Velikovsky's rapturous reception at Harvard seemed like the beginning of "the last phase" of the battle to have his work taken seriously by scientists. Indeed, the battle would soon end, but not as Velikovsky and his supporters had hoped. Later that year an astronomer named Walter Orr Roberts urged the American Association for the Advancement of Science (AAAS) to host a forum dedicated to specific questions raised by Velikovsky's work, such as the age of Venus. Many scientists feared that a forum would give legitimacy to Velikovsky's ideas, but such events were now so popular that the AAAS agreed. A symposium was scheduled for February 25, 1974, to focus on Velikovsky's claims about Venus. Velikovsky agreed to participate, but it soon dawned on his inner circle that the astronomers organizing the symposium aimed to refute, rather than debate, all of his work. Of particular concern was a scheduled speaker who had long urged the AAAS to interrogate Velikovsky's ideas: the planetary scientist Carl Sagan.[28]

Sagan, like Velikovsky, was a dreamer. Jewish like Velikovsky but born in a later generation, he thrilled at the wondrous future presented at the New York World's Fair in 1939, the same year Velikovsky immigrated from Palestine. Insulated by his doting parents from the horrors of the Second World War, Sagan lost himself in the Martian fantasies of H. G. Wells and Edgar Rice Burroughs (Chapter 15). Spending long hours in libraries, he pored over astronomy books and marveled at the possibility of extraterrestrial life. In 1960 he earned a PhD in astronomy and astrophysics from the University of Chicago. He then climbed to the forefront of planetary science and exobiology, emerging fields that sought answers to age-old existential questions and depended on the combination of previously isolated scientific disciplines. A gifted communicator, Sagan, like Velikovsky, was eager to reach the widest possible audiences. To that end he was willing—too willing, some colleagues complained—to speculate about ancient alien visitations to Earth, for example, or the persistence of microbial life across the solar system. In some respects he was similar to Velikovsky, but there was one crucial difference: he was firmly within the scientific establishment.[29]

When the AAAS convened its forum in February, a raucous crowd of perhaps fifteen hundred onlookers poured into the Grand Ballroom of the Hotel St. Francis in San Francisco. Velikovsky began by defending his mingling of "the humanistic heritage with the message of stones and bones," and then recounted the litany of successful predictions that, he said, had vindicated his theories. Other speakers explained why scientists had reacted to Velikovsky as they had; considered whether the celestial mechanics proposed by Velikovsky seemed plausible; and argued that Velikovsky had misread ancient texts. Finally, Sagan stepped to the podium.[30]

Sagan had taken his assignment seriously—so much so that he had drafted and revised a fifty-seven-page speech. It was too long to read in full, so he skipped through it as best he could. For Venus to have separated from Jupiter, he pointed out, would have required utterly implausible energies, and the debris it released would still be raining down on Earth. The near collisions between planets imagined by Velikovsky would be equally unlikely given the immensity of the solar system; nor could Earth's rotation stop and then restart at around the same speed during such a disaster. Indeed, when it came to Venus, Velikovsky's predictions were only vague and exaggerated versions of proposals originally made by astronomer Rupert Wildt, whom Velikovsky had neglected to mention. Unlike Wildt, however, Velikovsky had confused hydrocarbons with carbohydrates; misidentified the compositions of Venus and Jupiter; and failed to understand that reports of hydrocarbon clouds after *Mariner 2* were based on nothing more than one scientist's off-the-cuff speculation. In short, Sagan concluded, "where Velikovsky is original he is very likely wrong; and . . . where he is right the idea has been pre-empted by earlier workers."[31]

While Sagan stressed that "neither the critics nor the proponents of Velikovsky" had "read him carefully," his critique lifted arguments, calculations, and even phrasing directly from earlier reviews. In that sense, he was guilty of some of the same intellectual crimes of which he accused Velikovsky. He committed other such crimes by mischaracterizing and exaggerating aspects of Velikovsky's arguments. Some reporters viewed both men as attention-seekers, and on the following day both dominated the headlines. Velikovsky and his supporters soon detected a conspiracy to discredit his ideas in critical reports that mirrored one another by emphasizing Velikovsky's Judaism, "Slavic" appearance, and Russian accent.[32]

Velikovsky had had the benefit of an aggressively sympathetic audience. But it was clear even to some of his allies that the symposium had undermined, rather than legitimized, his theories. The scientists on the panel and Velikovsky would later publish dueling books, and Velikovsky's theories briefly retained a substantial following. News outlets, though, no longer gave them serious attention, and sympathetic journals collapsed in angry recriminations. As the countercultural moment ebbed, Velikovsky's theories faded into obscurity, and Velikovsky, who increasingly struggled with depression, died in 1979. In his blockbuster documentary *Cosmos* the following year, Sagan ridiculed Velikovsky's theories but criticized how scientists had attempted to suppress them.[33]

Velikovsky had been simultaneously behind and ahead of his time. Behind, in that scientists had long since abandoned efforts to find natural explanations for biblical miracles and no longer accommodated radical interventions by an outsider. Ahead, in that many scholars would only later recognize the benefits of integrating methods and data from the humanities, social sciences, and natural sciences, and only later accept that abrupt environmental changes, influenced by celestial bodies, really did imperil our distant ancestors. Today an entire field, geomythology, identifies environmental disasters in myths and legends, as Velikovsky attempted to do.[34]

It is telling that young academics who came of age on college campuses influenced by Velikovsky went on to incorporate within legitimate scholarship the ideas that he helped popularize. Yet Velikovsky's greatest contribution to history may have been to channel the discontent of a generation of college students, and some of their professors, against a form of science that no longer seemed to meet their needs. It is strange that a theory about ancient environmental upheavals, set in motion by Venus, had such a powerful impact on the counterculture. Yet the respect it granted to the humanities and social sciences; its rejection by the scientific establishment; and its apparent discovery of global catastrophes all helped it appeal to a generation increasingly concerned with where the megamachine was taking Western society. Velikovsky was wrong about the origins of Venus, but he was right about the precarity of life on Earth. Exposing that precarity would soon become the life's work of his most influential detractor, Carl Sagan.

8

Remaking Venus and Earth

When spacecraft confirmed that Venus is the hottest planet in the solar system, Carl Sagan benefited nearly as much as Velikovsky. As an unusually charismatic doctoral student at the University of Chicago, Sagan had speculated that an efficient greenhouse effect could heat the surface of Venus to such an extent that it would emit the microwaves radio astronomers had just detected on the planet. His theories about the atmospheres of Venus and other planets led him to help design instruments for the *Mariner 2* probe. Before the probe could reach Venus, he theorized that the growing luminosity of the early Sun had started to evaporate a primordial ocean that once covered much of the planet. Water vapor, a potent greenhouse gas, had trapped ever more heat near the planet's surface, creating a runaway warming that transformed an Earthlike world into a barren hellscape.[1]

Many scientists were initially skeptical of these ideas. Sagan later speculated that some would not accept that Venus could be so different from Earth. Indeed, such theories flew in the face of centuries of analogical reasoning about the planet. Worse, the discovery that the Venusian surface was clearly inhospitable to life seemed to weaken the case for exploring the planet. So committed were scientists on both sides of the Iron Curtain to the idea that Venus resembled Earth that the first Soviet landers carried aluminum busts of Lenin that would have survived on a temperate world, but immediately melted on Venus.[2]

By the end of the 1960s, robotic spacecraft had established beyond doubt that the surface of Venus was exceptionally hot, and that the cause was a carbon dioxide atmosphere that was thick enough to create a supercharged greenhouse effect. A scientific consensus around the climate of Venus had finally emerged, and Sagan's prestige correspondingly rose. Meanwhile, climatologists had started to grasp the peril posed by human-caused warming on Earth. Venus seemed to prove that the greenhouse effect could warm an Earthlike world until its surface became utterly inhospitable to life. That made it a potent symbol for climate scientists and activists, who in the 1980s raised widespread alarm at the prospect of global warming but could not secure an international agreement to limit greenhouse gas emissions. The discovery of the planet's great heat also motivated unprecedented plans for planetary engineering that briefly seemed to promise hope for the perceived crisis of overpopulation. Meanwhile, studies of Venus's dynamic atmosphere contributed to the discovery of another existential risk—the depletion of the ozone layer—that inspired a successful international response. The discovery of the Venusian environment helped reveal the threat posed by human alterations to environments on Earth, and spurred efforts to roll back some of the most dangerous changes.

Building Planet B

If a strengthening greenhouse effect had unleashed runaway warming on ancient Venus, could it be weakened to jumpstart runaway cooling? For Sagan, a possibility presented itself in the ability of microbial life to multiply exponentially, and thereby to alter planetary environments in short order. It was a solution that came naturally to him. While a doctoral student, his fascination with the possibility of life on other worlds led him to Joshua Lederberg, a pioneering geneticist at the University of Wisconsin at Madison. Lederberg had coined the term *exobiology* for a new, radically interdisciplinary science that would leverage the dawning era of interplanetary travel to compare organisms on different worlds, and so establish universal laws governing the emergence and evolution of all life.[3]

Prominent biologists and chemists ridiculed exobiologists for not having found alien life, the very subject of their science. Exobiologists replied that because little was known about the origins of life on Earth, it was better to assume that microbial life could evolve wherever conditions seemed suitable. In the Space Age, to do otherwise was to risk "forward contaminating" alien biospheres with terrestrial microorganisms and, worse, "back contaminating" Earth's biosphere with potentially dangerous extraterrestrial organisms.[4]

When Lederberg warned of these risks, the National Academy of Sciences (NAS) asked the International Council for Science to take action. In 1958 the council established an ad hoc Committee on Contamination by Extraterrestrial Exploration. This committee issued recommendations that informed the creation of a permanent Committee on Space Research to, among other things, encourage common standards of sterilization and quarantine in the Soviet and US space programs. Meanwhile, the NAS created a Space Science Board, which offered biological advice to newly formed NASA through a Panel on Extraterrestrial Life. It was a bureaucratic flowering that, in remarkably short order, established what would later be called "planetary protection" as a responsibility of any national space program—and, in that way, secured funding for experiments that aimed to detect extraterrestrial life. Owing to his partnership with Lederberg, Sagan now advised the US government on the issue that had kindled his love for outer space.[5]

Sagan's theories about the evolution of the Venusian atmosphere, and his work on experiments that reproduced the emergence of organic compounds on the early Earth, led him to speculate that microorganisms had evolved on the surface of Venus and escaped to the clouds as the planet warmed. He argued that those clouds could now be the last refuge of an Earthlike Venus. They were larger and longer lasting than those on Earth, but they swirled around Venus at an altitude where pressures and temperatures resembled those of Earth at sea level. Indeed, the density and complexity of Venus's lower atmosphere was so great that the planet's clouds in some ways resembled the floating, temperate shore of a vast ocean. In Venus's clouds, gravity is about the same as it is on Earth, and chemicals that can be used by life abound there.[6]

Sagan believed that if evidence were uncovered for life in the clouds of Venus, humans should never voyage there. It was not worth risking the contamination of an alien biosphere. Yet if exploration by sterilized robots established that Venus was lifeless, then a remarkable possibility presented itself.

Microscopic blue-green algae could be released into the atmosphere of Venus, Sagan theorized. The algae would perform photosynthesis and disassociate carbon dioxide into its carbon and oxygen components. The imported microbes would be light enough to reproduce before sinking into the scorching lower atmosphere. There they would be incinerated, but not before depositing their carbon onto the planet's surface. Algae introduced in small quantities would multiply exponentially, and quickly reach such numbers that carbon dioxide would rain out of the atmosphere, reducing the efficiency of the Venusian greenhouse and cooling the surface. Now landers could introduce plants to the surface, accelerating the process started by airborne algae until continued cooling allowed rain to sequester carbon dioxide in the planet's crust. Eventually Venus would be the verdant, watery world that astronomers had once compared to Earth: a planet ripe for colonization.[7]

In 1942 the American science fiction author Jack Williamson coined the term *terraforming* to refer to the process of transforming a planet to resemble Earth.[8] Today the concept is most commonly associated with audacious schemes for the future of Mars (Chapter 16). But Venus is the only world in the solar system with gravity comparable to Earth's, so in fact it is the one planet that could truly be terraformed. Sagan had suggested that the process could be irrevocably set in motion with a single robotic probe.[9]

The idea that Venus could be terraformed to quickly create a second world for human settlement and exploitation made international headlines when first proposed. Yet it was the growing influence of the environmental movement that inspired lasting interest in the scheme. The development of astronaut life support systems, which carried a part of Earth into the hostile vacuum of outer space, and images returned by the Apollo moon-landing missions, which framed Earth against the seemingly empty background of outer space, had profoundly influenced the movement. Leading environmentalists reimagined our planet as a closed system, a fragile and finite "spaceship Earth." In 1968 the American biologists Anne and Paul Ehrlich published a wildly successful bestseller, *The Population Bomb*, warning that rapid population growth would soon overwhelm the inescapable limits of the Earth-spaceship. Influenced by the alarming theories of Thomas Malthus, who held that population inevitably grew faster than food production, the Ehrlichs predicted that hundreds of millions would starve

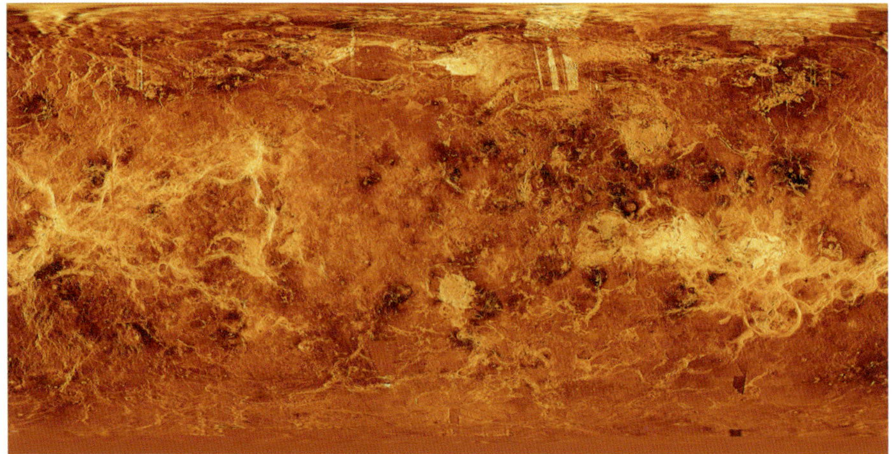

Terraforming Venus. Top: a radar map of Venus using data returned by the *Venera 13*, *Venera 14*, *Pioneer Venus*, and *Magellan* orbiters. Bottom: An imagined Earthlike Venus, with 70 percent of its surface covered by water.

in the 1970s and 1980s. Billions would follow, they insisted, without urgent efforts to control human reproduction.[10]

This bleak prospect naturally fueled interest in space settlement, even if the Ehrlichs stressed that colonizing new lands without managing reproduction would only delay a demographic collapse. Sagan's plan seemed to offer easy access to a second Earth, but it depended on the presumed existence of extensive water vapor in the Venusian atmosphere. A major feature in the *Daily Telegraph* announced, "It is no exaggeration to say that the whole future of human civilization" could depend on "remote controlled instruments that were sent in the past decade to Venus." If those instruments confirmed that there was enough water, the article concluded, then Sagan's calls to test for life before terraforming would fall on deaf ears. "Sooner or later somebody will use [Sagan's] plan without his permission."[11]

As waves of US and especially Soviet spacecraft departed for Venus, books and articles promoted and refined Sagan's plan. The physicist Freeman Dyson even proposed installing a network of motors on Venus that could accelerate its rotation and shorten its day, correcting the one difference between Earth and Venus that imported terrestrial life could not address. Sagan, together with the biophysicist Harold Morowitz, proposed genetically engineering algae to add tiny, hydrogen-filled sacks that would float in the Venusian atmosphere. Measurements returned by spacecraft, however, soon revealed that the Venusian atmosphere was far more massive than even Sagan imagined in 1961, and that it harbored less water than he had assumed. In 1974 two astronomers, James Hansen and J. W. Hovenier, showed that the clouds of Venus were probably composed of sulfuric acid rather than water ice. There was too much carbon and too little water in the atmosphere of Venus for Sagan's plan to work. Fortunately, the development and global diffusion of high-yielding crop varieties sharply increased terrestrial food production, preventing or perhaps delaying the apocalypse that the Ehrlichs had predicted.[12]

Even an unaltered Venus seemed to some like a "paradise planet" for human settlement. A balloon-habitat filled with Earth air would conveniently float in the bearable warmth of the planet's upper cloud deck. It would travel around Venus with the "super-rotation" of the planet's atmosphere, so that a day for its crew would last only four times as long as a day on Earth. Because the air pressure within the balloon would roughly match that of the surrounding atmosphere, the crew would have hours to repair any puncture. By contrast, explosive decompression is a mortal risk for astronauts in space, on the Moon, or on Mars. Already in 1971, Soviet scientists imagined colonizing Venus with settlers in cloud cities, who would oversee a terraforming project.[13] It was a tantalizing vision for the future of space settlement, one that was introduced to Western audiences with the 1980 release of *The Empire Strikes Back*.[14]

Floating over Venus. Left: An imaginary Venus landscape with a Soviet balloon drifting above a crater. Top right: A 1971 Soviet fantasy of a cloud city on Venus. Bottom right: A recent concept of a Venusian cloud colony by S. Zhitomorsky.

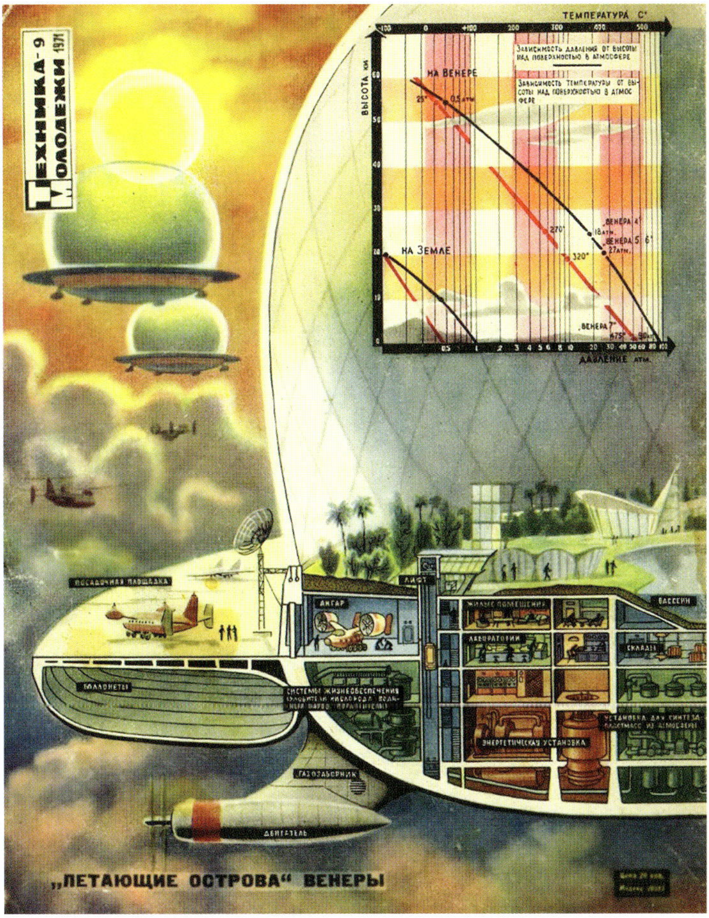
ТЕХНИКА-9
МОЛОДЕЖИ 1971
ТЕМПЕРАТУРА С°
Зависимость давления от высоты над поверхностью в атмосфере
Зависимость температуры от высоты над поверхностью в атмосфере
ВЫСОТА КМ
НА ВЕНЕРЕ
НА ЗЕМЛЕ
„ВЕНЕРА 4"
„ВЕНЕРА 5, 6"
„ВЕНЕРА 7"
ДАВЛЕНИЕ АТМ.
ПОСАДОЧНАЯ ПЛОЩАДКА
АНГАР
ЛИФТ
ЖИЛЫЕ ПОМЕЩЕНИЯ
ЛАБОРАТОРИИ
БАССЕЙН
СКЛАДЫ
БАЛЛОНЕТЫ
СИСТЕМЫ ЖИЗНЕОБЕСПЕЧЕНИЯ
УСТАНОВКА ДЛЯ СИНТЕЗА ПЛАСТМАСС ИЗ АТМОСФЕРЫ
ЭНЕРГЕТИЧЕСКАЯ УСТАНОВКА
ГАЗОЗАБОРНИК
ДВИГАТЕЛЬ
„ЛЕТАЮЩИЕ ОСТРОВА" ВЕНЕРЫ

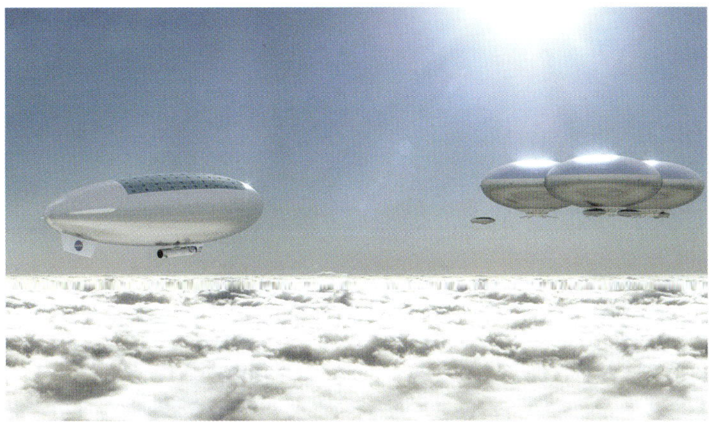

In 1985 the Soviet spacecraft *Vega 1* and *Vega 2* released French-designed balloons into the Venusian clouds, where they survived for two days. Since then, engineers around the world have worked to refine balloon technology for much longer stays on Venus. One recent proposal even imagines using genetically engineered bacteria to grow large balloon habitats in the atmosphere of Venus. Such an approach could have many benefits over current plans for settling Mars. The upper atmosphere of Venus is much easier to reach than the surface of Mars, for example. Spacecraft would not have to travel as long to reach Venus. They would use less energy to enter its orbit, given its stronger gravity, and could more easily exploit its atmosphere to slow down. Astronauts in Venusian cloud settlements would also not have to contend with the perilously thin atmosphere and toxic dust of Mars. If their balloons had a balcony, they could in theory venture outside with a thin protective garment and a mask. Habitats would be grown and supplied on site, rather than imported at great cost from Earth or extracted from previously pristine environments.[15]

Human settlement is invariably world-altering, and any plan that relies on bacteria could imperil an existing Venusian biosphere. Advocates for wholesale terraforming have not abandoned their dreams, however. Some still argue that cloud cities could deploy genetically engineered microorganisms to kick-start a process now requiring thousands, perhaps even tens of thousands, of years. Such plans would eventually involve herculean feats such as bombarding Venus with thousands of asteroids to accelerate its rotation or importing forty quadrillion tons of hydrogen from Uranus or Saturn. On Venus, runaway cooling appears out of reach for now—but settlement might not be.[16]

Changing Planets, Growing Risks

Altering a planet's atmosphere is a daunting challenge, but of course human greenhouse gas emissions are doing exactly that on Earth. The impact of these gases on Earth's atmosphere came into focus just as Sagan's terraforming plan fell out of favor in the 1970s. Over the following two decades a scientific consensus on global warming began to emerge—an agreement among the majority of climate scientists that Earth's climate was changing due to human greenhouse gas emissions and would continue to change until those emissions abated. The development of this consensus depended on the development of computer models, the interpretation of natural archives to evaluate how climate had changed, and missions to Venus that revealed just how profoundly the greenhouse effect could alter a planet.[17]

In the early twentieth century, the advent of aviation and then modern rocketry allowed scientists to study climate not just horizontally across Earth, but also vertically to the edge of space. It was a breakthrough that enabled them to measure patterns of atmospheric circulation, and thereby accelerate the ongoing process by which climatology transformed into a science of global change. Meanwhile, equations developed to describe atmospheric processes in every part of the world were, with military funding, translated to computers by the mathematician John von Neumann and meteorologist Jule Gregory Charney. Numerical weather predictions now became possible; so did crude computer simulations that accounted for global flows of energy and water to forecast decadal weather trends. Several groups in the United States experimented with these general circulation models in the 1960s, and by 1970 the climatologist William Kellogg urged that they be used to forecast Earth's future.[18]

This was a daunting task, and it was made more challenging by the fact that climatologists were working with a sample of one. There was, after all, just our Earth. To gain a clearer sense of how the planet might change in the future, scientists needed more samples. One way to get them was to analyze the different Earths of the past. As we will see, tree rings had begun to shed light on the planet's recent history of climate change (Chapter 15), yet the planet's deep past—when it had truly been a world distinct from our own—remained difficult to study with precision.

The American chemist Harold Urey proposed a way forward. Having won the Nobel Prize for his co-discovery of deuterium, the heavy isotope of hydrogen, Urey's prestige gave weight to his every proposal. In 1946 he suggested that ratios of heavy to light oxygen isotopes in the shells of fossilized, single-celled organisms could reveal how Earth's temperature had changed in the distant past. He encouraged Cesare Emiliani, a newly minted Italian geologist, to analyze deposits of these shells in cores extracted from ocean sediments. Emiliani believed that the shifts in the isotopic ratios of these shells could be attributed directly to ocean temperatures. While completing his

PhD in the 1960s, however, a young English geologist, Nick Shackleton, found that these shifts actually reflected changes in the isotopic ratio of the ocean water that surrounded the shells, which in turn registered fluctuations in global temperature (Chapter 1). With this insight, Shackleton discovered that there had been far more glaciations during the Pleistocene than previously imagined. Analysis of many sediment cores soon suggested that the glaciations had been paced by Milankovitch Cycles, and amplified by positive feedbacks (Chapter 1).[19]

Sediment accumulates slowly while layers in sediment mix together, and that is partly why microfossils could not reveal how quickly Earth had switched between glacial and interglacial conditions, or in other words how stable Earth's climate actually was. Yet scientists had long known that ice sheets contained layers corresponding to the annual accumulation of snow-turned-ice, and they suspected that close study of these layers could shed further light on the history of Earth's climate. Efforts to extract ice cores in the 1950s culminated, in 1960, with a project by the US Army to drill down to the bedrock at the bottom of Greenland's ice sheet. Detailed study of the core suggested that Greenland's glaciers moved too quickly for the Army to implement Iceworm, a top-secret project that would have hollowed out a four-thousand-kilometer network of nuclear missile silos in the ice sheet.[20]

Yet the core had another, more interesting story to tell. Scientists soon realized that because heavy oxygen isotopes evaporate less easily from the ocean in chilly weather, the layers in ice cores would have a greater quantity of light isotopes when Earth was colder. And because ice sheets provided a record of climate change dating back hundreds of thousands of years, they could yield a uniquely precise picture of the Pleistocene.[21]

After analyzing the core extracted by the Army, a Danish physicist named Willi Dansgaard found that temperatures during what previously seemed like stable glacials had actually oscillated, in just a few years, between exceptionally cold and relatively warm conditions. Analysis of bubbles trapped in ice cores, meanwhile, revealed that large-scale changes in climate correlated with shifts in the atmospheric concentration of carbon dioxide. Earth's climate, it seemed, was far less stable than previously imagined, and the same greenhouse effect that had superheated Venus appeared partly to blame.[22]

By the 1970s the different Earths revealed by ice cores and ocean sediments allowed climatologists to study more than one sample of a climate in flux. Now planetary scientists suggested that analyses of other worlds could yield even more, and more diverse, samples with which to improve forecasts of Earth's future. Besides Earth, the solar system offers three rocky worlds with substantial atmospheres: Mars (with an atmosphere approximately one hundred times thinner than Earth's); Saturn's moon Titan (with an atmosphere comparable in density to Earth's); and Venus (with an atmosphere roughly one hundred times denser). The American planetary scientist Yuk Yung argued that close study of such diverse atmospheres could clarify the basic principles by which atmospheres and climates functioned. Moreover, it now seemed that the climates of Mars and Venus had both changed to create worlds radically different from Earth. Studying the history of these changes could, in theory, give climatologists even more samples with which to hone their understanding of Earth.[23]

Few insights regarding Earth's climate were more profound, or influential, than those of the British chemist James Lovelock. In 1965 Lovelock joined a group at NASA's Jet Propulsion Laboratory that aimed to design life-detection experiments for the robotic Viking missions to Mars. As the historian of science Leah Arnowsky has shown, while other scientists focused their efforts on testing for microbes in the Martian soil, Lovelock considered the atmosphere. It dawned on him that, on Earth, methane and oxygen had persisted in their present concentrations for millions of years, even though they are combustible when mixed together. Because oxygen is primarily produced by photosynthesis and methane by anaerobic microorganisms, it seemed to Lovelock that life had maintained a chemical disequilibrium in Earth's atmosphere. He suggested that a similar atmospheric disequilibrium could be an indicator of life on Mars or Venus, because life inevitably "acquired the capacity to control the global environment to suit its needs." Earth could therefore be imagined as a single giant organism, which he called Gaia.[24]

The Gaia hypothesis was immediately controversial among scientists. It appealed to many environmentalists because it seemed to undercut claims of human supremacy over the natural world. It also aided efforts by oil companies to mislead the public on global warming, because it implied that the biosphere would act to prevent catastrophic climate change—"apart from a few brief ice ages," as Lovelock put it. Yet even as Lovelock developed his theory, atmospheric scientists came to believe that the potentially habitable climates of ancient Venus and Mars had in fact irreversibly deteriorated. To them, the lesson for Earth provided by both planets seemed to be that habitable climates were fragile, rather than resilient. Concerned, some would turn their attention back to Earth.[25]

James Hansen, an American atmospheric scientist, had until then focused his career on Venus. While he was an undergraduate at the University of Iowa in 1963, he and his friends had set up a telescope one night to view a lunar eclipse. Usually the Moon turns red when it passes into Earth's shadow, as sunlight is filtered and refracted through Earth's atmosphere on its way to the lunar surface. This time the Moon disappeared altogether. Hansen tracked down the cause: a dust veil lofted into the stratosphere by the recent eruption of Mount Agung, a volcano halfway around the world. He developed a method for using the magnitude of the Moon's darkening to calculate the amount of dust the volcano had created. After publishing his first scientific paper on the topic, he was accepted into graduate school at the University of Iowa, where he met the physicist James Van Allen. After Van Allen told him about the microwave radiation recently detected from Venus, Hansen realized he could merge his fascinations with astronomy and planetary atmospheres.[26]

In 1978 NASA launched the Pioneer Venus mission, the most ambitious US mission ever sent to Venus. The mission included two spacecraft: *Pioneer Venus 1,* an orbiter, and *Pioneer Venus 2,* a mothership that carried four atmospheric probes. By taking detailed measurements of the composition, structure, and circulation of the Venusian atmosphere, the probes would help scientists develop computer models for how the atmosphere had taken on its present characteristics. Hansen was working at NASA at that time and led an experiment on *Pioneer Venus 1* that would study how sunlight passed through and was absorbed by the planet's clouds.[27]

While working on the Pioneer Venus mission, Hansen joined with Yung and other scientists to study the effects of human activity on Earth's atmosphere. The scientists used a simple climate model to argue that the use of chemical fertilizers and fossil fuels had already increased the concentration in Earth's atmosphere of methane and nitrous oxide: greenhouse gases that do not stay aloft as long as carbon dioxide but induce more warming while they do. Hansen and his coauthors pressed for greater study of other planetary atmospheres, including that of Venus, to understand how Earth's climate was changing. Then, just as the Pioneer Venus spacecraft reached their destination, NASA tapped Hansen to lead a team that would design a new general circulation model to forecast Earth's climate. Here was an opportunity to apply lessons from Venus directly to the study of Earth. Within months Hansen's team had completed their first climate model, and they quickly used it to simulate how Earth would respond to human emissions. Because their model was partly based on Hansen's studies of the Venusian atmosphere, it accounted for positive feedbacks that could supercharge the greenhouse effect. As a result, it forecast far more warming in Earth's future than was predicted by another, less capable model that had just been developed by the climatologist Sykuro Manabe at Princeton.[28]

Both models indicated that human activity was changing Earth, but their different forecasts suggested different degrees of danger for human populations and the ecosystems on which they depended. To gain more clarity on the risks posed by future warming, the White House Office of Science and Technology Policy asked the National Academy of Sciences (NAS) to evaluate the state of climate science and determine to what extent Earth's climate was likely to change. In July 1979 an NAS committee chaired by Jule Charney released its assessment. It began by explaining that the temperature of every planet in the solar system reflected both the planet's distance from the Sun and the composition of its "atmospheric blanket." It concluded that if human emissions were to double the concentration of carbon dioxide in Earth's atmosphere, the planet would warm by an average of 3 degrees Celsius, give or take 1½ degrees—a figure precisely halfway between the values given by the Hansen and the Manabe models.[29]

Remarkably, this initial figure, based as it was on the most rudimentary climate modeling, turned out to be essentially correct. The Charney Report forecast that carbon dioxide concentrations could double by as early as 2030, given contemporary trends in the use of fossil fuels. Fortunately for human survival, by then the increase in global temperatures should actually be around half that total.[30]

As the Charney Report took shape, scientists digested a flood of data from the Pioneer Venus spacecraft. Even before the spacecraft reached their target, project scientists emphasized to reporters that measurements of the Venusian atmosphere would help them simulate how that atmosphere had grown so hot, and thereby determine whether Earth's could someday suffer the same fate. "Early results" from the spacecraft, the journalist Bill Densmore reported in December 1978, indeed allowed modelers to confirm "that the Venus atmosphere is an example of a runaway 'green house effect' which could eventually occur on Earth," as the burning of fossil fuels raised "Earth's temperature as on Venus." A wave of newspaper headlines now proclaimed that "hellish" Venus provided a warning for what might befall Earth. "A conservative man might shrink from suggesting that Earth could follow the evolutionary path of Venus," said the planetary scientist Thomas Donahue, quoted in the *Washington Post,* "but a cautious man might recognize the possibility." By the time the NAS published the Charney Report, coverage of the Pioneer Venus results had helped journalists communicate the threat of global warming by transforming Venus into a vision of the worst that could befall Earth.[31]

In subsequent efforts to mobilize public opinion and spur political action, Venus specialists took the lead. In 1981 Hansen paired his team's climate model with a reconstruction of global temperatures, along with an analysis of greenhouse warming on other planets, to establish that human greenhouse gas emissions were indeed heating Earth. This warming, the team emphasized, would become plainly obvious and destructive in the coming decades. However, it could be limited by avoiding "full exploitation of coal resources." The article inspired global headlines, but its policy prescriptions were anathema to Ronald Reagan, who had just been elected president on a platform that reversed the Republican Party's long-standing commitment to environmental protection. The Reagan administration had swiftly reshaped the Departments of Interior and Energy and gutted the Environmental Protection Agency (EPA) to increase coal and oil production in the United States. The Department of Energy, which due to NASA cuts had come to fund Hansen's lab, now pulled its support, and Hansen had little choice but to dismiss half his team. Hansen was the star witness, though, when a young congressman—Al Gore—began to hold congressional meetings on global warming. Under the weight of Hansen's testimony, it briefly seemed plausible that the United States, and in turn the rest of the world, might rally around a binding agreement to limit greenhouse gases.[32]

At a 1984 congressional hearing, Gore invited his friend Carl Sagan to offer objective commentary on the threat posed by greenhouse warming. "Suppose you were skeptical about the existence of a greenhouse effect or were skeptical about the possibility that it can really amount to a lot," Sagan began. "In this case it is very useful to look at other planets." Sagan's career had come full circle. He had established himself as a significant young scientist by identifying the greenhouse effect as the cause for extreme temperatures on Venus, and then established his celebrity as a science popularizer by leading the search for life on other worlds. Now he used the Venusian greenhouse to convince Congress that global warming was "absolutely something to take into consideration and to worry about." In subsequent years Sagan would become a leading voice for climate scientists by continuing to emphasize the example and warning provided by Venus.[33]

Widespread comparisons between Venus and Earth motivated diverse responses. To some, Venus offered reason for hope; Earth, after all, was not likely to resemble it any time soon. To others, it evoked darker emotions. A former biologist at the New York State Department of Environmental Conservation, for example, wrote Sagan—and the *New York Times*—after assuming that the accumulation of greenhouse gases in Earth's atmosphere would inevitably transform it into a world like Venus. The only solution would be to engineer a pathogen deadly enough to annihilate 99.9% of humanity, "eliminating many world problems in one quick and efficient manner." The "best part" was that the country that released such a pathogen could preemptively develop vaccines to protect its citizens as the rest of humanity succumbed. Two decades earlier, Sagan had suggested a scheme to solve overpopulation by transforming Venus to create living space for those who would otherwise crowd Earth. Now his comparison between the climates of Earth and Venus had suggested to one man another solution: a holocaust that would leave the United States as the last country standing.[34]

For less fatalistic thinkers, the example of Venus emphasized the increasingly urgent need to replace fossil fuels with renewable and nuclear energy. It also seemed to show, according to New York congressman Bill Green, that NASA's focus ought to be on the robotic exploration of other worlds—Venus in particular—rather than the new space station the agency was proposing. "There is little doubt among responsible scientists that our planet is curving" to more closely resemble Venus, with "carbon dioxide so prominent that life will disappear," one letter to the *New York Times* proclaimed. The more that could be learned about the history of Venus, the more humans could learn and perhaps control the future of Earth.[35]

By the time the self-proclaimed environmentalist George H. W. Bush was elected president in 1989, the warning from Venus seemed likely to be heeded. Within months, some thirty-two climate bills were introduced to Congress. Yet despite his claims to the contrary, Bush never regarded global warming as a pressing priority; worse, his chief of staff, John Sununu, distrusted climate modeling. Advertising by major oil and gas corporations, whose scientists had privately warned company leaders that warming could be catastrophic, weakened popular pressure for action. Climate policy in the United States went nowhere, and with Sununu's encouragement US diplomats helped derail international negotiations aimed at limiting global emissions. The world's lone superpower seemed intent on "veneraforming": turning Earth into a copy of Venus.[36]

For environmentalists, though, studies of Venus helped ensure that the 1980s were far from a lost decade. The reason had to do with a gas, ozone, that is present on Earth but largely absent on Venus. Following its discovery in 1839, scientists sought to measure ozone concentrations in places like European spa towns because they assumed that low-altitude ozone correlated with clean air. Laboratory experiments conducted in the late nineteenth century demonstrated ozone's ability to absorb energetic ultraviolet radiation. In 1921, the advent of aviation allowed physicists to discover that ozone accumulated at high altitudes, in the stratosphere. After scientists established the structure of DNA in 1953, they realized that without the protective layer of stratospheric ozone, the Sun's most energetic ultraviolet radiation would cause catastrophic damage to DNA on Earth's surface.[37]

Early in the 1970s, scientists discovered that the ozone layer was under attack. Clues came from two sources: Earth and Venus. Focusing on Earth, researchers including Lovelock found that chlorofluorocarbon (CFC) gases, which were widely used in refrigerators and spray cans, had spread through Earth's atmosphere. They also discovered that a component of CFCs, chlorine, could break down stratospheric oxygen. Concentrating on Venus, atmospheric scientists found that carbon dioxide remained stable in the Venusian atmosphere partly because of reactions between carbon monoxide and chlorine gases. They established the planetary scale at which chlorine could disintegrate the oxygen isotopes that constitute ozone.[38]

Both lines of evidence, from Earth and from Venus, enabled chemists to identify the threat posed by CFCs to Earth's life-giving ozone layer. Studies now revealed that a fast-growing hole had appeared in the ozone over Antarctica in the winter. Despite opposition from the chemical industry, in 1987 forty-six nations ratified the Montreal Protocol on Substances that Deplete the Ozone Layer, a roadmap to phase out CFCs. The Montreal Protocol has successfully reduced the quantity of CFCs in the atmosphere and prevented the ozone hole from spreading far beyond Antarctica. Without it, CFCs could have destroyed nearly two-thirds of the world's ozone by the year 2065. Loss on that scale would have caused a catastrophic collapse in the productivity and biodiversity of countless ecosystems, and a consequent surge in nitrous oxide that would have greatly amplified global warming. The Protocol likely prevented some 280 million cases of skin cancer in the United States alone.[39]

It would be too simplistic to conclude that activists, scientists, and far-sighted policymakers averted the apocalypse by restraining the chemical industry. The Montreal Protocol succeeded primarily because the industry could easily replace CFCs with hydrochlorofluorocarbons (HCFCs) and hydrofluorocarbons (HFCs)—two potent greenhouse gases. Major corporations went along with later amendments to ban these chemicals only because those amendments gave them a competitive advantage over smaller rivals that could not easily adopt alternatives. Yet when it comes to existential risk, motivations are less important than results. One warning from Venus went unheeded in the 1980s, but another spurred action—and, just barely, preserved a habitable planet for humanity.[40]

CONCLUSION

Earth's Once and Future Twin?

Perhaps it was no coincidence that in the early 1990s, as momentum for climate action faded in the United States and around the world, exploration of Venus all but ground to a halt. The reasons were diverse, but the most important may have been that Venus seemed a poor place to look for life. Still, there were Venusian mysteries that defied explanation. The ultraviolet-absorbing substance in the planet's atmosphere, for example, has no known cause, and its shifts in distribution across Venus seem to affect the breakneck speed of the atmosphere's superrotation. Planetary scientists speculated that bacteria, hiding, perhaps, in cloud droplets, could be absorbing ultraviolet radiation for photosynthesis, thereby reshaping Venus as fundamentally as life transforms Earth. At the very least, Venus's dynamic surface sustained chemical cycles that seemed capable of nurturing airborne bacteria.[1]

On September 14, 2020, scientists led by the astronomer Jane Greaves announced that they had detected the gas phosphine in radio emissions from the Venusian atmosphere. Phosphine had no known cause on Venus, but on Earth it is associated with microscopic ecosystems that do not require oxygen. Greaves and her team concluded that phosphine was a likely byproduct of life in the Venusian clouds. It was a sensational claim, and many astrobiologists were skeptical. Some uncovered telescope calibration errors that may have accounted for the phosphine signature. Others argued that Greaves and her team had misidentified a signature created by sulfur dioxide, or that the instrument used to detect phosphine could not have measured it in the temperate layer of the Venusian atmosphere. A few proposed that the phosphine had come from volcanoes, lightning, or chemical interactions within the Venusian atmosphere. As of this writing, Greaves maintains that her team did find phosphine, although less than originally claimed.[2]

Venus, therefore, may yet reveal whether life is unique to Earth. If so, it will once again tell us where we stand in the universe. Some four hundred years ago the movements of Venus were used to calculate the basic dimensions of our solar system, even though attempts to get precise measures of its atmosphere have been frustrated. Changes in its atmosphere, driven perhaps by volcanic eruptions, strengthened widespread belief in the existence of life on other worlds, and even early proposals to make contact. Studies of the atmosphere supported pseudoscientific ideas that blamed its violent formation for the course of human history; animated the counterculture movements of the 1960s and 1970s; and anticipated new, radically interdisciplinary fields of study. Dreams of artificially changing its atmosphere suggested that space exploration could solve the problems of Earth, and further studies of the atmosphere then helped reveal that those problems were far worse than previously imagined. Indeed, by demonstrating the magnitude of the risk associated with the ozone hole, the exploration of Venus helped save all of us on Earth, at least for now.

The variable environments of Venus, therefore, call on us to change—and maybe not just on Earth. If we learn to think less like surface-dwellers, we may yet find on Venus the most comfortable and interesting environment for human settlement beyond Earth. And even without our help, Venus' surface may someday resemble that of Earth, as closely as astronomers once assumed. When the volcanism that likely sustains today's Venusian atmosphere subsides, in a billion years or so, sulfuric acid and eventually carbon dioxide will fall from the sky, and a much cooler world with a nitrogen and argon atmosphere may then emerge. Unless we are around to stop it, the brightening Sun will at the same time boil away the oceans of Earth, launching the same runaway greenhouse that might have transformed ancient Venus. For millions of years, both planets—one cooling, the other warming—would again be the closely matched siblings they might have been, had primordial Venus been covered with water. In the end, they could share the same fate, as the Sun balloons into a red giant and consumes the inner solar system. Ultimately, the twins would perish together.[3]

PART III

Moon

A drawing of the Moon and its maria.

“I saw . . . the Moon’s white cities, and the opal width / Of her small glowing lakes, her silver heights . . . And the unsounded, undescended depth / Of her black hollows.”
—Alfred Lord Tennyson, “Timbuktu” (1829)

“The most significant thing that we ever found on the whole goddamn Moon was that little bacteria who came back and lived and nobody ever said shit about it.”
—Pete Conrad in Brian Duff, “The Great Lunar Quarantine” (1994)

It was just past 9:00 p.m. on a crisp autumn night in 2019. A full Moon rose over the glowing monuments of Washington, DC, as I hunched over a telescope high up on a rooftop. The Moon commanded my full attention. By then I had been preparing to write this chapter for over a year. Every big crater, every major mountain range and lowland plain, had stories to tell me. For a few long minutes, I peered through my telescope, lost in the lunar landscape. And then I saw something that should not have been there.

A little circle—perfectly black, perfectly round—appeared over the Moon's western limb. Looking for all the world like Venus transiting the Sun, it advanced across the brilliant surface of the Moon. I racked my brain: What could it be? Not a planet, of course. An asteroid? Impossible. To look like that black circle, an asteroid would have been inconceivably big or close to Earth.

My thoughts abruptly changed when the circle reached the eastern limb of the Moon. Astonished, I gawked as the circle reversed course and started traveling first southwest, and then southeast. Nothing in space—nothing that obeyed Newtonian physics—could behave that way. My mind raced through more improbable possibilities. Could it be a nearby balloon, caught in an updraft? Of course not. This circle moved too steadily and seemed to follow the Moon.

Soon I had the uncanny feeling that I was seeing something I should not be seeing—something I was not supposed to see. My imagination ran wild. I remembered the black monoliths in Arthur C. Clarke's *2001: A Space Odyssey.* I checked the lenses of my telescope. Could it be dust? But my telescope would never focus on something so close, and in any case, it was perfectly clean. When I returned my attention to the Moon, the circle was gone. I would have doubted my eyes or my sanity, had I not taken plenty of pictures.

Back inside, I obsessed over that circle. I typed out a report and sent it off to the Central Bureau for Astronomical Telegrams at Harvard University. Within minutes, comet hunter Michael Rudenko responded by drawing my attention to a recent sighting of the same phenomenon. The culprit seemed to be a giant balloon, floating twenty kilometers above the Earth, made by Google to connect rural communities to the internet and provide backup connectivity in event of a disaster.[1]

My unidentified flying object apparently had little connection to the actual Moon. Yet for one evening, at least, I lived the history you are about to read. I grasped how a change in the appearance of the Moon enthralled, confused, and in some cases terrified observers from antiquity to the present. How do you react when one of life's great constants—the Moon, undergoing its regular phases—is suddenly anything but constant?

Unexpected changes in the Moon's appearance inspired and supported revolutionary new ways of understanding the universe, from the first efforts to validate a Sun-centered model of the solar system in the early seventeenth century, to the emergence of the new astronomy nearly two centuries later. Between the eighteenth and twentieth centuries, observations of changing lunar environments, and fears that those environments could change Earth, encouraged fresh ways of thinking about the possibility of extraterrestrial life, with profound consequences for the rise of mass media, amateur astronomy, science fiction, mountaintop observatories, and today's efforts to protect Earth from alien contaminants. Changes to the Moon shaped the great ambitions and projects of the Space Age, from the Space Race that culminated in the Apollo program to present-day efforts to expand human settlement beyond Earth. It is no exaggeration to say that changes in the appearance of the Moon's environments repeatedly changed the course of human history on Earth—and beyond.

That is fitting, because the Moon emerged out of the greatest physical change in Earth's history. Computer models suggest that the Moon coalesced out of detritus left over from a cataclysmic collision between Earth and at least one other world in the primordial solar system. What that collision looked like, precisely, is still up for debate. Most scientists favor a so-called giant impact theory, which supposes that a planet roughly the size of Mars smashed into the young Earth some 4.5 billion years ago. So violent was the collision that the two planets fused together, though some of their combined plasma whipped out into space to form the Moon. The Earth's core may have accumulated extra iron, and its rotation may have sped up, activating the powerful magnetic field that shields our atmosphere from the solar wind. The collision also could have triggered the engine of plate tectonics, which cycles rock from Earth's mantle to its surface. Newly exposed rock absorbs atmospheric carbon dioxide before being pulled back underground, so the creation of the Moon might have protected us from the runaway greenhouse warming that overwhelmed Venus.[2]

Because the Earth rotates a bit faster than the Moon orbits it, the gravity of the tides raised by the Moon pulls the Moon forward ever so slightly, while the Moon pulls Earth backward. For eons, this tug of war pushed the Moon farther from us, at a rate of just over an inch per year, slowing Earth's originally breakneck rotation and stabilizing the axial tilt that is responsible for its seasons. The comparatively gentle seasons that eventually took hold would be conducive to life; so were tides, which mixed ocean waters with nutrients and oxygen, all the while creating dynamic coastal habitats. Humanity may owe its very existence to the environmental changes that created the Moon. Those changes continue to influence us, for lunar environments have had a surprising influence on the past five centuries.[3]

9

Muses of the Moon

Although they were systematically excluded from the revolutions in natural philosophy that unfolded in the seventeenth and eighteenth centuries, women nevertheless made important, and long overlooked, contributions to emerging scientific disciplines. In astronomy, the men who made discoveries, proposed theories, and invented new practices or technologies in fact depended on the often-uncredited labor of men and women who prepared or operated instruments, sketched observations, kept logbooks, made calculations, and discussed, or even suggested, interesting ideas. In the right conditions—when allied to an open-minded spouse, parent, or sibling, for example, in a culture or economy that permitted women's participation in public life—some women could emerge as astronomers in their own right.[1]

Maria Clara Eimmart was such a woman. She was born in Nuremberg in 1676, a time when the city governed its own affairs within the Holy Roman Empire and could prevent guilds from controlling participation in professional life. In Nuremberg, families controlled the means of production, and that provided an opportunity for women to participate in paid work. In the Eimmart family, this work involved the heavens. Maria's father, Georg, was a well-known engraver and an amateur astronomer who poured his substantial wealth into a home observatory. Maria trained as his apprentice, and from him learned that sketching with pastels could evoke the subtle shadings of flowers, or celestial objects. It was the Moon that most interested her. Between 1693 and 1698, she made some 350 illustrations of its surface, all on a rich blue cardboard background. With realistic detail that no other seventeenth-century depiction of the Moon could match, Maria captured how the Moon's appearance changed as its cycled through its monthly phases.[2]

Maria Clara Eimmart died in childbirth, like so many women of her time. Her career was fleeting, but her drawings endure as the most scientifically accurate and visually striking examples of the seventeenth-century explosion of interest in selenography, the cartographical method of mapping the lunar surface. Selenography was an expression of a mapmaking and publishing enterprise that supported, and was supported by, the commercial and territorial ambitions of European states; it was also a reflection of the growing importance of observation in the culture of natural philosophy on the continent. By mapping the Moon with painstaking detail, many selenographers hoped above all to find that its environments could change—and with that, to discover extraterrestrial life. It was a dream that would fade in the eighteenth century, only to reemerge with even greater popularity when it seemed that the Moon's surface really had changed in the nineteenth century.

Mapping the Changing Moon

Scientists long believed that most of the Moon's current features formed in a roughly 150-million-year surge of comet and asteroid impacts, a "late heavy bombardment" that came and went some 3.9 billion years ago. New evidence, however, suggests that the Moon withstood continuous collisions from the time of its formation until the supposed bombardment, when it ebbed enough for the Moon's bright highlands to freeze into place. When the impacts finally abated, scorching basalt repeatedly broke through the Moon's crust, spilled into giant, hollowed-out impact basins, and froze to form flat, dark lowland plains: the *maria* that give the Moon its distinctive appearance.[3]

The Moon is enormous: five times more massive than Pluto, twenty-five times more massive than the entire asteroid belt. Its highlands cover roughly the same area as Africa, while its maria cover an area as large as the continental United States. Its size, proximity to Earth, and bifurcated appearance make it the only celestial body in the night sky whose surface is clearly visible to the unaided eye.[4]

Yet not a single realistic depiction of the Moon predates the fifteenth century. Different cultures viewed the Moon as a deity, a symbol, an omen, an influence perhaps on Earth, but, with rare exceptions, not a fully realized world. To the likes of Aristotle and Ptolemy, the Moon was a perfect sphere that divided the immutable heavens from the flawed and changeable realm of Earth. There was no need to see in its complex surface the intricate environments of another world. Beginning in the fifteenth century, a series of breakthroughs—in navigation, shipbuilding, mathematics, and printing—catalyzed a revolution in the quantity and quality of European maps. By the end of the sixteenth century, maps had become tools that aided conquest, commerce, and extraction on the frontiers of European empires; they were also a source of entertainment for a public that craved depictions of the fantastical and exotic. In that context, English natural historian William Gilbert, who later served as Queen Elizabeth's physician, drafted the first map of the Moon.[5]

Gilbert was a Copernican; his map, he hoped, would let him catch the Moon's major features in the act of changing. Any change would undermine the Aristotelian idea that the heavens were perfect and immutable. Gilbert had established a precedent that would endure for nearly three centuries. Realistic depictions of the Moon, including maps, were subsequently designed to reveal environmental changes on another world.[6]

Gilbert believed that because Earth orbits the Sun while the Moon orbits Earth, observers on Earth would see the Moon as slightly swaying as their viewpoint continually shifted. By detecting landscapes along the Moon's eastern or western hemisphere as they rocked into view, Gilbert hoped to gather more evidence for a Sun-centered reimagining of the solar system. He never detected this swaying, called libration, but Galileo did in 1632. Sometimes, libration happens because the Moon moves fastest through the part of its elliptical orbit that is closest to Earth, revealing slivers of the far side that otherwise go unseen. Other times, libration zones emerge because the Moon is tilted ever so slightly on its axis, or because the rotating Earth allows observers to see subtly different parts of the Moon at different times.[7]

More intriguing than libration to Gilbert's contemporaries was the Moon's "ashen glow": the dim light that can expose parts of the Moon not illuminated by direct sunlight. Some took it as evidence that the Moon had an internal source of light; others correctly reasoned that light from the Sun bounced off Earth to brighten the lunar night. This earthlight enabled Copernicans to promote the idea that Earth and Moon were similar worlds, illuminating and illuminated by each other, both orbiting the Sun. Johannes Kepler, for example, imagined how a traveler to the Moon might look back at Earth from the lunar surface. By drawing on the latest reports of exotic environments beyond Europe, Kepler described a Moon that was distinct from Earth, with a porous surface riddled with mountains and caves and subject to extremes of heat and cold, but similar enough to have air, seas, vegetation, and animals of "monstrous size" that had adapted to thrive. Three centuries later, spectroscopic measurements of earthlight would reveal Earth to be blue as seen from space—and therefore very different from the Moon.[8]

By spurring breakthroughs in cartography and transporting sailors to far-flung environments, newly seaworthy sailing ships had played a key role in the beginning of a new conception of the Moon in European scholarship (Introduction). Then, in 1609, Galileo used his light-ship—the telescope—to reveal in the Moon a world "not unlike the face of the earth, relieved by chains of mountains and deep valleys," utterly unlike the smooth sphere imagined by Aristotle. Galileo stressed that these environments betrayed their nature by gradually increasing "in size and brightness" with the advance of lunar day, as "more and more peaks shoot up as if sprouting now here, now there, lighting up within the shadowed portion" of the Moon. He supposed that the larger appearance of the daylit side of the Moon, compared to the darkened side, hinted at a thick "vaporous globe" that surrounded the Moon: an atmosphere, perhaps, akin to Earth's.[9]

After they were printed and widely distributed, Galileo's descriptions and sketches of a rugged, changeable Moon astonished Europe's literate classes. Although Galileo had exaggerated the Moon's seemingly terrestrial features, he cautioned against the assumption that "earth and water" accounted for the contrast between light and dark on the lunar surface. Still, the English poet John Donne expressed the opinion of many when he declared that Galileo had "instructed himself of all the hills, woods, and cities in the new world, the Moone." Admirers across Europe compared Galileo to Columbus, and some judged Galileo to be a superior explorer, for his discoveries had come without bloodshed.[10]

Seventeenth-century intellectuals authored the first recognizably modern works of science fiction to imagine just how close the resemblance between Earth and Moon might be. Many assumed that because colors on Earth seemed to fade in the distance, the Moon was really more colorful than it appeared—and that Earth would take on a monotone hue as viewed from the Moon. English bishop Francis Godwin imagined flocks of migratory birds, insects, and "Moon Men" traveling between similarly vibrant lunar and terrestrial environments. Perhaps Moon and Earth were part of one vast, dynamic ecosystem?[11]

The question was far from idle speculation, for Galileo's discoveries challenged the authority of the Roman Catholic Church, widening the schism in Europe that had been ripped open by the Reformation. Between 1545 and 1563, the Church had responded to the Reformation by clarifying its doctrine and authority in the Council of Trent. The council indirectly reinforced the synthesis of Aristotelian philosophy and Church doctrine, a synthesis that scholars like Galileo undermined with their support for a Sun-centered solar system and discoveries that collapsed distinctions between terrestrial and celestial environments. More importantly, the council confirmed the Church's supreme authority to interpret scripture. Heliocentrism seemed to clash with specific passages in the Bible. And worse, its proponents attempted to justify their cosmology by offering their own interpretations of the Bible. While the positions of Church leaders were complex, contradictory, and evolved with time, the early seventeenth-century Church increasingly opposed efforts to prove the superiority of the heliocentric model by identifying changing environments on the Moon and other worlds. As the religious wars provoked by the Reformation swept across Europe, scholarship and literature that promoted changes in celestial environments, including those on the Moon, served as political propaganda by undermining Catholic claims to authority.[12]

Changes in the Moon's appearance seemed to have practical applications for the agents of all European empires, Catholic and Protestant. Contemporary sailors had no way to accurately calculate longitude, which meant that maps were riddled with inaccuracies and voyages were longer and more dangerous than they had to be. In 1598 Philip III, Habsburg king of the Spanish Empire, offered a handsome reward to anyone who could solve the problem, and two years later the States-General of the Dutch Republic offered their own prize. Key to the effort would be a method for determining the difference between local time and time in a city of known longitude—without using a clock, since the pendulums in the clock designs of that time could not be used to accurately tell time aboard a moving ship. One proposed method was to precisely record the position of the Moon relative to nearby stars, which could then be compared to the predicted position in locations of known longitude. The trouble was that the Moon's movement was not known with great enough accuracy to allow precise calculations of longitude.[13]

Some of Europe's leading scholars suggested that sailors time the advance of shadows into lunar craters, which could then be compared to predictions of the timing as viewed from cities of known longitude. This solution required more accurate lunar maps. Across Europe, cartographer-astronomers raced to complete the first and finest maps, sometimes with the enthusiastic support of state patrons. Their efforts culminated in a series of competing maps with contradictory naming schemes that, to varying degrees, depicted the Moon as a world similar to Earth. Yet none enabled sailors aboard moving ships to use a telescope accurately enough to observe the subtlest details in lunar craters.[14]

The new selenographers found that lunar landscapes frustrated the most meticulous efforts to draw them, so that even the new maps were far from accurate. One thing they did not know was that the Moon's soil, or regolith, is strewn with glass forged in the heat of cosmic impacts. This glass only reflects sunlight directly back to its source, rather than off to the side, which means that lunar features invisible from Earth, when illuminated at a low angle, suddenly snap into view when the Sun is directly overhead. Because the distribution of glass is uneven across the lunar landscape and does not reflect the Moon's actual topography, the face of the Moon appears to change in bewildering ways as it waxes and wanes. Added to that is the extra complexity of the Moon's libration, which subtly changes the appearance of most lunar features over an eighteen-year "Saros" cycle, and the Earth's atmosphere, which can abruptly obscure or reveal parts of the lunar surface.[15]

The extent of these changes slowly came into focus as telescopes gradually improved over the seventeenth century. Glassmakers learned how to forge larger lenses, and telescope makers learned to lengthen their refracting telescopes to compensate for both spherical and chromatic aberration, two defects that can keep refractors from gathering all rays of incoming light at the same focus. The best telescopes in Europe now presented the viewer with somewhat brighter, sharper images at higher magnification, although the largest—such as the

150-foot-long monstrosity used in Danzig by self-styled "Muse of the Moon" Johannes Hevelius—were impractical when focused on the fast-moving Moon.[16]

Improved telescopes led lunar observers to wonder: Did features on the Moon blur and wink into and out of view solely because of Earth's restless atmosphere, or did the Moon have an atmosphere, too? Most believed it did. But in 1637 the Moon passed over the Pleiades, a cluster of scorching blue-white stars, brilliant and easily visible to the naked eye. Using a telescope, Jeremiah Horrocks noted that the cluster's stars instantly vanished when the Moon obscured them. He deduced that the Moon could not have an atmosphere, since starlight passing through an atmosphere would gradually fade before disappearing behind the Moon. Other lunar observers agreed. The Dutch polymath Christiaan Huygens now reasoned that without air, the Moon could not possess water, let alone life. Still, changes on the Moon continued to fuel suspicions that there was life on its surface. The French astronomer Giovanni Cassini, for example, wrongly thought he had espied a "kind of whitish cloud" drifting across the lunar surface. It was actually bright debris unearthed by the creation of a crater, but it suggested to Cassini that the Moon's environments might after all be similar to Earth's, and capable of harboring life.[17]

Newton's theory of gravity gradually won over skeptics, however, weakening the Copernican impulse to compare the Moon to Earth. Gravity explained and therefore supported the Sun-centered model of the solar system far better than any observation of dynamic, Earthlike environments on the Moon ever could. With Copernican cosmology on surer footing and the quest for longitude no closer to resolution, selenography devolved into a niche pursuit for a small group of scattered observers, each equipped with telescopes woefully unsuited to the task. For much of the eighteenth century, most astronomers considered the Moon only to measure its movement, and thereby to verify Newtonian physics.[18]

Cities on the Moon

Toward the end of the eighteenth century, the appearance of change on the Moon's surface inspired another profound shift in understandings of the cosmos. Just a few years after Caroline Herschel arrived in England, she and her brother

William began to study the Moon. The first telescope they crafted revealed a wealth of lunar detail that had been hidden from other observers. Their logbooks indicate that William adopted the analogical method, the principle behind his new astronomy, after detecting what he thought were environmental changes on the Moon. Before long he noticed "a *forest*"—he underlined the word—with clearings and trees seemingly big enough to identify from Earth.[19]

These observations convinced Herschel that the Moon and Earth also shared common engines of environmental change. In the autumn of 1780, Herschel described "one of the most extraordinary Phenomena I ever saw": a thin, glowing ribbon, resembling "the Lava of a Vulcano," apparently flowing from the Moon's brightest crater, Aristarchus. That ribbon seemed to come and go in the ensuing month, perplexing Herschel. Three years later, he and his sister again noticed "channels created by lava" around Aristarchus. They also observed a "beautiful treble halo" surrounding the Moon, caused perhaps by dust from the ongoing volcanic eruption of the Laki fissure in Iceland. Laki's "dust veil" may well have given a fiery hue to the otherwise colorless lines around Aristarchus.[20]

In 1787 the Herschels tracked apparent volcanism around Aristarchus, showing it to dignitaries who peered through their telescope on royal invitation. Three years later, during a ruddy lunar eclipse, brother and sister spotted no fewer than 150 active "volcanoes" on the Moon. At Aristarchus, they had undoubtedly detected shifting patterns of sunlight striking a "rille"—a lunar feature carved by ancient lava—winding from a crater that, at just 450 million years old, was too young to have faded under the bombardment of the solar wind. "Perhaps conclusions from the Analogy of things may be exceedingly different from truth," Herschel wrote. But, he went on, "we have no other way to come at knowledge," so "we must allow great weight for arguments taken from this source." On the Moon—as in the rest of the cosmos—he was partly right.[21]

Of course, it is humans who carry out some of the most profound environmental transformations on Earth. The most tantalizing application of analogical reasoning to the Moon's environmental changes suggested that intelligent builders were also at work on the lunar surface. Kepler, reflecting on the implications of the craters described by Galileo, immediately found it curious that the Moon should be littered with circular shapes. A circle would conveniently provide shade throughout long lunar days. Perhaps some craters, Kepler speculated, were the work of intelligent builders who had carved underground cities along crater rims. Herschel took that idea and reversed it. In a rare atmosphere, he mused, circular buildings would allow anyone within to enjoy either direct or reflected sunlight. Whatever their use for alien architects, craters seemed to offer the key to proving the existence of intelligent life beyond Earth at a time when most assumed that God would not create barren, thus useless, worlds.[22]

But how to confirm that some craters, at least, were the work of extraterrestrial architects? William decided that he and his sister would watch for "the erection of any new small circus"—that, is, any new crater—"as the Lunarians may [monitor] the Building of a new Town on the Earth." First, the Herschels needed to plot all craters already in existence—"an exact list of those Towns that are already built"—across a "perfectly smooth" part of the Moon, where uneven terrain would not confuse the matter. They had to observe these "towns" during every lunar phase, since "experience shows that what is not visible in one phase or one light of the Moon will be so in another."[23]

As the months passed, William and Caroline undertook their crater survey. It was grueling work. They spotted canals and pyramids, turnpike roads and thickly forested mountains—or so they thought—but their survey seemed no closer to completion. There were simply too many craters. In his log, William Herschel initially categorized craters by size—"Metropolis, Cities, Villages." Later, disillusion setting in, he replaced those words with less evocative terms: "Large Places, Middling Places, Small Places." In 1779 Herschel's first published paper appeared only after editors had scrapped his speculative passages on the Moon's city builders. By then, he had lost his zeal for finding new craters.[24]

Herschel's reports of lunar volcanism led Johann Hieronymus Schröter to examine Aristarchus in 1783, the year of the Laki eruption on Earth. His initial observations with a modest refractor so intrigued him that he purchased reflectors crafted and shipped by the Herschels themselves. Using the high magnifications that could be achieved with those reflectors in good weather, he scoured the Moon for signs of environmental change that could have biological, rather than geological, causes.[25]

As he lingered over individual mountains, rilles, and craters—he was the first to systematically apply the word *crater* to circular formations on the lunar surface—Schröter grew convinced that he could detect topographical transformations that, while subtle through the telescope, would have to be truly immense in scale. The rugged walls and floor of the crater Alhazen and the straight clefts within the crater Hevelius, for example, seemed to change with repeated observation. The shifting appearance of the lunar surface helped convince him that the Moon had an atmosphere, and although he never found direct evidence for life, he did adorn his sketches with features that resembled terrestrial forests.[26]

The bulk of Schröter's work was lost when Napoleon's soldiers sacked his observatory in April 1813. All the same, Schröter managed to publish seventy-five engraved plates of the Moon. The apparently Earthlike details that his telescopes had revealed led other observers to revive the seventeenth-century dream of finding environmental changes on the Moon that revealed its true nature, and possibly its inhabitants. Many would adopt the method pioneered by Schröter and the Herschels of sketching distinct lunar features, then comparing sketches to determine whether the features changed, and what the changes might reveal.[27]

The Holy Roman Empire would emerge as the hotbed of this new selenography. In a time of strengthening ties between Napoleonic France and the imperial state of Bavaria, French and Bavarian surveyors launched a long effort to draft more accurate maps of Bavaria, which would allow for more lucrative taxation. Surveying required fine optics and detailed optical knowledge, so Bavarian officials scrambled to import British lenses, which at that time were the best in the world. The impact of the French Revolution and the Napoleonic conquests in Europe had also set in motion a process of secularization across the Holy Roman Empire. In Bavaria this resulted in the closing of the abbey of Benediktbeuern. The abbey's Benedictine monks, known for their technical expertise, were now enlisted to establish an optical institute that would reduce Bavaria's dependence on British imports. The effort gained urgency in 1806 when the debut of Napoleon's Continental System closed European ports to trade from the hostile British Empire, and the British government responded by blockading the French Empire.[28]

That year the self-made optical pioneer Joseph von Fraunhofer joined the institute, where he learned to craft lenses of unprecedented quality. With these he perfected a new, "achromatic" refractor design that sharply reduced optical distortions. Now, for lunar observation, small refractors could rival the performance of the bigger reflectors designed by the Herschels. In his laboratory, Fraunhofer invented the science of spectroscopy and launched an optical trade that eventually spanned Europe. For a while, however, German astronomers had special access to telescopes that brought even subtle lunar environments within easy reach.[29]

One such astronomer was Gruithuisen. Using no fewer than three small Fraunhofer refractors, Gruithuisen compared Schröter's sketches of the lunar surface to his own observations. Any discrepancy, he believed, would reveal environmental changes characteristic of lunar life. After detecting what seemed like overwhelming evidence for forests and lakes, channels and polders, rising mist and migrating animals, Gruithuisen concluded that one's "conviction of the existence of organized and living beings" on the Moon could only grow "as one becomes acquainted with . . . the many changes on its surface." "Selenites" would no doubt build massive structures to endure the Moon's thin atmosphere and long days and nights. Therefore, Gruithuisen suspected, their buildings might just be visible on the lunar surface. They would reveal themselves, he speculated, by changing their appearance during ongoing construction projects, by ejecting garbage, or by releasing vapors from chimneys. In the supposed lunar countryside, within the crater Alhazen, he soon found a promising clue: a recurring oscillation between light and dark that he assumed must signify the cultivation and harvesting of crops.[30]

Then, at 3:30 a.m. on the night of July 12, 1822, with the Moon in its last quarter and the atmosphere over Germany perfectly calm, Gruithuisen happened across a sight that, he later claimed, moved him to cry out, "O, Schröter, there it is, what you always sought in vain!" In the darkest and ostensibly wettest and warmest spot on the Moon, near its supposedly verdant equator, Gruithuisen spied what he took to be a latticework of gigantic structures, sharply defined in the lunar sunset, big enough to "mock our pyramids," radiating out in "mathematically regular" angles along a north–south axis (supposedly to exploit local winds). Here, it seemed, was incontrovertible proof of intelligent lunar life—a "bird's-eye view of a city." Gruithuisen carefully observed the area in subsequent months. When he noticed small changes, owing, he believed, to smoke rising

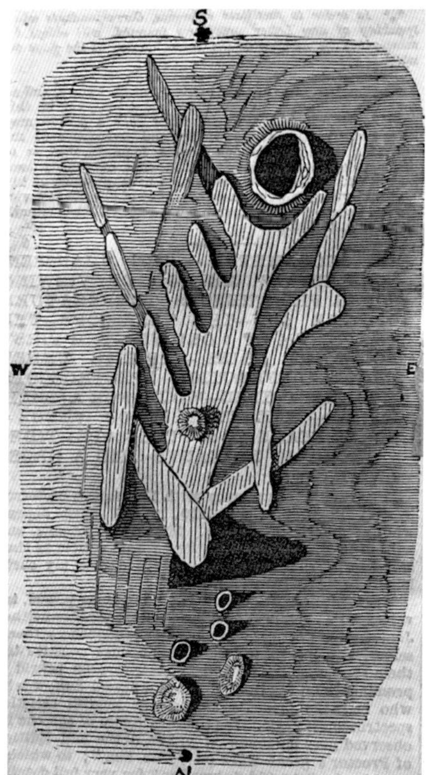

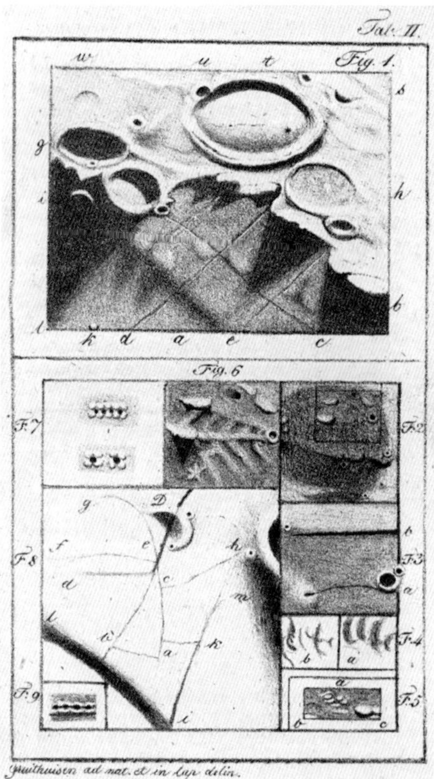

A lunar city? Left: Wallwerk, sketched by French astronomer Casimir Marie Gaudibert in 1873. Right: Gruithuisen's sketches of Wallwerk and its surroundings.

from invisible chimneys, he decided "this could not be a mirage." He named the city Wallwerk, meaning "Wall-Work." It was a rare lapse in imagination.[31]

After two years of painstaking preparation, Gruithuisen published his observations of plants, animals, and civilization on the Moon. He then embarked on a promotional tour across Europe, using telescopes to win over astronomers and dignitaries alike. European newspapers captivated readers with stories of Gruithuisen's observations of changes to his city, which seemed to him like evidence of ongoing activity. He proposed a plan to carve immense geometric shapes across Earth's surface that would alert the Lunarians to the presence of another intelligent species, and perhaps open a channel for communication. In 1828 he even announced the discovery of a second lunar city. Yet by the

1840s, astronomers armed with telescopes far more capable than Gruithuisen's generally agreed that Wallwerk was not, after all, an obviously artificial latticework of straight lines.[32]

The Moon and Mass Media

Even after Wallwerk was discredited, intellectuals and the general public maintained the dream that there was life on the Moon, and the hope of finding it by detecting changes on the Moon's surface. In August 1835 the *New York Sun* published a series of articles announcing "Great Astronomical Discoveries" by John Herschel, William's son, whose methodological work and study of the southern skies had made him perhaps the world's best-known astronomer. On August 25, the first article described a collaboration between Herschel and David Brewster, a Scottish scientist, to build a telescope that used its own light to augment the dim glow of celestial objects. Whereas the maximum magnification of Gruithuisen's largest telescope was at best 200x, Herschel's supposedly could reach 42,000x. If small refractors had uncovered signs of life on the Moon, readers surely wondered, what would Herschel's telescope reveal?[33]

On the following day the *Sun* began publishing Herschel's "observations" of the Moon. The paper established its credibility by reporting that Herschel had disproven earlier, fanciful accounts of lunar pyramids, roads, and canals. Then came the sensational news: sightings of trees, birds, and bizarre animals with a "hairy veil" that protected their eyes from lunar extremes of light and dark. New Yorkers flocked to purchase copies of the *Sun*. On August 27 they discovered that Herschel had sighted creatures that looked like beavers and used fire; on August 28 they learned he had glimpsed sentient life: the winged "*Vespertilio-homo*, or man-bat." On August 30 they found that the Moon had racial hierarchies resembling those imposed by Europeans and European settlers on Earth. Herschel introduced another race, "less dark in color, and in every respect an improved variety," and then at last, in a valley at the foot of a mountain, the highest form of man-bat, "in our eyes scarcely less lovely than the general representations of angels."[34]

With cholera epidemics, riots, and political unrest spurred in large part by debates over slavery, the 1830s were a frightening decade in New York. Not surprisingly, New Yorkers embraced the otherworldly diversion of Herschel's lunar observations. "It was too wonderful not to command instant belief," the author and astronomer Richard Proctor later observed; "the excitement spread to other countries, and, finally, the infatuation became . . . universal and complete." Many journalists and academics took the observations seriously. In an era of unprecedented technological progress, Herschel's fictional telescope seemed implausible only to specialists. More importantly, reports of lunar life seemed perfectly sensible in the wake of Gruithuisen's apparent discoveries. The *New Yorker* announced "a new era in astronomy and science generally," and many other publications followed suit. "Not one person in ten discredited" the reports, the poet Edgar Allan Poe later exaggerated, "and it is especially noticeable that the doubters were, for the most part, those who doubted without being able to say why."[35]

A month later, newspapers continued to print the reports, theaters began to dramatize them, and museums started to design exhibitions around them. Eventually, though, journalists unveiled the hoax for what it was and traced its origins to Richard Adams Locke, a reporter known thereafter as the "man in the Moon." The *Sun* boasted that it had increased its circulation by means of the hoax, but in all likelihood that claim was also false. At a time when newspapers sold for six cents, the *Sun* had set its price at a single penny since its 1833 launch. It was able to turn a profit only by using steam-powered printing presses to mass produce copies of its paper, and then by hiring newsboys to sell copies across New York. These techniques would be hallmarks of a new "penny press" that targeted the burgeoning urban working class ignored by six-cent papers, albeit with sensational stories and fraudulent advertising. The unique capabilities of the *Sun* ensured that just about everyone in New York heard about Herschel's discoveries at the same time. But the haste with which other papers reprinted the hoax ensured that the *Sun*'s circulation scarcely increased.[36]

The *Sun* did earn a handsome profit by selling lithographs of fanciful lunar environments and pamphlets with Locke's complete narrative. Yet the true profit for the *Sun* lay in revealing that six-cent papers, whose editors derided penny dailies for publishing low-brow nonsense, would reprint the same false stories that appeared in the cheaper rags. The "Moon Hoax" accelerated an ongoing shift toward the penny system of newspaper publishing, and with it a lasting democratization of news media. It also revealed that newspapers could earn handsome profits exploiting popular interest in outer space by promoting astronomers as swashbuckling explorers—a lesson that, we will see, later inspired some of the most sensational news stories of the nineteenth century. "The sensation created by this immense imposture, not only throughout the United States, but in every part of the civilized world," the American showman P. T. Barnum reflected, "will render it interesting so long as our language shall endure." Ironically, for millions the hoax had, for a few days, realized the centuries-old ambition of serious lunar scientists: to reveal a world like Earth in the changeable environments of the Moon.[37]

10

The Promise and Peril of Lunar Life

Like scientists in other fields, by the 1830s selenographers were beginning to doubt that observing an exceptional state of their subject—a change in the lunar surface, for example—would reveal its true nature. When astronomer Johann Heinrich von Mädler met wealthy banker and astronomy enthusiast Wilhelm Beer in Berlin, they agreed that Schröter, Gruithuisen, and other selenographers had been fooled by tricks of light. Using a Fraunhofer refractor in Beer's private observatory, they drafted spectacularly detailed maps that, when published in 1834, seemed like the last word in lunar cartography. Devoid of contrast or features resembling terrestrial environments, the maps published by Beer and Mädler depicted the Moon as a homogeneous wasteland. "The light contained in [their] work was . . . a 'dry light,'" Agnes Clerke later reflected, "not stimulating to the imagination." It recast the Moon as an airless, waterless exception to the plurality of worlds. Few now perceived any reason to search the Moon for signs of change. Schröter's former assistant, Friedrich Wilhelm Bessel, even condemned efforts to seek "traces of industry of lunar inhabitants."[1]

But ironically, maps of unprecedented accuracy actually provided a powerful tool for detecting change: they were detailed, reliable baselines that could be used to identify the subtlest variations in the lunar surface. Before long the new maps sparked a revived interest in lunar life that culminated, in the twentieth century, in a massive and ultimately futile effort to protect Earth's biosphere from lunar contamination.

Changing views of the Moon. Top left: The astronomer Johannes Hevelius charts a Moonlike Earth (complete with regions exposed by libration), 1647. Top right: The romanticist artist John Russell drafts a mysterious, living Moon, 1805. Bottom left: The Beer and Mädler map of an inert, lifeless Moon. Bottom right: A 1966 USAF map of a Moon to be explored and exploited.

Reinventing an Earthlike Moon

On October 16, 1866, J. F. Julius Schmidt, director of the National Observatory of Athens, Greece, suddenly realized that the crater Linné, clearly marked on the Mädler-Beer maps, had vanished, leaving behind a small white patch. Schmidt was not the first to announce that a part of the lunar surface mapped by Beer and Mädler had changed—but his reputation made all the difference.[2]

The news electrified the scientific community and spread like wildfire through the popular press. Were alien builders at work on the lunar surface, after all? For months every major telescope in every major observatory across Europe and the colonial world followed the Moon. Energized by the possibility of lunar life, thousands of people across Europe and North America bought small telescopes, catapulting amateur astronomy from being a niche amusement for eccentric gentlemen to a serious interest for the scientifically minded, with lucrative profits for the makers of fine optics and telescopes. Millions of amateurs have kept the solar system under constant surveillance ever since.[3]

Astronomers initially concluded that Linné really had vanished. Their most powerful telescopes later revealed that the crater plainly seen by Beer and Mädler was now visible only by its shadow, at local daybreak or sunset. Apparently it had shriveled to a fraction of its former size and surrounded itself with white debris. Astronomers also found that in 1788 Schröter had reported a sudden darkening of the crater, which suggested that a force had been at work around the crater for nearly a century. Then in 1878 Hermann Klein, a German selenographer in Cologne, announced the appearance of a new crater near Hyginus, a lunar caldera. Amid the sensation caused by these apparent environmental changes, California's richest man, the aging land baron James Lick, abandoned his dream of building a gigantic pyramid to cover his casket when he died. Instead he chose to finance the construction of the world's largest telescope, atop the world's first permanently installed mountaintop observatory. The Lick telescope, he decided, would use the thin and steady mountain air to detect lunar change and thus lunar life, immortalizing his legacy better than a pyramid could. That discovery never came, of course, but astronomers received a windfall of more than $1.2 billion (in 2024 US dollars), and Lick was buried under his observatory.[4]

In response to the commotion caused by Linné, William Radcliff Birt, an English amateur astronomer who had performed meteorological work for John Herschel, organized an international campaign to track changes on the Moon. From the pages of the *English Mechanic,* a widely read newspaper that published exchanges between laymen and experts, Birt recruited "workers" who would follow his instructions in observing lunar environments. Writing under pseudonyms, women and men reported their lunar observations, theories, and suggestions to Birt, who published a selection with his reflections and directions. Far from being a niche pursuit, in Britain this study of the Moon belonged to a populist reinterpretation of science as a practical activity for ordinary people. By the end of the nineteenth century, it was a reinterpretation that unfolded in museum exhibitions, public lectures, scientific societies, and popular publications. By scouring the Moon for signs of change, Birt's workers contributed to a broad movement that gradually integrated science within British popular culture.[5]

Readers of the *Mechanic* searched for the "minutest objects" new telescopes could reveal on the Moon, because the lunar surface now seemed to change on relatively small scales. Enthusiasts reported that craters and bright patches winked in and out of existence around the crater Picard in Mare Crisium, for example, while the size and shape of the twin craters Messier and Messier A relentlessly fluctuated. They detected changes in the "walls" and "ribs" of Wallwerk; in mountains along the floor of the crater Frascatorius; in the extent of winding rilles charted by Beer and Mädler; in the shape of the central peaks of the giant crater Gassendi. Of special interest was the floor of the ancient and enormous crater Plato. There, observers agreed, tiny craters appeared and disappeared—a phenomenon that would long sustain theories of lunar vegetation. Enthusiasts also noticed flashes that seemed to be a product of weather in the Moon's atmosphere.

To support observations by readers of the *English Mechanic,* in 1878 Birt cofounded the Selenographical Society. Its members, all of them elite lunar observers, shared their sightings and collected them in a tattered notebook. They reported striking changes to complex lunar environments around the Hyginus caldera, for example, where a new crater seemingly appeared; along the east wall of the great crater Archimedes, where a triangular spot emerged; and at the even larger crater Cleomedes, where a "bright streak" suddenly came into view. Birt reported lunar changes identified by the Society in the pages of *English Mechanic,* alongside descriptions of new techniques and technologies that serious observers should now use to record the subtlest lunar details.[6]

Reports of lunar change dominated columns in the *New York Times* and inspired letters to local newspapers. Meanwhile the Moon Hoax had paved the way for penny dailies to launch what the historian of science Joshua Nall calls an era of "event astronomy" that exploited the expectations of colonial readers.

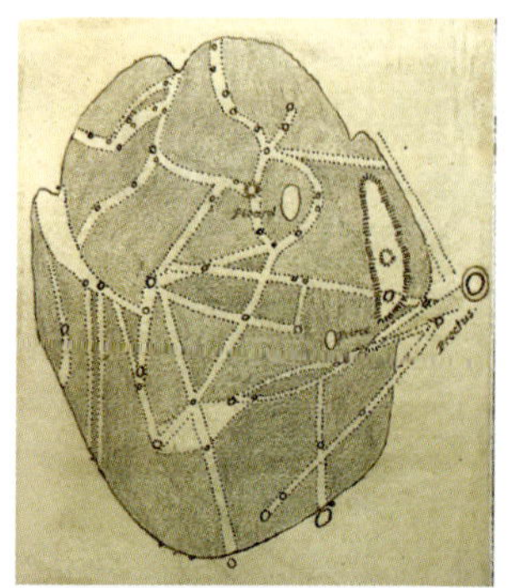

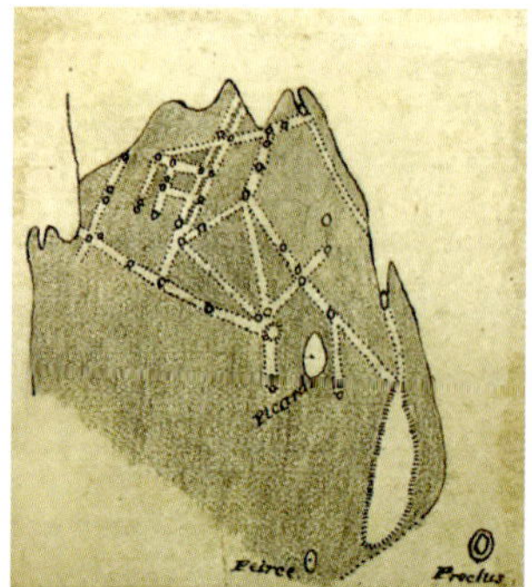

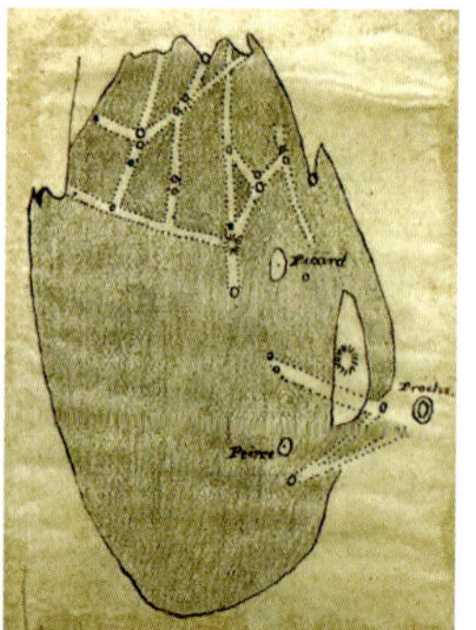

Canals on the Moon? Top: Straight lines crisscross sketches of Mare Crisium by A. Stanley Williams, in the Selenographical Club logbook. Bottom: rilles surround the central peaks of Gassendi, in a photo taken during *Apollo 16*.

By building and operating telegraph lines to far-flung observatories, penny dailies like the *New York Herald* ran sensational stories in which swashbuckling astronomers on the frontiers of empire uncovered cosmic environmental changes that had never been observed before. Few fared better in this new media environment than William Pickering of the Harvard College Observatory. In 1891, William's brother, the observatory's director Edward Pickering, dispatched William to establish a humble mountaintop observatory in Arequipa, Peru. William was supposed to supply Harvard's all-woman team of "computers" with detailed spectroscopic and photometric measurements of Southern Hemisphere stars. Instead William spent lavishly to construct a handsome retreat devoted to observing change in specific environments on Mars and the Moon. From Peru, he would send reports to the *Herald* that inspired an explosion of popular interest in canal-building Martians, while offering nothing of value to Harvard's computers (Chapter 14).[7]

Well before theories of Martian canals entered the scientific mainstream, observers like Pickering used increasingly capable telescopes to uncover more and more rilles on the Moon. Some took these for canals. Pickering explained that "the lunar canals are much smaller and perhaps broader, in proportion to their length, than those of Mars, yet on account of the nearness of the Moon its canals are much more readily seen than the Martian ones." Paying attention to the rilles prepared observers to identify and map similar but subtler features on Mars. Percival Lowell, for example, drew straight lines along the fractured floor of the lunar crater Atlas while charting canals on Mars in 1894.[8]

For Pickering the significance of such observations lay not in training the eye but in explaining the causes of canal formation on both Mars and the Moon. Changes in canals on the Moon seemed to unfold with every month-long lunar day, which to Pickering hinted at the growth and decay of vegetation. Canals on the Moon and Mars, Pickering concluded, revealed "the tenacity with which life will exist throughout the universe in situations that seem to us, from our ignorance, most unfavourable and most unsuited to it." Later Pickering blamed the migration of insect swarms for changes in the appearance of the crater Eratosthenes.[9]

Pickering's observations of what he called "real, living changes" on the Moon were taken seriously by contemporary astronomers, because very little was definitively known about the lunar surface at the turn of the century. "The theory of our satellite being a burnt-out cinder," the president of the Royal Astronomical Society of Canada assured an audience in 1897, "will no longer hold." Pioneering silent films by Georges Méliès and Fritz Lang soon thrilled audiences by imagining fanciful (in Méliès's *Le Voyage dans la Lune*) and seemingly realistic (in Lang's *Frau im Mond*) voyages to an Earthlike, habitable Moon. Professional astronomers, though, would soon lose interest in the Moon. Earth's turbulent atmosphere limited the lunar detail visible with even the biggest and newest telescopes, and so there seemed to be little hope of resolving the controversies raised in the age of selenography. Breakthroughs that revealed the expansion of the universe, the existence of other galaxies, and exotic physics of the most extreme environments in deep space drew most astronomers' attention away from the study of planetary and lunar environments.[10]

Surviving Lunar Life in the Space Age

Scientists returned to the Moon only as advances in rocketry allowed competition between the United States and Soviet Union to spill beyond the confines of Earth's atmosphere. A popular "space craze" underway since the 1920s had inundated children in particular with books, comic strips, movies, and toys that gradually raised popular enthusiasm—and expectations—for space travel. In this context, the launch of *Sputnik* in 1957 seemed to reveal that achievements in space had a unique capacity to signal the superiority not only of a nation's military-technological capabilities, but also of its political system and potency as an ally. In the United States, a web of politicians, bureaucrats, engineers, and scientists laid out careful plans to match Soviet triumphs in space. Yet the perception of a widening gap between the United States and the Soviet Union launched a devoted Cold Warrior, John F. Kennedy, to the presidency and encouraged him to upend those plans. In 1961 he called for the United States to land a man on the Moon by the end of the decade. This would be the finish line to a new "Space Race" that, NASA administrator James Webb assured Kennedy, was spectacular enough to signal victory, near enough to reach within politically expedient time frames, and distant enough to give the United States time to leapfrog the Soviet Union. In response, Soviet first secretary Nikita Khrushchev eventually authorized multiple Soviet efforts to land cosmonauts on the Moon before NASA could get there. Suddenly lunar environments became matters of national security—and renewed scientific scrutiny.[11]

Soviet and US spacecraft quickly established that primitive vegetation did not grow in sheltered cracks on the lunar surface, as many scientists had suspected. Yet although exobiologists had developed a device to test for the presence of microbial life, robotic missions were not designed to look for it and could not rule out its presence on or below the lunar surface. Because it seemed entirely plausible that microorganisms had evolved on the Moon as on Earth, exobiologists pointed out that spacecraft could forward contaminate a lunar biosphere, and back contaminate the terrestrial biosphere by returning microbes from the Moon (Chapter 8).[12]

Once NASA established its Apollo program to land astronauts on the Moon, the risk of back contamination elicited widespread concern within the US federal government. In 1962 scientists convened by the National Academy of Sciences concluded that "the introduction into the Earth's biosphere of destructive alien organisms could be a disaster of enormous significance to mankind," a "tragically ironic consequence of our search for extraterrestrial life" that NASA "should do all in its power" to prevent. Carl Sagan, who had, as an undergraduate student, written Secretary of State Dean Acheson to ask how the federal government would respond to hostile UFOs, now catapulted to prominence partly by warning that microbes, eking out an existence on the Moon, could multiply explosively in Earth's benign environment.[13]

NASA administrators ignored these warnings until congressional hearings in the spring of 1963, when senators grilled agency officials over their preparations for preventing the contamination of Earth. Regulators at the National Institutes of Health and the Department of Agriculture soon threatened to prohibit the return of astronauts from the Moon if NASA did not take steps to prevent back contamination. In 1965 the Public Health Service joined NASA in establishing the Interagency Advisory Committee on Back Contamination (ICBC). The committee was supposed to ensure that everything returned from the Moon, including astronauts, would be quarantined "until shown to be free of possible pathogenic or infectious materials and otherwise not harmful to man's biosphere." Experts privately agreed that

NASA could only delay, rather than prevent, the release of microbes introduced from the Moon.[14]

Nevertheless, it was now clear that the success of the entire program hinged on how well NASA could appear to protect Earth from lunar life. NASA officials drafted plans for the most complex biological facility ever built, an 86,000-square-foot "Lunar Receiving Laboratory" (LRL) that would prevent lunar microbes from escaping, and, to preserve the purity of samples returned from the Moon, keep terrestrial microbes from entering. Engineers and scientists designed the facility's procedures, sterilization techniques, and biological barriers to contain *Yersinia pestis,* a pathogen that was impressively deadly—it causes bubonic plague—but so attuned to a specific host organism that nothing like it could have evolved on the Moon. Worse, when the facility finally came online in late 1968, its equipment repeatedly failed in rounds of testing that alarmed ICBC representatives. Officials also struggled to hire technicians who would monitor the astronauts for sickness, grind lunar samples to dust, and introduce the dust to plants, animals, and living tissue. All were required to sign agreements that forbade them from leaving the facility if exposed to lunar contaminants, and prevented their relatives from claiming their bodies.[15]

In the months leading up to the *Apollo 11* Moon landing, tensions worsened between ICBC representatives who prioritized low-likelihood, high-consequence risks to Earth, and NASA managers who prioritized high-likelihood, relatively low-consequence risks to astronauts and equipment. Quarantine advocates insisted that the *Apollo* command module be sealed as soon as it returned to Earth, that astronauts stay within until they reached a mobile quarantine facility, and that astronauts remain quarantined even in the event of a medical emergency. They proposed using *Coxiella burnetii,* the pathogen responsible for Q fever, to test the LRL and its personnel. *C. burnetii* forms spores that could protect it on the Moon's surface, unlike *Y. pestis,* and would resemble a dangerous lunar microbe by spreading easily through suspended lunar dust. NASA managers rejected these suggestions, which could have killed astronauts or delayed the first lunar landing. It was easier for them to accept a less likely risk to the entire Earth.[16]

The nature of that risk largely escaped public attention until a deadly influenza pandemic killed one million people globally in 1968. Late that year, major US newspapers and magazines published a series of exclusives that compared the potential consequences of lunar contamination to the horrific toll in Indigenous communities of smallpox, measles, and other diseases introduced by European settlers. Then, just months before *Apollo 11,* Michael Crichton published *The Andromeda Strain,* a wildly popular book that imagined efforts to contain an extraterrestrial pathogen in a facility inspired by the LRL. NASA officials rushed to placate public fears by assuring journalists that their quarantine protocol used technology of unparalleled sophistication that had been rigorously tested. Privately, however, they fretted over the unpreparedness of the LRL and attempted to ignore the ICBC's recommendations.[17]

In July 1969, after Neil Armstrong, Buzz Aldrin, and Michael Collins returned to Earth aboard the *Apollo 11* command module, the public followed along as journalists reported perplexing and at times frightening news from the quarantine wing of the LRL. Shortly after the astronauts arrived, for example, six technicians were exposed to Moon dust while opening film from the *Apollo* capsule. They immediately stripped, showered, and joined the astronauts. Meanwhile, botanists found to their surprise that plants exposed to Moon dust grew unusually quickly. Luckily, the cause turned out to be mundane: elemental nutrients in the dust had never encountered water and so were easily absorbed by plants.[18]

That November, Alan Bean and Pete Conrad touched down on the lunar surface within sight of *Surveyor 3,* a robotic lander that had preceded them by about two and a half years. After the astronauts returned to Earth with the robot's camera housing, LRL technicians found bacteria within its interior. It was not immediately clear whether the bacteria posed a threat to the terrestrial biosphere. Nevertheless, the discovery seemed to justify the entire LRL facility. Technicians soon found that the bacteria had no effect on a laboratory mouse, and then identified it as alpha hemolytic *Streptococcus mitis,* a common pathogen in the human respiratory system. Although Crichton had feared that microorganisms in inadequately sterilized spacecraft would return to Earth "different," LRL technicians determined that the bacteria had originated on Earth and survived on the Moon without changing. One year later, microbiologists raised fears of a different kind, after reporting that some terrestrial bacteria died after just ten hours of direct exposure to Moon dust.[19]

Many other accidents and alarms were never communicated to the public. *Apollo* astronauts, for example, found it impossible to avoid breathing in the

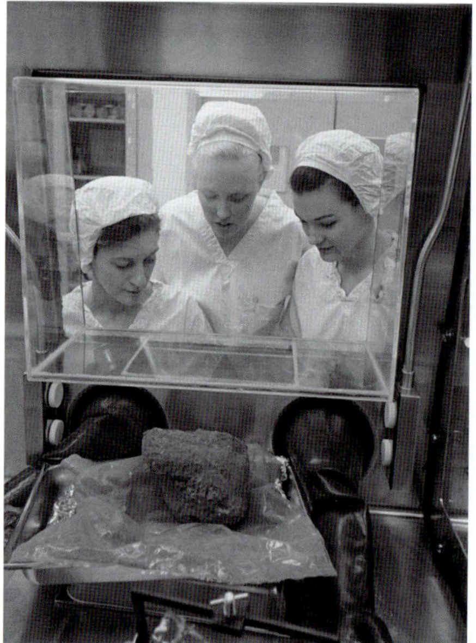

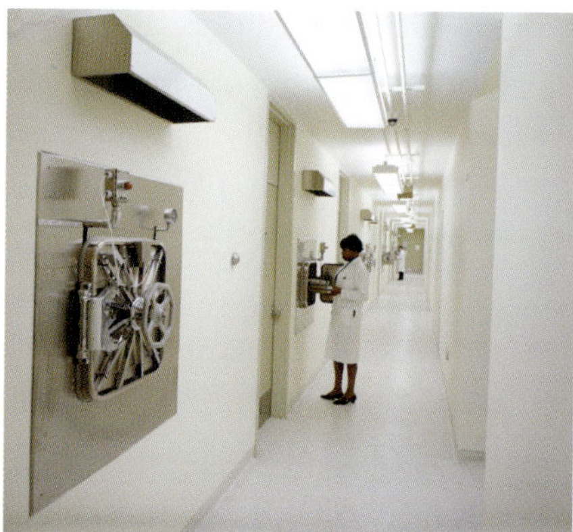

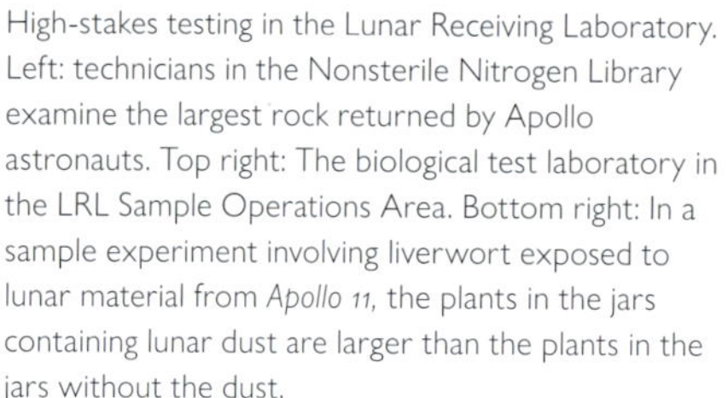

High-stakes testing in the Lunar Receiving Laboratory. Left: technicians in the Nonsterile Nitrogen Library examine the largest rock returned by Apollo astronauts. Top right: The biological test laboratory in the LRL Sample Operations Area. Bottom right: In a sample experiment involving liverwort exposed to lunar material from *Apollo 11*, the plants in the jars containing lunar dust are larger than the plants in the jars without the dust.

fine Moon dust, which mysteriously smelled like gunpowder. Summary reports drafted by LRL officials reveal routine breaches involving punctured gloves, spills, plumbing problems, and even exposed personnel. Rumors circulated that technicians had spotted cockroaches entering and exiting the facility. Lunar microbes, had they existed, would in all likelihood have breached the LRL.[20]

At least some of the technicians and doctors who implemented the quarantine protocol realized that its procedures were flawed, and suspected that NASA did not take the threat of back contamination seriously. The astronauts' doctor, an Army officer named David Yates Graham, called me shortly before the publication of this book. In 1969 NASA had asked him to help with preparations for the return of the *Apollo 11* astronauts. Their capsule was slated to splash down in the Pacific Ocean, whereupon divers would extract them and a helicopter would airlift their capsule to a waiting aircraft carrier. When Graham realized that Navy personnel planned to use disinfectant to wash the capsule and the divers, it dawned on him that "the whole thing is a farce," because the disinfectant would spread potentially dangerous microorganisms without killing them all. He insisted that he was "not going to be involved in this," and protested to his superiors at NASA, who promptly asked the Army to remove him. To keep him from talking to reporters, they gave him a long vacation. "So I said, fine, good, I'm out of here," he recounted, "and that was the end of that."[21]

Like physicists in the Manhattan Project who tested the first atomic bombs despite lingering suspicions that the explosions would ignite Earth's atmosphere or oceans, NASA managers were willing to risk a great deal to help America defeat its enemies. And like those physicists, they gambled alone, without adequately informing elected policymakers. Had the Moon actually harbored life—as its supposedly changeable surface once seemed to suggest—then the consequences could have been catastrophic. It was a good thing that the quest for life on the Moon uncovered nothing. Humanity might not be as fortunate if samples are brought to Earth from other worlds in the solar system.[22]

11

Lunar Changes, Human Ambitions

In 1936 a disgruntled graduate student informed the People's Commissariat for Internal Affairs—the Soviet secret police—that counterrevolutionary activities were afoot at Pulkovo Observatory in St. Petersburg. Thirty-one-year-old observatory astronomer Nikolai Kozyrev was sentenced to ten years in the gulag. In 1941 he was sentenced to another ten years. As he trudged to work in Siberia, guards commanded him to keep his eyes on the ground, rather than the heavens. He endured solitary confinement in an unheated cell, shared a room with a madman, temporarily went deaf, and nearly died. He "saved himself," Aleksandr Solzhenitsyn wrote in *The Gulag Archipelago,* "only by thinking of the eternal and infinite: of the order of the Universe—and of its Supreme Spirit; of the stars; of their internal state; and what Time and the passing of Time really are."[1]

In January 1947, lobbying by his colleagues won Kozyrev an early release. For a decade he struggled to rebuild what had been a promising career. Then, on a November night in 1958, he made what seemed like the discovery of a lifetime. Peering into the eyepiece of a big reflector at the Crimean Astrophysical Observatory, he noticed a red flare in the lunar crater Alphonsus. He quickly obtained what seemed like clear spectroscopic evidence of lunar volcanism. The director of Leningrad Observatory, Aleksander A. Mikhailov, announced that this appeared to inflict a "crushing blow" to the "conception of the moon as a dead body."[2]

It was just over a year since *Sputnik* had launched the Space Age, and it would be a little more than a year until Yuri Gagarin became the first human to orbit Earth. Already the Moon seemed a promising target for Soviet cosmonauts and US astronauts. Scientists and engineers scrambled to determine whether the Moon's environments really did change, in ways that could threaten human explorers and maybe dictate the outcome of the Space Race. Selenography had become a Cold War science.

The Hidden Figures of Project Alphonsus

New lunar maps and photographs did not reveal any evidence for the changes in the shape, size, or elevation of lunar environments that many nineteenth- and early twentieth-century selenographers believed they could see. For centuries, though, observers had also reported strange illumination effects over a small selection of bright lunar craters: lights that seemed to come and go; red or blue spots that seemed to appear and disappear; patches of the surface that seemed suddenly hazy before snapping back into focus. In 1956 the US astronomer Dinsmore Alter noticed that ultraviolet photographs he had taken of the crater Alphonsus showed a "definite haze."[3] With much greater fanfare, Kozyrev soon reported spots that he attributed to volcanism. Then, on two nights in 1963, a group of astonished Air Force mappers at Lowell Observatory reported a wave of blue and red flares over another crater, Aristarchus. The news made international headlines, and the strange anomalies received a name: transient lunar phenomena, or TLPs.[4]

Officials at NASA and the US Department of Defense sought to determine whether TLPs existed, because the phenomena seemed capable of destroying lunar spacecraft, potentially causing catastrophic damage to US prestige. They were also intrigued by the possibility of exploiting the phenomena. TLPs hinted at a potential source of energy on the lunar surface that could, for example, refuel rockets for travel across the solar system. This might justify the soaring cost of the Apollo program. In 1964, NASA representative Alan Rice argued it was "vital to our national as well as scientific interests" that scientists uncover the truth about TLPs. "The timeline for the exploration of our solar system can be greatly influenced by the date this information is obtained."[5]

Leading the search for lunar changes. Astronomers Jaylee Burley (left) and Winifred Cameron (right) spearheaded NASA's push to explain TLPs.

Pioneering women soon took the lead in organizing the effort to demystify TLPs. In 1965, astronomers Jaylee Burley of the Goddard Space Flight Center and Barbara Middlehurst of the University of Arizona assembled the first chronological list of observations of lunar changes. Out of more than 200 reports, they found 159 to be credible. They dated them to periods in which Earth's gravity tugged on the Moon with particular force, owing to especially close approaches of the Moon to Earth. Because these events appeared to correlate with tidal forces that could plausibly force gas through weak points along the lunar crust, TLPs seemed even more likely to be real.[6]

To unravel the TLP mystery before its astronauts landed on the Moon, NASA signed a contract with Trident Engineering Associates to manufacture a strange new device: an alternating red and blue filter that, when attached to telescope eyepieces, would reveal otherwise subtle TLPs as flashing black dots. At NASA's Goddard Space Flight Center, astronomer Winifred Cameron coordinated with

Trident engineers to distribute this Moon-Blink contraption to observatories across the United States. Cameron no doubt knew that amateurs had fallen far from the cutting edge of lunar science, yet military surveillance programs had successfully mobilized thousands to monitor the skies for Soviet satellites and bombers. Perhaps with that in mind, Cameron eagerly recruited amateurs to her network.[7]

The outcome was a "global visual patrol," as Trident engineers put it, that kept the Moon under "constant surveillance." In 1966, the network—now named Project Alphonsus—expanded when some two hundred amateurs in the British Astronomical Association (BAA) voted to form a team that would use makeshift Moon-Blink devices to hunt for changes on the lunar surface. "Had anything of the sort been suggested ten years earlier," reflected seasoned BAA lunar observer Patrick Moore, "it would have invited ridicule." Interest quickly spread across Canada and into Japan, where Shotaro Miyamoto led a Moon-Blink team at Kwasan Observatory. Newspapers in the United States reported when amateur astronomy clubs in their cities joined the international effort to search for, as one Houston-based journalist put it, "explosions, eruptions, or volcanic action" that "could endanger the lives of America's astronauts and spoil a multi billion dollar [*sic*] space probe."[8]

By the middle of the 1960s, Cameron had established perhaps the largest observing campaign ever directed at a celestial environment. She set up a hotline that enabled up to sixteen US observatories to coordinate their TLP observations in real time through an operator in Washington, DC, and had them report TLP sightings on standardized datasheets. She also gave free rein to amateur associations that, collectively, directed the global observations of thousands of seasoned lunar observers. Results were not long in coming. NASA press releases repeatedly announced when observers in the network seemed to detect TLPs, praising its Moon-Blink devices for making the sightings possible. "TLPs do really occur," Moore concluded in 1967, "and may be less uncommon than has been thought." All the while, speculation mounted about the causes of transient lunar phenomena, in debates that spilled beyond the pages of the world's most prestigious scientific journals and assumed increasing urgency as the first lunar landing approached.[9]

Transient Lunar Phenomena and the Apollo Mission

In January 1965 Lyndon Johnson informed Congress that NASA's new program to detect phenomena on the Moon hinted at the presence of "extraterrestrial resources," including "energy sources on the lunar surface which could be exploited." The obvious next step was to dispatch robotic spacecraft to Alphonsus and Aristarchus, the craters where TLPs seemed most common. By now NASA engineers had worked out the sequence by which their robots would pave the way for the crewed lunar landing that would, ostensibly, prove the superiority of the US political and economic system. *Ranger* probes had already smashed into the Moon, taking high-resolution photographs of select sites during their descent. Beginning in 1966, *Surveyor* landers would test the Moon's surface—its regolith—to ensure that it could support the weight of an astronaut or two. Meanwhile, *Lunar Orbiter* spacecraft would assemble high-resolution maps of the Moon to permit detailed mission planning. When the robots had done their work, the *Apollo* craft would land in a "safe zones"—flat, accessible areas of the Moon's near side where astronauts could maintain round-the-clock radio contact with Earth. Parallel stages of exploration unfolded in the Soviet space program, with *Luna* probes at first outpacing their US counterparts, but the Soviet crewed program eventually fell far behind.[10]

The purpose of NASA's robotic exploration campaign was to scout the least dangerous (and for many scientists, least interesting) landscapes on the Moon, rather than to prospect for resources in what seemed like the Moon's most volatile, unpredictable environments. Prominent scientists, though, ultimately convinced NASA administrators to incorporate science, especially geology, in its robotic and crewed lunar exploration programs. In 1965 NASA allowed *Ranger 9* to explore a "scientific" target: Alphonsus, with Aristarchus as the backup. In late March *Ranger 9* smashed into Alphonsus, broadcasting to US television live images that revealed small craters along volcanic fractures.[11]

To geologists, Alphonsus did not look like an especially safe site to land an *Apollo* craft, but it did seem more interesting than any other lunar environment visited to date. NASA's Office of Space Science and Applications now suggested an ambitious plan. It involved robotic missions not only to Alphonsus and Aristarchus, but also to that other locus of supposed change, the crater Linné. Soon after the Surveyor program got underway in 1966, Benjamin Milwitzky, its program manager at NASA headquarters, argued that a new wave of robotic landers should be sent to Aristarchus and Alphonsus. In the summer of 1967 scientists suggested both craters as landing sites for *Surveyor 6*, but the demands of the Apollo program redirected the spacecraft to the relatively dull plains of Senius Medii. Still, that summer scientists did manage to devote the end of the *Lunar Orbiter 5* flight to studying Aristarchus and its immediate vicinity, where they had hoped, but failed, to uncover an explanation for TLPs.[12]

The summer of 1967 was high noon for NASA's lunar exploration program. The agency's annual budget had declined since 1966 but still hovered at nearly $50 billion (in 2024 USD), nearly double its current total. That was more than enough to envision ambitious schemes for lunar science and settlement. A proposal from the Aerospace Systems Division of Bendix Corporation, a NASA contractor, suggested deploying instruments on the Alphonsus crater floor to monitor volcanic activity. In August the Manned Spacecraft Center (MSC) in Houston, which later became NASA's Lyndon B. Johnson Space Center, organized a major summit and study in Santa Cruz that resulted in extraordinary plans for two simultaneous landings at Alphonsus, and two more at Aristarchus. In both missions, astronauts would stay on site for at least six days. Upon departing they would release a giant, robotic rover to explore new destinations hundreds of kilometers away. Although some geologists worried about the risks, the US Geological Survey subsequently proposed Alphonsus and the Aristarchus Plateau as possible landing sites for later Apollo missions. NASA's byzantine process for selecting the sites weighed both options, along with the Aristarchus crater itself. In March 1968 the Apollo Site Selection Board considered Alphonsus as a destination for the second or third Apollo landing, while MSC planners pored over Lunar Orbiter photographs to see if astronauts could land there, or at Aristarchus. Yet another committee, the Group for Lunar Exploration Planning, met in July to consider landings at both Alphonsus and Aristarchus.[13]

Later that year, *Apollo 8*, the first crewed flyby of the Moon, promised to revolutionize the study of TLPs. Barbara Middlehurst, the astronomer who had earlier helped compile a list of all reported TLPs, realized that observations of

TLPs from terrestrial telescopes could be verified in real time by astronauts orbiting just a hundred kilometers above the lunar surface. She approached Robert Citron, an engineer who had earlier helped establish a program, named Moonwatch, that organized volunteers, including amateur astronomers, to observe Soviet satellites. Citron now established a communications network centered on the newly established Center for Short-Lived Phenomena, an office at the Smithsonian Institution that focused on the rapid study of meteorite impacts, volcanic eruptions, oil spills, and other sudden environmental changes. The Lunar International Observers Network (LION) would link the center to 125 observing stations in twenty-one countries, all of which would keep the Moon under constant surveillance on every clear night. Reports of transient phenomena on the Moon would be relayed to the center, and then to a dedicated LION desk in the Science Support Room at the MSC in Houston. After receiving LION reports, the director of the Support Room might decide to contact Mission Control, and the *Apollo* astronauts might then be asked to observe the TLPs on location. NASA contracted with Lockheed Electronics Company to manage the network; Lockheed in turn chose Middlehurst to help organize LION and interpret its observations. Though Middlehurst was listed last in rosters of the committee that oversaw LION, the network depended on her expertise and on many of the same observatories and astronomers that she and Cameron had previously connected in Project Alphonsus.[14]

When Frank Borman, James Lovell, and William Anders finally rocketed to the Moon on *Apollo 8* in December 1968, LION kept watch. By the following May, as Thomas Stafford, John Young, and Gene Cernan journeyed to the Moon in *Apollo 10,* LION connected more than three hundred professional and amateur astronomers in forty-six US observing stations and another 130 stations in thirty-one countries. Citron developed a code whereby observers could communicate TLP reports without speaking English; Soviet astronomers even used the network to send comments and congratulations to the *Apollo* astronauts.[15]

During the Apollo 10 mission, "bad weather was reported from many parts of the world," according to Lockheed, thwarting TLP observations in Tokyo, much of New Zealand, Kazakhstan, parts of Brazil, and large swaths of Europe. Nevertheless, observers transmitted fifty-four TLP reports through LION, seventeen of which involved Aristarchus. Twice the LION support desk contacted the director of the Science Support Room at MSC, who in turn asked Mission Control to relay requests for TLP confirmation to the *Apollo* astronauts. Mission Control patched through the first request, but the astronauts reported "no sign of activity in Aristarchus." They were asleep when the second request reached Mission Control, so it was not sent through.[16]

Just two months later, in July 1969, Neil Armstrong, Michael Collins, and Buzz Aldrin crowded into *Apollo 11* and embarked on their historic quest to land on another world. As they hurtled to the Moon, LION received six reports of TLPs at Aristarchus, four of which were independently verified. The LION desk passed on a request for astronaut observations to A. W. Patteson, the science missions manager, who relayed it to Lou Wade in the Apollo Mission Control Center just before the crew initiated the all-important engine burn that would bring them into the Moon's orbit. The LION request was denied.[17]

On July 19, capsule communicator Bruce McCandless finally transmitted LION's request as the crew approached Aristarchus on their first, elliptical orbit around the Moon. As the minutes ticked by, observers on Earth reported TLPs in the crater and relayed the news to LION. Eventually Collins spoke up. "Hey, Houston," he began, "there's an area that is considerably more illuminated than the surrounding area. It just has—seems to have a slight amount of fluorescence to it." Aldrin agreed. "That area is definitely lighter than anything else that I could see out this window," he reported.[18]

In their first debriefing after returning from the Moon, the astronauts endured question after question about the TLP they had apparently sighted at Aristarchus. "In this general area, there was a generally higher level of illumination," Armstrong reflected while in quarantine at the LRL, but added, "I don't mean to imply that it looked like it was self-illuminated." Still, it was then two weeks after the fact, and Middlehurst lamented that a LION representative had not been present to ask Armstrong further questions. US magazines and newspapers emphasized the crew's apparent sighting of the TLP in their first, sensational articles on the lunar landing. "Many scientists believe the glows are caused by lunar eruptions, complete with fire fountains and lava flows," a story in *Time* magazine reported on July 25.[19]

Brilliant Aristarchus. *Apollo 15* photograph of the Cobra Head volcanic vent, with Aristarchus to its right.

That same month the Site Selection Board debated whether *Apollo 12* should risk landing at Aristarchus. Committee members eventually chose another site, but TLP hotbeds remained attractive targets for mission planners. That autumn the Group for Lunar Exploration Planning considered Alphonsus as a primary or perhaps a backup landing site for *Apollo 13*. However, its astronauts never made it to the Moon's surface, and their fortunate return to Earth in a crippled spacecraft underscored the risks inherent in the Apollo program. By the spring of 1970 it was abundantly clear that the soaring cost of the Vietnam War, the skepticism of Richard Nixon, and perhaps especially the growing public sentiment that lunar landings represented misplaced federal priorities were all conspiring to bring the Apollo program to an untimely end. The repeated failure of the Soviet N-1 moon rocket and the growing spirit of détente between the Cold War superpowers meant that there was no geopolitical threat to galvanize public or political support for US lunar exploration. Indeed, what was there to explore? NASA still could not confirm that TLPs were real, still less that they revealed a source of energy or profit on the lunar surface. The Apollo missions had seemingly revealed the Moon to be a dead world, an unchanging wasteland that could never justify the expense required to repeatedly travel to it.[20]

NASA planners had agreed that the Apollo 18 mission would unravel the mysteries of transient phenomena by visiting the Aristarchus region, despite the risk. Yet by 1970 it was clear that the Apollo 17 mission would be the last in the program. Because the program had been planned around an orderly sequence of progressively more capable missions to progressively more interesting sites, it would now "fall short of minimal recommended scientific exploration" and would be all but useless for supporting what was still NASA's long-term goal—constructing a lunar base. NASA's administrator, Thomas Paine, sent a series of letters that pressed senators and representatives to support the final Apollo missions. Prominent scientists argued that it would be absurd to waste the expensive vehicles and equipment that had already been built. Yet the Apollo program had lost the support of politicians and the public. Astronauts had never visited the mysterious TLP hotbeds on which NASA had once pinned its hopes for human expansion into the solar system.[21]

Instruments ferried to the Moon by Apollo astronauts, however, did hint that volatile gases periodically vent into space from the lunar surface, particularly in and around Aristarchus. On at least three occasions astronauts reported sudden flashes on the Moon: bizarre phenomena plausibly caused by electric discharges on the lunar surface. Was the Moon a dynamic place, after all? When it came to most aspects of lunar science, the aborted Apollo program offered more questions than answers. Still, because Apollo astronauts had uncovered no clear proof of lunar change, scientists firmly disavowed the existence of transient phenomena. Most assumed that TLPs were nothing more than products of wishful thinking, more examples of lighting effects misinterpreted by eager but naive observers.[22]

Most historians of the Apollo program have ignored the women and amateur astronomers who sought evidence for a dynamic Moon. Yet Barbara Middlehurst, Jaylee Burley, and Winifred Cameron anticipated some of the most innovative and important areas of present-day lunar science. A new wave of research suggests that transient phenomena really do occur around Aristarchus and Alphonsus. There is a rush to confirm the phenomena because the growing pace of robotic and crewed missions to the Moon may soon alter the tenuous lunar exosphere and overwhelm trace gases released by TLPs. Evidence for change on the lunar surface once seemed to threaten NASA's plans for the Moon. Now what may be imperiled is that evidence itself.[23]

12

Exploiting the Moon

Soon after the launch of *Sputnik* in late 1957, Joshua Lederberg (Chapter 8) and geneticist J. B. S. Haldane met in Calcutta during a lunar eclipse. It was forty years since the October Revolution set the stage for the formation of the Soviet Union. Lederberg and Haldane calculated that, during a lunar eclipse, the explosion of a nuclear bomb on the Moon's surface might be visible as a "red star." It would be a "second coup" for the Soviets after *Sputnik,* said Haldane, for it would provide a truly worldwide demonstration of the power of Soviet rocketry.[1]

Now Lederberg and Haldane considered the depressing possibility that radioactive fallout would imperil lunar microorganisms, and thereby foil efforts to better understand the origin of life. This concern led Lederberg to call, for the first time, for efforts to protect "the moon and other planets from inadvertent radioactive or biological contamination." Before long the newly formed Committee on Contamination by Extraterrestrial Exploration (Chapter 8) also warned against detonating nuclear bombs on Mars or Venus. Exobiology, with its emphasis on the existential risk posed by space travel to terrestrial and cosmic environments, was entering the deliberations of the US and Soviet space programs. What Lederberg did not know was that his protégé Carl Sagan had secretly been advising the US Air Force on its own plans to detonate a nuclear weapon on the Moon.[2]

Indeed, the Anthropocene—the name proposed for a geological epoch distinguished by human dominance over our planet's great systems and cycles—may have reached the Moon not long after it started on Earth. Owing to military schemes, space exploration programs, and, lately, the ambitions of corporations and nonprofits, human action has disturbed at least fifty-eight locations on the lunar surface, and it is transforming the Moon's tenuous atmosphere. The environmental changes wrought to date pale in comparison to those that never left the drawing board, but it is all too clear that a lunar Anthropocene has begun.[3]

Bombing the Moon, and Bombs in the Moon

The Moon's atmosphere, like Earth's, may soon change irrevocably under the weight of human industry and ambition. But its surface, like Earth's, has long been scarred by human machinery. Between 1958 and 1976, the Soviet Union and United States dispatched seventy-eight robotic spacecraft to the Moon.[4] First, audiences around the world followed by radio or television as successful *Luna* and *Ranger* spacecraft smashed into the Moon, with flashes and impact plumes that some (mistakenly) claimed to see from Earth. After *Luna 5* failed to fire its retrorockets and slammed into the crater Copernicus at full speed, raising a cloud of dust some two hundred kilometers across, Soviet and US space agencies prioritized controlled landings that could be survived by a human crew. In January 1966 *Luna 9* became the first spacecraft to achieve that feat. In its wake, gleaming landers slowly populated the lunar landscape, some with robotic arms to scratch and scoop the Moon's age-old regolith.[5]

For a while, violent impacts only occasionally altered the Moon's surface. This usually happened on purpose (in the case of all five *Lunar Orbiter* spacecraft), but sometimes by accident. In July 1969, for example, *Luna 15* rushed to the Moon on a desperate mission to return with lunar samples before *Apollo 11* astronauts could bring back a much larger horde of rocks. It was the Soviet Union's last chance to salvage a propaganda victory in the Space Race, by showing that it could accomplish with robots what the United States could only achieve with a far costlier and riskier crewed program. Yet *Luna 15* spent an extra day in lunar orbit—ensuring that it would reach Earth only after the *Apollo* astronauts did—then rammed into a mountain just hours before the astronauts lifted off.[6]

The Apollo program heralded a new wave of lunar bombardment. Astronauts left behind seismometers to measure "Moonquakes," then calibrated the instruments with a test tremor by deliberately steering their ascent stage—the part of their landers that returned to the orbiting command module—into the lunar surface. Beginning with *Apollo 13*, the giant upper stages of the Saturn V moon rockets slammed into the Moon as well, at speeds of nearly three kilometers per second. One impact after another left gaping craters on the lunar surface, creating the artificial changes to the Moon's surface that selenographers had sought in past centuries. During the Apollo 16 and Apollo 17 missions, astronauts even fired grenades into the lunar surface to test their seismometers with smaller explosions than those generated by crashing spacecraft. Even that bombardment, however, paled in comparison to what defense planners in both the Soviet Union and the United States once had in mind.[7]

In 1919 the pioneer of US rocketry, Robert H. Goddard, had already suggested detonating on the lunar surface an explosion large enough to be seen from Earth, as a way to prove that the United States could achieve the impossible. At the time the idea invited ridicule. But in 1956 the prominent meteorologist William Welch Kellogg wrote an unclassified report at RAND Corporation that proposed a nuclear assault on the Moon. Shortly thereafter, another William Pickering—this one the influential head of NASA's Jet Propulsion Laboratory (JPL)—suggested that an atomic bomb detonated on the Moon by a US "Interspatial Ballistic Missile" could "shower the Earth with samples" of scientifically valuable lunar dust and yield "beneficial psychological results," such as proving the superior reach and capability of US missiles. At Lockheed, the contractor responsible for the upper stage that would carry the bomb to the Moon, an engineer named Saunders Kramer, calculated that a nuclear warhead would be able to explode on the lunar surface before being crushed by impact.[8]

In 1959 the popular periodical *Aviation Week* circulated rumors that the Soviets had equipped at least some of their *Luna* spacecraft with nuclear warheads that would create an explosion visible from Earth. Air Force intelligence officers secretly warned their superiors that exploding a nuclear weapon on the Moon would be an "extremely attractive" option to Soviet leaders, as the flash of the explosion "could demonstrate the Soviet claim to superiority and be supported by the personal experience of every observer." In 2006 Boris Chertok, once a prominent engineer in the Soviet space program, confessed that his colleagues had not only proposed a version of the *Luna* spacecraft that could nuke the Moon, they had in fact designed one. The physicist Yakov

Zeldovich, who had played a major role in developing the Soviet atomic bomb, allegedly suggested the scheme to provide "irrefutable proof of our hitting the Moon." According to Chertok, Soviet planners did indeed assume "that when the atomic bomb struck the Moon, there would be such a flash of light that all observatories capable of observing the Moon at that moment would easily record it." Engineers "fabricated mock-ups of the lunar capsule with a mock-up nuclear warhead"; the warhead was riddled with detonators to ensure it would blow up when it hit the Moon, no matter its orientation.[9]

Chertok claimed that Mstislav Keldysh, known as the chief theoretician of the Soviet space effort, asked the chief designer, Sergey Korolev, to hold off on informing Khrushchev about the scheme "until we had discussed everything." When Korolev hesitated, Chertok approached him "on behalf of all the guidance specialists" to stress that the nuclear-armed *Luna* should be launched only if there was no chance of an accident. Keldysh agreed. "Imagine the furor if this thing were to come down on foreign territory," he told Korolev, "even if it didn't explode." In any case, Zeldovich appears to have changed his mind after calculating that a nuclear explosion in the vacuum of space might not even be visible from Earth. If Chertok's story is true, then the Soviets avoided detonating a nuclear bomb on the Moon only because the risk no longer seemed worth the reward.[10]

Serious preparations for a nuclear attack on the Moon definitely progressed in the US Air Force. Since 1949 the US physicist Leonard Reiffel had directed projects at the Armor Research Foundation of the Illinois Institute of Technology that explored the environmental effects of nuclear explosions. In the late 1950s the Air Force approached Reiffel to investigate how a nuclear detonation near the Moon would affect the lunar environment, and whether it could be seen from Earth. In 1959 a classified report, written by Reiffel and a small team of scientists for an Air Force organization that focused on the development, testing, and evaluation of nuclear weapons, considered the scheme. From the military's perspective, Reiffel wrote, the purpose of nuclear explosions near the Moon was clear, "since information would be supplied concerning the environment of space, concerning detection of nuclear device testing in space and concerning the capability of nuclear weapons for space warfare." It seemed obvious, too, that the United States would benefit from such a display of technological mastery, provided "the climate of world opinion were well-prepared in advance." Measuring the effects of a nuclear explosion might even reveal the origin of the elusive TLPs. There were some scientific obstacles to a nuclear attack on the Moon, however, including "environmental disturbances, biological contamination, and radiological contamination" of the lunar environment.[11]

The scientific and military usefulness of nuking the Moon depended on preparation, and not only of popular sentiment. On Earth, derelict ships and mock towns were used in weapons tests to measure the destructive potential of nuclear weapons. Similarly, to estimate the effect of a nuclear explosion on the Moon, a series of giant white spheres could be deployed as targets above the Moon. Ideally, spacecraft would also deliver instruments to the Moon's surface that would revolutionize lunar science by measuring the environmental consequences of the explosion. Seismometers, for example, would provide data on the Moon's composition by measuring tremors created by the detonation. The instruments could release a series of reflective balloons that, when pushed by the Sun's light, would create a line visible from Earth. That would allow scientists to pinpoint the location of the instruments. It all made a great deal of sense—from a certain point of view.[12]

The foremost ambition of the US national security establishment, concerning the Moon, was to establish a lasting presence on the lunar surface. As with most security issues in the early Cold War, competition raged between branches of the Department of Defense: in this case, the Air Force and Army. On June 9, 1959, a secret Army report proposed an effort with "authority and priority similar to the Manhattan Project" to urgently establish an armed outpost on the Moon. "Once having been second best in the eyes of the world's population," the report argued, referring to *Sputnik*, "we are not now in a position to afford being second on any other major step in space." Indeed, "the results of failure to first place man on extra-terrestrial, naturally-occurring, real estate will raise grave political questions and at the same time lower United States prestige and influence." The Kennedy administration would later use the same argument to support NASA's Apollo program. Yet the Army insisted that it was not enough to merely land an astronaut or soldier on the Moon. Everything hinged on building a base, not only for the prestige it could give, but also because world opinion might well conclude that the country with the first lunar base owned the Moon. There were (at the time) no laws to preclude such a claim or prevent the construction of military infrastructure on the Moon. And in any case, a common view was that "force is the ultimate sanction of the law."[13]

The Army imagined an initial base of up to twenty people that would be expanded over time to conduct surveillance of Earth, provide a communications relay, serve as a base for exploration of the solar system, undertake military operations in space, support the commercial exploitation of lunar resources, and, incidentally, lead scientific investigations into the nature of the lunar environment, including the possibility of lunar life. The plan would depend on a series of gigantic rockets and landing vehicles, the largest potentially powered by nuclear bombs. Launch sites would be built at four (apparently annexed) equatorial locations, and communications stations established around the world and in orbit. The expense would be staggering, although perhaps not by the standards of the Department of Defense: well over $600 billion in 2024 dollars, according to the Army's optimistic estimate. Essential to the program, therefore, would be the "industrialized expansion" of the lunar outpost to give it "a capability of self-regeneration" through the wholesale exploitation and transformation of the lunar environment. If the outpost could be made sustainable in an environment "more hostile than any heretofore encountered by man," and especially if it could be made profitable—provided lucrative resources were identified on the Moon—then the initial expense would pay for itself.[14]

When remembered at all, the Army's Project Horizon is often portrayed as an absurd fantasy from the heady days of the early Cold War.[15] Yet when it came to exploiting the Moon's environment, Army engineers anticipated problems that continue to guide plans for lunar settlement today. Given the intense radiation and routine micrometeoroid bombardment at the lunar surface, they correctly reasoned, it was logical to cover bases with lunar regolith—which meant building them in "holes or caves," or else hollowing them out of the lunar landscape.[16] Army planners figured that inflatable or folding rooms would be easily transportable by rockets and potentially invaluable on the Moon. Power shovels and dump trucks would be useless in the Moon's low gravity, but the Army designed lunar-grade suits and vehicles to dig trenches and haul regolith. Using solar or nuclear power, "oxygen and water may be extracted from the natural environment," then obsessively recycled. Transplanted environments from Earth would aid in these efforts. Algae farms, for example, would not only provide officers with fresh salad—a valuable source of morale, apparently—but also convert carbon dioxide into oxygen. Just a few years later, NASA abandoned a plan to process astronaut waste using algae—partly because it seemed to conflict with the assumption that humanity was separate from and superior to nature.[17]

The Army drafted a list of possible locations for this lunar military base, including several near Alphonsus, that would be easily accessible from Earth and afford swift movement across the lunar surface. When established, "the lunar base will be operated under the control of a unified space command," the report stressed, because "space . . . will be considered a military theater." Specialized weapons, suits, and combat vehicles would allow soldiers to control lunar environments so they could be mined and industrialized. At the same time, the Army report acknowledged, more research was needed on just a few lingering issues before the base could be set up. These included "food and oxygen, clothing, chemical, biological, radiological, bio-medical, vacuum conditions, weightlessness, meteoroids, lunar-based systems, moon mapping, explosives in lunar environment, power generation, material and lubricants, liquid hydrogen production and handling, and lunar 'soil' mechanics." Urgently needed, it seemed, was a "major technical facility for environmental research"—a Lunar Environment Research, Development, and Training Center that, in Army plans, would sprawl around a giant sphere resembling the Moon.[18]

Army planners felt entitled to take the lead in developing a lunar base because Army soldiers would be ideally suited to operate an installation that controlled territory and resources. Getting to the Moon required flying, though, and that provided an opening for the Air Force to develop its own schemes. The Army could scarcely justify its participation in a lunar expedition if there were no soldiers permanently stationed on the ground, so Army officers downplayed the significance of merely landing on the Moon. Air Force scientists and engineers, by contrast, secretly drafted extensive plans that began with simply landing Air Force personnel on the Moon and bringing them back to Earth: an achievement, they stressed, that would be more than sufficient to "excite the whole world." Army plans featured rovers and suits. The Air Force imagined spaceships that would look and fly like aircraft. On May 26, 1961—the day after Kennedy committed NASA to landing a man on the Moon—the Air Force undercut NASA, a civilian agency, by releasing detailed plans for Project LUNEX (LUNar EXpedition). Air Force planners concluded that this project was an "economical," "reliable," and above all quicker alternative to whatever NASA could propose.[19]

Project LUNEX also imagined a second expedition, in 1968, to begin construction of an Air Force base on the Moon. The Army had concentrated on controlling lunar territory and resources. In contrast, the Air Force, which managed most of the US nuclear arsenal, proposed building missile silos on the Moon. Soviet and US officials feared that the other side could develop the capacity for a nuclear first strike that would annihilate their arsenal or decapitate their leadership before they had a chance to retaliate. This, it was believed, would invite aggression—and it was precisely why the Kennedy administration risked Armageddon to prevent the construction of nuclear missile silos in Cuba, which (for a missile) was just minutes from Washington, DC. To the US Air Force, a Moon base offered the perfect solution—a Lunar Based Earth Bombardment System would survive any first strike.[20]

This logic was what had motivated the Army to attempt Project Iceworm in Greenland's ice sheet (Chapter 8) and the Navy to launch a fleet of nuclear submarines. In theory, a Moon base offered even more security. Yet it did not seem to occur to Air Force planners that the Soviets could nullify this ultimate deterrent simply by stationing their own missiles on or near the Moon. Then the diabolical logic of mutually assured destruction would simply replicate itself on a new world, and the extraordinary expense of tunneling a missile base into the lunar regolith would go to waste. In the wake of Project LUNEX, Soviet base-building on the Moon would therefore invite an American attack, precisely the kind of nuclear holocaust that lunar missile silos were supposed to deter. The Air Force acknowledged that decisions about the nature of strategic assets on the Moon could wait for a few years, but directed that plans for a base should "be started *immediately* if maximum military advantage [is] to be derived from a lunar program."[21]

Lunar Water and Lunar Life

Like the Army, the Air Force dreamed up creative and prescient ways of exploiting the lunar environment. Water could provide fuel and support life, Air Force engineers concluded, and water ice could accumulate in permanently shadowed craters. Gruithuisen had originally proposed that a primordial bombardment of comets had formed the Moon's craters, and that the comets had brought water that survived in the eternal shadows of polar crater rims. In June 1961 the chemist Harold Urey suggested that astronauts land near such a crater rim to determine whether it preserved enough water to sustain human settlement. The poles had been well beyond the safe zone established for Apollo mission landings, and rocks returned by astronauts seemed to reveal a Moon bereft of all water, and therefore hostile to human settlement.[22]

Soviet and US spacecraft photographed just about all of the Moon in the 1960s and 1970s. However, the angle of sunlight left one narrow strip extending from the south pole, nicknamed Luna Incognita, too dark to map. In 1972 it dawned on amateur astronomers that there would be no forthcoming mission to complete the lunar mapping project. The Association of Lunar and Planetary Observers, one of the observing networks mobilized by Winifred Cameron to detect transient lunar phenomena, took action. Over nearly two decades, the group had mapped Luna Incognita when the Moon's libration brought it within view. Their completed map revealed, among other new features, two polar craters—later named Shackleton and Shoemaker—that were largely shielded from direct sunlight. In 1994 the *Clementine* spacecraft, a joint effort by NASA and the Ballistic Missile Defense Organization, clearly observed both craters for the first time. Using radar to look for the telltale reflection of radio waves on ice, the probe obtained evidence of water in the vicinity of the lunar south pole.[23]

If water really did exist on the Moon, the dead world of the Apollo missions would suddenly be a much more appealing target for human settlement. Just four years after *Clementine* reached the Moon, scientists and engineers started to bombard the Moon's polar craters with spacecraft to search for water in the impact plumes. The orbiter *Lunar Prospector* detected hydrogen at the Moon's poles but could not confirm the presence of water. Then in 2008 the Indian Space Research Organization's *Chandrayaan-1* orbiter did detect water ice in Shackleton after smashing a probe into the crater.[24]

One year later NASA's *Lunar Reconnaissance Orbiter* and its companion mission, the *Lunar Crater Observation and Sensing Satellite* (*LCROSS*), verified the presence of shaded water ice. The Centaur rocket upper stage that had launched *LCROSS* to the Moon smashed into the crater Cabeus, at the Moon's south pole, with the force of a small atomic bomb. *LCROSS* subsequently hurtled through a plume of debris that rose some sixteen kilometers above the crater. Before it too crashed into the Moon, the probe employed its spectrometers and other instruments to record the composition of the plume. High above the Moon's surface, the orbiter performed further analysis of the plume. The results revealed the presence of water, along with potentially lucrative light metals and volatile chemicals. NASA's *Lunar Atmosphere and Dust Environment Explorer* (*LADE*) subsequently identified trace water in the Moon's tenuous atmosphere, suggesting the existence of an active water cycle at the poles.[25]

Water is essential to human life, of course, but perhaps even more important in space settlement are its constituent elements, oxygen and hydrogen. Oxygen can provide breathable air. And when chilled and combined with hydrogen, it creates the same rocket fuel often used as liquid propellant on Earth. Water can therefore provide a powerful source of fuel for flights to and from outposts beyond Earth. And the lunar south pole has another, equally important source of energy. While the floors of some craters at the Moon's poles never see sunlight, crater rims at the poles only rarely see shade. Solar panels stationed there would receive far more solar radiation than similar installations on Earth, where sunlight is filtered through a thick atmosphere. They also would provide power nearly continuously, rather than sporadically as on Earth. Given these resources and its close proximity to Earth, the Moon's south pole may be the most attractive target for human settlement in the solar system, although robotic missions will need to confirm that its water ice can be extracted easily from lunar regolith. India's *Chandrayaan-2* rover would have hunted for water at the pole, but it crashed; forthcoming Chinese and joint Russian and European missions, however, plan aggressive prospecting.[26]

Today the US and the Chinese space programs aim to exploit the same environments on the lunar south pole, and the logic that informed Project Horizon seems back in vogue. It is likely, though, that the most ambitious plans for the Moon's south pole are being drafted by corporations, not governments. Technological advances descended from the Apollo program, bipartisan political support, rising popular interest in outer space, fears of environmental degradation on Earth, and perhaps especially the soaring wealth of tech industry billionaires in an era of booming socioeconomic inequality have all laid the groundwork for a second Space Age shaped in part by the towering ambitions of private, "NewSpace" corporations. Many such companies, large and small, have plans to survey, mine, and settle the Moon, thus contributing to the emergence of a lucrative economy in cislunar space—the expanse of space between and around Earth and the Moon. The best-known NewSpace company, SpaceX, aspires to build a Moon base that would generate propellant for interplanetary colony ships bound for Mars.[27]

Less well known, for now, is Blue Origin, the company founded and funded by former Amazon CEO Jeff Bezos to expand the human frontier into outer space. Bezos imagines a limitless future for humanity in which space cities multiply beyond Earth to exploit the inexhaustible resources of the solar system, and thereby ease the unsustainable pressure imposed on the terrestrial biosphere by the consumptive capitalism that made him rich. Yet that future will demand an initial outpost in space where machines and fuel can be manufactured on site, without the tremendous cost in energy and money of hauling them out of Earth's deep gravity well. As early as 2017 Bezos engaged in private discussions with NASA about developing a distribution system to land supplies at the Moon's south pole in support of a public-private Moon base. NASA ultimately developed its own plans for a return to the Moon—at present, headlined by an orbiting Lunar Gateway space station and a lander designed by SpaceX—but a base in Shackleton crater remains a focus for Blue Origin. If the company has its way, the environment of the Moon's south pole could soon look radically different—and perhaps a bit more like Earth's.[28]

Confirmation of water on the Moon and hints of a water cycle at its poles suggest a more complex and dynamic lunar environment than scientists believed they had uncovered in the wake of the Apollo missions. High-resolution photography and careful measurements of anomalies in the Moon's gravity have, moreover, revealed enormous tubes, hollowed out by primordial lava flows, lurking just beneath the lunar surface. These underground caverns could be ideal sites for human settlement, and newly uncovered mineral deposits may even make such colonies profitable. In particular, lunar stockpiles of Helium-3—a potential fuel for pollution-free fusion reactors—dwarf those on Earth.[29]

The renewable energy potential of the Moon is even greater, as solar panels manufactured from lunar regolith could transmit limitless clean, microwave energy to Earthbound receivers. We might be at the start of a new space race, between corporations and space agencies, to develop a cislunar economy that sprawls across the region of space dominated by Earth's gravity and centers on the plentiful resources of the Moon. If so, lunar landscapes may change more than they have in billions of years. It would be an extraterrestrial extension of the Anthropocene that could, ironically, ease the environmental footprint of humanity on Earth.[30]

Yet humanity's most profound transformations of the Moon may be the hardest to detect, because they could involve Earth's smallest organisms. When, in 1958, Joshua Lederberg warned that the lunar surface could soon be irrevocably forward contaminated, the newly formed Space Science Board at the National Academy of Sciences asked him to establish an ad hoc committee to recommend procedures for spacecraft sterilization. Within months the Lederberg committee's uncompromising recommendations were endorsed by the Board, and in turn by NASA's inaugural administrator, Thomas Keith Glennan.[31]

For the physical scientists and engineers responsible for launching the first Moon-bound spacecraft, the devil was in the details. Sterilization procedures used in hospitals, which relied on steam chambers called autoclaves, were clearly inadequate for interplanetary probes. Among other things, their steam could not reach spacecraft interiors that would be splayed across the lunar landscape in the event of a collision. The Lederberg committee recommended using procedures designed by the US Army Biological Warfare Laboratories, which used ethylene oxide, a sterilizing gas. The gas was able to penetrate the interiors of complex machines, but it could not work its way through the plastics or ceramics that protected spacecraft electronics, which were known to harbor microorganisms. Meanwhile, exposing spacecraft to enough heat or radiation to kill microorganisms also risked severe damage to sensitive equipment. Nor was it clear what percentage of bacteria would have to be killed before engineers could consider a spacecraft sterile; it was, after all, impossible to eliminate them all. Officials could not even decide whether they should target viruses as well as bacteria. Viruses seemed less likely to contaminate an alien biosphere because they could only reproduce within cells they had evolved to infect. Yet it was conceivable that those cells could include bacterial or fungal spores capable of surviving on another world.[32]

The Jet Propulsion Laboratory opted to use dry heat to sterilize the components of its *Ranger* spacecraft before they were assembled, tested, and reassembled. Mission staff cleaned each assembled spacecraft with alcohol, and then, after loading it into its rocket, doused it with ethylene oxide. The spacecraft was repeatedly recontaminated during assembly, testing, and transportation, though, so it would still have harbored abundant microorganisms when it was placed in its rocket. The sterilization procedure therefore depended entirely on ethylene oxide gas, which could not penetrate the interior of the spacecraft electronics. The *Ranger* probes were designed to crash into the lunar surface. Their contaminated interiors repeatedly scattered microorganisms across the face of the Moon.[33]

The repeated failure of the *Ranger* spacecraft led to widespread calls to abandon efforts to decontaminate Moon-bound missions. But it was the Kennedy administration's goal of sending astronauts to the Moon that doomed the sterilization program. Human bodies are miniature biospheres, each teeming with some forty trillion bacterial cells; the decision to send them Moonward played a key role in convincing NASA officials to abandon efforts to prevent lunar forward contamination. As of 1963, NASA aimed only to avoid "widespread or excessive contamination" of the lunar surface—and then only until NASA judged that it had gathered enough information about the Moon. In practice the Apollo mission astronauts made no meaningful effort to avoid microbial contamination. When they unloaded their spacecraft for the ascent from the lunar surface, for example, they left particularly rich, and protected, microbial communities on the Moon—inside urine tanks and diapers they called "maximum absorbency garments." In 1970 NASA administrator Thomas Paine acknowledged to the *Chicago Tribune* that "there has been some" forward contamination of the lunar surface. Given that NASA's ongoing quarantine protocol rested on the possibility that the Moon could harbor its own biosphere, replete with indigenous microscopic life, it was a remarkable admission. NASA officials assumed that terrestrial microorganisms could not survive for long on the Moon. But as we have seen, the contaminated camera housing returned to Earth by the *Apollo 12* astronauts seemed to prove otherwise.[34]

It is possible that microorganisms delivered to the Moon by the Apollo missions clung to life until, on April 11, 2019, the first privately funded Moon lander disintegrated upon impact with the lunar surface. Operated by Israel Aerospace Industries and the Israeli nonprofit SpaceIL, and named *Beresheet* ("In the beginning") after the first word of *Genesis,* the spacecraft amounted to a half-ton Noah's Ark. The Arch Mission Foundation, a nonprofit devoted to seeding the solar system with human knowledge and biota, loaded *Beresheet* with a "lunar library," a capsule that included thousands of tardigrades: microscopic "water bears," organisms capable of surviving for years in the harsh environment of outer space. Engineers had reinforced the library capsule with resin, which in turn carried human DNA. It may have survived the crash. Yet it is equally likely that it ruptured, and that thousands of tardigrades today survive on or under the lunar surface.[35]

Just over a month later, the China National Space Administration landed the *Chang'e 4* robotic lander on the Moon's far side: a first in lunar exploration. The spacecraft included the Lunar Micro Ecosystem: a sealed container filled with plant seeds, yeast, and fruit fly eggs. As they sprouted and hatched together, the seeds and eggs were supposed to create a simple, self-contained or "closed" ecological system not unlike the algae waste-processing systems that NASA had pioneered at the dawn of the Space Age. The fly larvae would produce carbon dioxide, the plants would release oxygen, and the yeast would regulate quantities of both gases while decomposing waste. The idea was to test the viability of a terrestrial ecosystem on the Moon, which in turn would pave the way for lunar settlement. Some of the plants did grow as planned, but the experiment terminated abruptly when its spacecraft could not provide sufficient warmth with the onset of lunar night. Nevertheless, there is a small chance that its microorganisms survive on the lunar surface today.[36]

Since the dawn of the Space Age, dozens of spacecraft have scattered trillions of terrestrial microorganisms across the surface of the Moon. They are not the first microbial visitors from Earth. Millions of years ago, for example, rocks blown off Earth by the asteroid or comet that helped doom the dinosaurs must have carried other organisms to the lunar surface. Still, the magnitude of today's migration may have no precedent. It might be the most important, but least visible, expression of a lunar Anthropocene.

CONCLUSION

Navigating the Lunar Anthropocene

On an ordinary autumn night I crouched over a delicate refractor, waiting in breathless silence. Peering through my eyepiece, I admired faint stars glittering in the gloom, diamond-dust on black velvet. Gradually, a milky glow appeared in my peripheral vision. Subtle at first, it brightened quickly, drowning out the stars: a white fire to herald the coming of a world. An undulating horizon wheeled into view: mountains, cliffs, peaks and valleys, entangled, intersecting. A bewildering complexity. Infinite shades of grey and yellow, framed by brilliant snow-white and utter blackness. A Moon-world that, to our modern eyes, is immediately, obviously alien. Bereft of nearly everything that sustains life on Earth, but human nonetheless—a collection of environments that have shaped our culture for three hundred thousand years.

If the Anthropocene today encompasses at least two worlds, what should be preserved in the Moon's austere beauty? What kinds of environmental transformation are permissible on the Moon, and who should reap the benefits of lunar exploitation? At present, answers are elusive. Some argue that space agencies and companies should abandon all plans for further alterations of the lunar surface. Conservation of the Moon, according to this view, would provide a needed symbol of humanity's capacity to build what the conservation biologist Rick Steiner calls a "more compassionate, sustainable, cooperative future" in an age of environmental crisis. Of course, if lunar exploitation can alleviate parts of that crisis, then symbolism might have limited value. Representatives of some Indigenous communities, such as Buu Nygren, president of the Navajo Nation, argue that the Moon's sacred place in Indigenous cosmologies should preclude some ways of using the lunar surface, including as a burial ground for human remains. Space archaeologists, meanwhile, call for the preservation of historic landing sites on the lunar surface—not only to protect human cultural heritage on the Moon, but also to enrich future study of the ideologies and technologies of the early Space Age. Few have considered whether some lunar environments have their own cultural heritage, and therefore should be conserved: the crater walls of Plato, for example, or the rilles winding out from Aristarchus.[1]

Efforts to conserve the Moon's environments and artifacts are both complicated and aided by existing space law. On the eve of the Apollo landings, the UN Committee on the Peaceful Uses of Outer Space drafted the Outer Space Treaty in order to begin to regulate human activities on other worlds. Once it became international law in October 1967, the Treaty forbade states from claiming sovereignty over any part of the lunar surface. Because it precluded states from acquiring territory on the Moon, however, it also made it impossible for them to regulate access to, for example, the Apollo landing sites (even though spacecraft themselves remain under the control of the state that launched them). It was also unclear whether the Treaty prohibited countries and companies from owning resources extracted from other worlds. In 1979 the Moon Treaty attempted to clarify this point by spelling out that lunar resources are the "common heritage of mankind" and cannot become the property of a state, nongovernmental organization, corporation, or person. Because that Treaty would have prohibited or at least discouraged any development of the Moon, no space-faring nation signed onto it. National laws now stipulate that companies do have the right to own and sell whatever they mine on other worlds, unless it harbors extraterrestrial life.[2]

Today, forty-three countries have signed onto Artemis Accords, a nonbinding agreement, drafted by NASA and the US State Department, that provides a roadmap for the pending return to the Moon by America and its allies. Based on the Outer Space Treaty, the Accords emphasize that no state may own lunar territory and promote the "sustainable" extraction and ownership of lunar resources. They articulate an intent "to preserve outer space heritage," including "historically significant human or robotic landing sites, artifacts, spacecraft, and other evidence of activity on celestial bodies." Yet many of the most important players in the coming age of lunar exploitation—not least the China National Space Administration—have not signed the Accords, and in any case, it is difficult to know how effectively any aspect of space law can be enforced far from Earth.[3]

Across today's Moon, continuity is at least as interesting as change. It was not always so. Perceived fluctuations in the Moon's appearance strengthened Copernican cosmology, motivated efforts to improve navigation, and inspired the first search for extraterrestrial life—with profound consequences for everything from the development of mass media to the emergence of plans to mitigate existential risk. Change on the Moon shaped the ambitions of the early Space Age, which in time made our species the dominant influence on the age-old lunar landscape. As we will see, the history of human engagement with Mars followed a roughly similar trajectory. Yet the windswept environments of Mars—more dynamic by far than those of the Moon, and quite possibly home to indigenous microbial life—played an even larger role in the story.

PART IV

Mars

A drawing of Mars, with dark regions and white clouds.

"Over the face of [Mars] sweep changes that show it to be not a dead but a living world, like ours in this, and luring curiosity by details unknown here to further exploration of its unfamiliar ground."

—Percival Lowell, *Mars and Its Canals* (1906)

"There is something fascinating about science. One gets such wholesome returns of conjecture out of such a trifling investment of fact."

—Mark Twain, *Life on the Mississippi* (1883)

If some part of Earth's surface is an analogue for another world, it is Antarctica. Long rumored to exist, in 1773 its ices were first recorded by Captain Cook and his crew; its land was sighted by sealers, decades later. In the twentieth century, it became a frontier of exploration and imperial competition—and then, in the twenty-first century, Earth's most spectacular site of planetary change. Antarctica is also the place where Earth's connection to other planets may be easiest to discern. Gathered by the movement of ice are accumulations of scorched rock: meteorites originally blasted into space by the impact of asteroids or comets and the eruption of volcanoes, on other worlds.[1]

By comparing pockets of gas within these rocks to the known composition of the Martian atmosphere, scientists have discovered that some were blown off Mars. For thousands or millions of years, they had tumbled through the Sun's immense gravity well, falling in and out of smaller, planet-sized depressions in space-time. At last, they burned through Earth's atmosphere. When they landed, most of them disappeared from view. But in the Antarctic they stood out against the brilliant ice, so scientists found them easily.[2]

Just after Christmas 1984, Roberta Score of the Antarctic Meteorite Laboratory happened across a large rock in the Allan Hills of Eastern Antarctica. Rock ALH84001 was a piece of primordial Mars, which made it interesting. To date, only about three hundred of the roughly sixty thousand meteorites found on Earth have come from Mars. That rock became even more interesting in 1996, when a team of scientists, led by the planetary geologist David McKay, announced they had found fossils of bacteria within the meteorite. "Rock 84001 speaks to us across all those billions of years and millions of miles," President Bill Clinton reflected from the South Lawn of the White House. "It speaks of the possibility of life."[3]

Research into the origins of life and the extremes that life could endure had come a long way since scientists and engineers debated the Apollo quarantine protocol. In 1996 it seemed plausible that microorganisms had evolved on Mars billions of years ago and survived in the hollows of a rock as it tumbled through space. The implications were startling. Had life first evolved on Mars, then settled on Earth? Were Earth and Mars twin poles in a common ecosystem? Or had life evolved independently on the two worlds, implying that it emerged wherever conditions were favorable? If so, the galaxy should be teeming with living things.[4]

Scientists soon argued that nonbiological processes unknown on Earth could have tunneled the structures that McKay and his colleagues interpreted as fossils. Explanations that did not involve life were assumed to be simpler and thus more likely to be true, so ALH84001 was reinterpreted as an ordinary meteorite. Still, the controversy helped scientists clarify what evidence would clearly reveal the presence of extraterrestrial life, and where on Mars they should look for it.[5]

The furor over ALH84001 was nothing new. For centuries astronomers, journalists, politicians, and interested laypeople made sense of the red planet by interpreting a complex web of fluctuating environments, not only on Mars but also on Earth. Beginning in the seventeenth century, astronomers eager to observe detail on the surface of Mars contended with instability on different scales in two dynamic atmospheres: one on Earth, and the other on Mars. Often, local weather and small-scale atmospheric turbulence on Earth, occasionally shaped by the causes or consequences of climate change, either prevented or distorted observations of Mars. Astronomers responded by adopting the practices and language of Earthbound explorers. They sought out remote environments where the atmosphere was thin, dry, and stable. At the same time, Martian dust storms, by abruptly altering the shape of the planet's dark regions, convinced most astronomers that Mars, like Earth, had dynamic oceans and clouds. That in turn led astronomers, journalists, and ordinary people to conclude that life, quite possibly intelligent life, thrived on Mars.

In the 1870s Mars made two particularly close approaches to Earth just as technical innovations increased the power of telescopes, and Western empires embarked on an unprecedented wave of canal construction. Sharp-eyed observers spotted mysterious, apparently linear details on Mars that seemed to resemble artificially constructed canals. By mapping these canals, observers came to believe that supposed seas on Mars were actually dry land. The red planet now seemed like a world that was older and drier than Earth, which many thought would also someday lose its seas. Popular and scientific discourses held that Martians were ancient too, clinging to life on a dying world by using canals, the most impressive technologies known to observers on Earth. Canal enthusiasts justified their theories by arguing that humanity had begun to transform Earth's environment in ways that would eventually lead it to master nature as the Martians evidently had. The progress of industry at the expense of nature seemed to confirm the existence of Martian canals, while the canals proved the inevitability of such progress.

The natural dynamism of both terrestrial and Martian environments also encouraged new impressions of Mars as a future Earth, and of Martians as future humans. While observers sketched and popularized canals on a Mars that suddenly seemed largely arid, a series of severe El Niño events led to catastrophic droughts on Earth, drawing attention to the social consequences of persistent water shortages. On Mars, past climate changes had left a ring of dark dunes beneath the planet's northern ice cap that emerged every spring as the polar ice retreated. Influential observers wrongly concluded that a short-lived sea of meltwater accounted for that dark band around the warming poles. They argued that canals siphoned water from this sea to irrigate farmland that persisted in the dark patches of Mars. Then, in the final decades of the nineteenth century, observers began to see lights around Mars's terminator (the boundary area between day and night). Enthusiasts mistook those lights as being either Martian attempts to contact humans or the exhaust plumes of Martian spacecraft bound for Earth. The mysterious flashes were actually caused by sunlight catching the tops of Martian dust clouds; but all the same, popular paranoia about an imminent Martian invasion reached its peak. Entrepreneurs, meanwhile, hatched ambitious plans to respond to Martian messages. Environmental changes on Mars had inspired new conversations about existential risk, and opportunity.

By the early twentieth century, astronomers increasingly doubted the existence of canals on Mars. The dust storms that had helped inspire the canal theory, however, provoked breakthroughs in the study of climate change on Earth. When dust storms slowed efforts to map the surface of Mars using spacecraft, scientists began to consider how suspended dust could cool Earth's climate. Their research inspired worldwide calls for nuclear disarmament when it seemed to reveal that dust thrown into Earth's atmosphere during a nuclear war would cool the globe so much that it would doom the last remnants of humanity. Meanwhile, scientists and science fiction authors envisioned grand projects to transform the Martian climate and thereby create the Earthlike world once imagined by astronomers. Such a world seemed capable of ensuring that humanity would survive an Earth-shattering catastrophe, such as a nuclear war. Today, terraforming Mars into a second home for humanity has become the stated ambition of companies that aim to colonize Mars. The dynamic environments of the red planet have long inspired both utopian and apocalyptic visions of humanity's future—with profound implications for its present.

13

Unveiling a New World

In April 1672, England and France declared war on the Dutch Republic. Under the leadership of Louis XIV, the Sun King, the largest army in Europe exploited drought to ford rivers that otherwise defended the Republic. By early June the invaders neared Holland, the richest and most populous province of the Republic. In desperation, Dutch engineers did the unthinkable. They breached the dikes and opened the sluices that usually kept the sea from flooding their country. Despite the resistance of farmers, water surged into a vast crescent around Holland, transforming the province into a fortified island. At great cost the Dutch had stymied the invaders—for the moment. As autumn turned to winter, everything depended on the weather. If temperatures dipped too low for too long, the floodwaters would freeze, the road to Holland would open, and the Republic would fall. In the Little Ice Age (Chapter 1), this was a frighteningly real possibility.[1]

As French troops prepared for winter, Mars, the planet of war, glittered with disturbing brilliance in the night sky. Although a Martian day is about as long as ours, a year on Mars is twice as long. Every two of our years, therefore, Mars makes a close approach to Earth. At these oppositions, Earth is exactly between Mars and the Sun, and Mars can then be fewer than sixty million kilometers from Earth. At conjunctions, when the Sun is directly between Mars and Earth, Mars can be more than four hundred million kilometers distant. The apparent size of Mars in our night sky is up to ten times bigger during oppositions than it is in conjunctions, and its brightness varies accordingly. In 1672 an opposition brought Mars especially close to Earth, and the red planet outshone everything in the night sky save the Moon.[2]

The Republic's most esteemed scholar, Christiaan Huygens, lived in Paris, where he had helped establish the French Academy of Sciences. With his homeland imperiled by his adopted country, it must have been a distressing summer. Yet Huygens was an optical pioneer. He had invented a revolutionary telescopic eyepiece, and he knew the opportunity that awaited in the sky. War or no war, he took advantage of the opposition. Struggling with his unwieldy telescope, he noticed a never-before-seen bright patch at the Martian south pole. Precisely as the fate of his homeland depended on the coming of ice, he was one of the first humans to discern the presence of ice, and thus water, on another world. It seemed to confirm the Copernican idea that Earth was just another planet, orbiting the Sun, and it strengthened the emerging concept of the plurality of worlds. In the imagination of scholars and, eventually, ordinary people, Mars would become a world whose changing, familiar features revealed it to be a planet like Earth.[3]

Close Approaches of a Windswept World

The two-year opposition rhythm provided a regular schedule for observations of Mars from antiquity to the present. However, Mars is tilted on its axis, its orbit is much more elliptical than Earth's, and Mars is inclined differently than Earth relative to the Sun's equator, so not all oppositions are alike. The differences between oppositions took on new significance after the first telescopic observations of the planet convinced most natural philosophers that Mars was a world, like Earth. During every opposition, Mars presents a different angle of its globe to terrestrial observers. This means that the features of its surface that are easy to observe in one opposition may be hidden in another. Astronomers soon grasped that the most important cycle repeats itself every fifteen to seventeen years. Once during every such interval—as in 1672, for example—Mars reaches perihelic opposition, the point in its orbit that brings it closest to Earth at the same time that it is nearest the Sun. (Aphelic oppositions occur when Mars is closest to Earth at the same time that it is farthest from the Sun.) As seen from Earth, Mars can look nearly twice as large in diameter during perihelic oppositions as during aphelic oppositions. And owing primarily to distant

Jupiter's gravitational pull, the distance between Mars and Earth is never quite the same in every perihelic opposition.[4]

The perihelic opposition cycle is about much more than the size of Mars through the telescope. From Earth's Northern Hemisphere, Mars is high in the sky only during aphelic oppositions, whereas from Earth's Southern Hemisphere, it is high in perihelic oppositions. When Mars is near the horizon, the planet's light passes through more of Earth's turbulent atmosphere than when it is overhead, making it much harder to observe. In fact, the large telescopes that were constructed in domed observatories during the late nineteenth century could not even be turned to objects that were just above the horizon. The very best views of Mars can be had from Earth's Southern Hemisphere, during perihelic oppositions.[5]

Even during the oppositions most favorable for observation, the first astronomers armed with telescopes found Mars a difficult subject. Dark, or low-albedo, regions across the planet's surface and bright ice caps are, in theory, visible using even modest instruments. Subtle details and shadings on the planet's disk, however, are easily overwhelmed by chromatic aberration, an optical distortion that was common to refractors until the twentieth century (Chapter 9). They are also readily obscured by waves and convection cells in Earth's atmosphere. These, as the historian, psychiatrist, and amateur astronomer William Sheehan puts it, cause Mars to resemble "a quarter shimmering at the bottom of a swimming pool," on which "uncertain details dart and then are gone like doubtful visions."[6]

Astronomers, moreover, reckoned with changes not only in Earth's atmosphere, but also in that of Mars—because, in some respects, that planet really is strikingly similar to ours. Its rotation is only slightly slower than Earth's, and its axial tilt is just marginally greater. Mars has seasons like Earth, and days that last only a little longer than ours. Its surface area is nearly equal to the total area of all exposed land on Earth, and it has ice caps. Its atmosphere is much thinner than Earth's, but it churns with complexity, as intricate convective and radiative processes register the flow of solar energy and respond to the gentle pull of two small moons. Carbon dioxide and water ice clouds are visible through even modest telescopes, obscuring or distorting surface features for observers on Earth.[7]

A Martian dust storm. The European Space Agency's *Mars Express* orbiter photographed dust clouds welling up around the Martian north polar cap in April 2018.

Billions of years ago Mars may well have had oceans, a thick atmosphere, and a magnetosphere similar to Earth's. Its likeness to our world has waned since then. Owing perhaps to its small size, its core cooled more quickly than Earth's, and that may partly account for the demise of the swirling dynamo that formerly sustained its magnetosphere. Its crust froze into place. The solar wind gnawed at its atmosphere, which drained into space until it was too thin and cold for water to linger as liquid on the Martian surface.[8]

The soil, or regolith, of Mars is now so bereft of liquid water that even the slightest breeze can loft dust high into the sky. Temperature gradients near the edge of the planet's ice caps create air currents that can lift heavy dust particles, creating regional dust storms. Because these storms create their own winds, they can spread until they cover the entire planet. On average, that happens once every three Martian years. Dust storms most often occur in the southern hemisphere of Mars when the planet makes its closest approaches to the Sun. As a result, perihelic oppositions, which otherwise offer the best views of Mars, frequently coincide with dust storms that obscure much of the Martian surface.[9]

Martian Changes and the Limits of Analogy

Dust storms likely impeded seventeenth-century astronomers from observing surface features on Mars. Only in 1659, after years of observing, did Huygens identify the vast, heart-shaped volcanic plain that is today called Syrtis Major Planum. By observing the region as Mars rotated, Huygens accurately deduced that the planet has a day roughly as long as Earth's. Huygens had proved that environmental features were discernible on Mars, but dust storms and the modest pace of optical improvement prevented him from making further discoveries. More importantly, the advent of Newtonian physics led early eighteenth-century astronomers to preoccupy themselves with studying celestial movements, rather than environments. During the perihelic oppositions of 1704 and 1719, however, the Italian astronomer Giacomo Filippo Maraldi noticed that the "glittering" patch at the Martian south pole changed in size, and that another bright patch had appeared at the planet's north pole. He also glimpsed previously undetected dark patches and bright features that scientists have now identified as seasonal cloud formations near the Martian equator and north pole. "We see, then," Maraldi concluded, "that great changes occur on the surface of Mars."[10]

Maraldi had discovered that Mars is a variable world, but it was William Herschel who began to transform how astronomers, and eventually much of the public, understood the planet. His logbook reflects his wonder at the apparently Earthlike world that his unrivaled telescopes brought within view. On March 25, 1781, for example, he detected a dark patch—almost certainly Syrtis Major—that changed dramatically in subsequent days until a second patch appeared, followed by two bright spots at the Martian poles. On July 22 of that year, Herschel mused that "the poles being surrounded with snow and ice may occasion the white appearance" of these spots. Given that "we know that our planet [Earth] has polar regions which are covered by mountains of ice and snow, which melt when exposed alternately to the rays of the Sun," he later wrote, "it is permissible to suppose that the same causes probably produce the same effects on Mars."[11]

Herschel complained incessantly that the "undulation of the air" interfered with his attempts to discern detail on the planet. However, in 1783, when the eruption of the Laki fissure killed thousands in Iceland and contaminated much of Europe with sulfuric gases, he benefited from the suspended dust (Chapter 9). By removing "every troublesome ray" from his optics, the terrestrial dust unveiled a Martian surface that at that time was mercifully free of dust storms. Now Herschel concluded that the bright patches of Mars were ice caps waxing and waning with the passage of seasons, and that the dark patches were clouds. "The analogy between Mars and the earth," he wrote, seemed "by far the greatest in the Solar System."[12]

Most astronomers accepted Herschel's explanation for the bright patches on Mars. The real controversy involved the nature of the variable dark regions. Some observers believed they were oceans, because bodies of water look dark when viewed from afar. But how could oceans change shape overnight? After climbing Brocken mountain in central Germany shortly after sunrise, Johann Hieronymus Schröter watched as fog settled into the lower valleys and took on a dark grey appearance. The contrast between the mist and the bright mountain, he reflected, "is very much as the dark patches on Mars are to the brightly illuminated surface." Martian winds, he determined, blew vast cloud formations across the surface of Mars to create the planet's shifting dark patches, a sure sign that the planet closely resembled Earth.[13]

Winds do transform the dark expanses of Mars, but by redistributing dust, not water vapor. Because these environmental dynamics are alien to Earth, astronomers armed with relatively small telescopes had little chance of accurately perceiving them by seeking analogies for Martian environments on Earth. Still, occasionally environmental changes on Earth did inspire genuine insights about the Martian environment. During the Martian opposition of 1809, for example, the French astronomer Honoré Flaugergues speculated that the dark regions of Mars were either clouds or "immense fogs" similar to those released by Laki in 1783. The volcanoes of Mars are largely inert, and therefore not responsible for major changes to the dark expanses of Mars. Particles suspended in the atmosphere, though, do cause both volcanic dust veils on Earth and alterations in the appearance of dark regions on the Martian surface. A volcanic eruption on Earth had led Flaugergues to a more accurate view of the Martian environment than any other astronomer would imagine for decades to come.[14]

The notion that clouds or fogs alone accounted for Martian dark patches, however, could not survive the first truly rigorous observations of the red planet. Using the small refractor that had helped them map the Moon with unprecedented precision, the Prussian astronomers Wilhelm Beer and Johann Mädler observed Mars during the close opposition of 1830. Stable atmospheric conditions over Berlin helped them confirm earlier observations by another of the city's astronomers, Georg Carl Friedrich Kunowski, which suggested that at least some of dark patches were permanent features. At the next opposition, with Mars a little farther from Earth and the atmosphere less favorable over Berlin, and with dust storms likely raging on Mars, Beer and Mädler nevertheless managed to identify the same patches they had observed in 1830.[15]

Yet some patches did seem to change dramatically. During the opposition of 1837, for example, even Syrtis Major seemed indistinct, in all probability because of a dust storm on Mars. Some of the dark patches of Mars, Beer and Mädler speculated, were probably marshes that waxed and waned with the arrival and evaporation of meltwater from the Martian ice caps; others seemed to resemble terrestrial clouds. The men who had discredited fantasies of an

Earthlike Moon now wrote that Mars "appears as an image of the Earth in the firmament seen from a great distance."[16]

During the opposition of 1862, the Italian astronomer Antonio Secchi believed he had glimpsed oceans on Mars, and perhaps a "great squall." In the optical illusion called the contrast effect, neutral colors appear blue or green when surrounded by orange or red. So the dark regions of Mars can indeed look like oceans through the eyepiece. Changes in the distribution and thickness of Martian clouds can strengthen the effect by lending a blue-green hue to those regions, but only in the absence of a major dust storm. Those conditions seem to have prevailed in 1862, convincing some astronomers that Mars that was truly a mirror image of Earth.[17]

Others were less certain. Owing in large part to the work of Beer and Mädler, most astronomers had abandoned the idea that the Moon's dark regions were seas. In 1863 the English geologist John Phillips cautioned that the supposed oceans of Mars could be as dry as those on the Moon. Only the telltale reflection of the Sun on waves, he wrote, could prove that water covered much of Mars. Still, he acknowledged, the waning and waxing of the Martian polar caps created "an obvious requirement of water somewhere on Mars."[18]

There was also the troublesome matter of the lack of resemblance between drawings of the Martian surface by different observers, who used different instruments at different times, in different places. To some observers, discrepancies between drawings seemed to suggest that Martian changes were at least in part a consequence of differences between observers, telescopes, and terrestrial environments. Most, however, blamed changes in the Martian clouds and seas. Camille Flammarion wrote that the appearance of Mars varied owing to not only shifting cloud formations but also major inundations "flooding vast plains and changing their coastlines." The Reverend William Rutter Dawes, an English astronomer, argued that river deltas on Mars seemed to fluctuate with the movement of Martian seas. According to Flammarion, some dark regions varied from "dark and blue" to "yellow and bright" because the "water of these seas retreats or evaporates."[19]

The image of a watery, cloudy, and clearly variable Martian environment dominated scientific and popular understandings of Mars by the middle of the nineteenth century. The changing features of Mars provided apparent proof that water in all its states could create dynamic, Earthlike landscapes on other worlds. It also shaped assumptions about what Earth might look like from space. "It may be doubted," John Herschel wrote in 1849, "whether, in their perpetual change, the outlines of our continents and seas can ever be clearly discerned" by hypothetical inhabitants of the Moon. Above all, the seeming similarity of Mars to Earth led many observers to conclude that, as Phillips wrote in 1864, "we should regard Mars as habitable."[20]

By appearing to suggest that Mars was a world like Earth, changes in the dark regions of Mars also led astronomers and enthusiasts to a new consensus on the nature of the bright red continents that seemed to surround the blue-green seas. In the early nineteenth century, observers debated whether the atmosphere, soil, or vegetation of Mars accounted for the color of its continents. In 1830 John Herschel expressed a popular view by concluding that these continents owed their color, "no doubt," to "an ochrey tinge in the general soil." Camille Flammarion countered that soil or rock could only explain the color of the continents in the absence of vegetation, "a suggestion that can hardly be accepted, since there is air, water and sunshine" on Mars. Moreover, he argued, "the color is not everywhere red; it is usually a warm yellow, which can best be compared with ripe wheat." Over the course of the 1860s and 1870s, the assumption that Mars had vegetation similar to Earth's gained credibility. In 1873 the French physician and science writer Jean Chrétien Ferdinand Hoefer summarized what had become a mainstream opinion when he wrote that, because the "natural forces of fertility" were at work on Mars as much as on Earth, "it is probable that the dominant color of Mars is due to some kind of vegetation."[21]

In 1867 Richard Proctor drafted the first map of Mars. He patriotically named some of the planet's most prominent features after his fellow Britons, including relatively obscure English astronomers. Of course, astronomers in the French and newly formed German empires protested. Still, few could dismiss the characterization of Mars in Proctor's map. Mars, the map said, was a "miniature of our Earth" with oceans, continents, and ice caps that could be discovered, charted, and claimed for the glory of terrestrial nations. In fact, for nearly a century the alien dynamism of the Martian environment had created the perfect trap for William Herschel's new astronomy (Chapter 3). Astronomers armed with increasingly capable telescopes and observation techniques could not firmly establish the identity of the environments they believed they could discern on Mars. Using the analogical reasoning championed by Herschel, and influenced by the enduring popularity of the theory of the plurality of worlds, they all too readily agreed that changes in Martian environments they could barely discern revealed Mars to be a world similar to the obviously changeable Earth. In subsequent years this belief in an Earthlike Mars would encourage one of the most sensational (false) discoveries in the history of science: an extraterrestrial civilization, clinging to life on a dying planet.[22]

14

Aliens on an Aged Earth

On the night of September 5, 1877, a perihelic opposition brought Mars closer to Earth than it had been in decades. Dust storms seem to have obscured Mars during the decade's other oppositions, but now astronomers hoped that capable new telescopes would offer an unprecedented view of the planet's apparently Earthlike surface. On the night of the opposition, clear skies prevailed over much of Europe and eastern North America. Yet much to the frustration of astronomers, a planetwide dust storm again rendered Mars a featureless globe.[1]

Astronomer and hydraulic engineer Giovanni Schiaparelli had been observing the planet for weeks, well before the onset of the dust storm, and he continued his vigil long after the storm cleared. From his rooftop above Milan, he used his celebrated eyesight and capable refractor to sketch countless studies of Mars as it neared Earth. Previously observers had drawn only the most obvious features on the planet. In moments when atmospheric cells over Milan converged, the Martian surface was suddenly revealed with such clarity that Schiaparelli discerned a host of subtle details that others had missed. The challenge of observing Mars was to accurately translate to paper the intricate features that could be glimpsed in such "revelation peeps," as Percival Lowell later called them. Schiaparelli attacked the problem at scale. By drafting countless sketches in the immediate wake of innumerable peeps, he recorded far more detail than others had. Then, combining his drawings, he identified sixty-two enduring features on Mars and used the Mercator projection to arrange them on the first micrometrical map of the planet. It seemed like a breakthrough.[2]

His training as a hydraulic engineer convinced Schiaparelli to triangulate Martian features with, as he put it, "precise measurements!" As he synthesized hundreds of sketches to draft his map, his zeal for precision led him to convert subtle features, drawn with delicate pencil in different ways on different nights, into hard, sharp outlines firmly delineated with ink. A strange web of lines came to crisscross his map of Mars. Because they seemed to connect the planet's dark oceans, Schiaparelli called them *canali* (Italian for channels), a word infamously mistranslated into English as "canals."[3]

Schiaparelli's status as a respected professional with renowned eyesight lent authority to his new vision of Mars, as did his use of cartographical methods from the burgeoning colonial science of geography. In the late nineteenth century, expeditions by geographers thrilled publics in world-straddling empires with descriptions of seemingly exotic peoples and landscapes that appeared ripe for imperial competition and conquest. The creation of maps with new features distinguished the successful from the unsuccessful explorer-geographer, and now the same was true for astronomers of Mars. The definitive lines, new features, and cartographical format of Schiaparelli's map of Mars gave the impression that areography had become a rigorous science, with Schiaparelli its foremost practitioner.[4]

By adapting to the constraints of weather on Earth and Mars, Schiaparelli had developed observational and cartographical techniques that enabled him to reinvent how astronomers perceived the red planet. Over the next five decades, environmental transformations on both worlds inspired others to refine his vision of the Martian surface, and then to imagine an integrated climatology, hydrology, biology, and sociology of Mars—a Martian "world system"—that seemed to explain the planet's latticework of straight lines. At the heart of that system was a hypothesized civilization that had survived the gradual dying of a planet by developing infrastructure to reshape the planet's environment with every season. The development of this vision of Mars amounted to one of the most influential blunders in the history of science. Even though there were no canals on Mars, in all probability thousands believed they saw them, fooled by fluctuating environments on two worlds.

The Canal Controversy and the Mars Boom

Schiaparelli added so many features to his map of Mars that he decided to scrap Proctor's naming scheme and replace it with one that drew on the classical education then in vogue among European scholars. Rather than a familiar, terrestrial planet with place names that commemorated the glories of British astronomy, Schiaparelli's Mars was a romantic and above all mysterious world named after gods and goddesses. For all its revolutionary and evocative detail, though, Schiaparelli's map did not yet resemble the watercolor world of shifting hues and shadings that most astronomers saw through their eyepieces. Proctor, his naming scheme discarded, joined other British astronomers in insisting that "no one who has ever seen Mars through a good telescope can accept the hard and unnatural configurations depicted by Schiaparelli." Even Schiaparelli acknowledged that his published maps did not entirely conform to his drawings, and blamed engravers for the discrepancy. As for the channels, he wrote, it was as "impossible to doubt their existence as that of the Rhine on the surface of the Earth."[5]

In 1878 Schiaparelli purported to believe that his channels were natural waterways that merely added to the evidence for Earthlike environments on Mars. During the opposition of 1879, with Mars bereft of dust storms and clouds, he drafted drawings that were even more detailed, using them to revise the *canali* on his maps into a network of straight, occasionally parallel lines that bore little resemblance to natural waterways. British astronomers, whose names no longer adorned Mars in Schiaparelli's naming scheme, were outspoken skeptics. Yet the discourse had changed. Now, most argued that Schiaparelli had merely misrepresented features that were genuinely present on the Martian surface. British amateur astronomer William Denning acknowledged that ideal weather permitted sightings of what looked like canals—but they lacked the "straightness of direction and general uniformity of tone" that Schiaparelli had depicted. Solar astronomer Walter Maunder similarly admitted that "there certainly seemed to be a number of spider-like markings" on Mars, but he preferred to depict these canals with delicate shadings, for they did not always seem to be in exactly the same place from night to night. Others pointed out that because canals would need to empty into stable shorelines, they would not make sense on a planet with oceans apparently prone to "marvelous 'inundations.'"[6]

Just about every Mars observer was talking about canals by the 1880s. As a result, many competed to identify canals when they studied the planet. Some of the most reputable astronomers outside of England strengthened their reputations as explorers on a par with Schiaparelli by announcing that they, too, had spotted canals. The Belgian astronomer François Terby even argued that previous observers had glimpsed canals without knowing what they were. "Now that their existence is known," the astronomer William Pickering concluded, "there should be no difficulty for anyone with a little practice and moderate eyesight in seeing the more conspicuous ones."[7]

Even if the existence of canals was a "generally accepted fact" among Mars observers, as Pickering put it, the question remained whether they were natural formations or artificial constructs. Proctor insisted that they were illusions created by mist blanketing riverbeds. A suggestion that they were cracks in enormous ice sheets convinced few astronomers because the ice caps of Mars clearly did not cover the entire Martian surface and melted dramatically in the spring, which appeared to indicate that Mars was warmer than Earth. A clue seemed to come in 1881 when Schiaparelli reported that a canal he had named "Nilus" in 1877 now formed two parallel lines. Many observers subsequently noticed the same phenomenon, later called "germination." Because it was difficult to imagine a natural cause for the sudden appearance of parallel lines on an otherwise Earthlike world, some believed that germination revealed the presence of alien builders. Flammarion, for example, proposed that the Martians had constructed overflow canals that filled in case of flooding, to protect surrounding farmland.[8]

Such ideas captured the attention of newspaper publishers and the public, but the canals only became an actual sensation during the exceptionally close perihelic opposition of 1892. Three years earlier William Pickering had started writing for the *New York Herald,* the penny daily that had become perhaps the most influential newspaper in North America. In 1891 he arrived in Arequipa, Peru, where he derailed his brother's plans for a stellar observatory just as the *Herald* secured exclusive control over telegraph lines between North and South America (Chapter 10). Mars does not rise high above the Northern Hemisphere horizon during perihelic oppositions, so Pickering would have the best view of any astronomer, including those armed with more capable telescopes. The *Herald* promised exactly the kind of breathless coverage-by-telegraph that it had used to launch the era of event-astronomy, and in Pickering it had an ideal protagonist. As Mars approached, he saw himself as an explorer following in the footsteps of Columbus, "discovering a new country," as he put it, and gradually beginning to "feel quite at home upon the planet."[9]

Although the *Herald* teased that Pickering might uncover "the nature of the Martian canals," Pickering actually doubted that the lines crisscrossing Mars were canals at all. "Think of the labor involved in covering over, and then reopening, a canal, say, sixty miles wide by three thousand miles long," he wrote in 1888, "and all in the space of a few weeks." The changes that the supposed canals seemed to undergo, Pickering argued, revealed them to be strips of vegetation, rather than waterways. Mars, for Pickering, was not an enormous hydraulic machine created by an alien civilization, but instead the more Earthlike world that many astronomers had believed it to be before the opposition of 1877. In 1892 Pickering's telegrams to the *Herald* therefore echoed earlier descriptions of Mars by emphasizing the staggering scale on which "melted snow" from the planet's southern ice cap "apparently transferred to seas across land." The *Herald* revised his reports into a vivid travel narrative that, when splashed across front pages by syndicates and rivals alike, popularized a vision of Mars as the ultimate imperial frontier, one that skilled adventurers were beginning to explore.[10]

Astronomers in observatories across the Northern Hemisphere had publicly downplayed the coming opposition, not only because Mars would be low in their night sky but also because they disdained the sensationalism with which the press covered the prospect of canal-building Martians. These astronomers faced widespread ridicule for their inability to see the environmental changes described by Pickering and reported by the *Herald*. None fared worse than Edward Holden, who was now director of the Lick Observatory. Holden had scorned "the very uncritical attitude of the public to what is printed," especially regarding Mars, a subject "in which the whole intelligent world is interested." Now, to preserve his reputation and that of the Lick Observatory, Holden informed the *Herald*'s chief rival, the Associated Press, that astronomers at Lick had confirmed Schiaparelli's observations of germination. It was a striking reversal for Holden; germination had been controversial precisely because it seemed to indicate that the canals were artificial constructs. When Pickering described only "single" canal sightings, the *Herald* and other newspapers played up the controversy.[11]

The popular expectations and rapid exchange of views that characterized event-astronomy, itself an outcome of the penny press and the telegraph, had led two canal skeptics to debate the appearance, rather than the existence, of canals on Mars. Neither Pickering nor Holden mentioned the possibility that aliens had constructed the canals they claimed to see. Yet the media frenzy had sparked a "Mars boom," an explosion of popular interest in the latest updates about canals and canal builders on Mars.[12]

William Pickering had been sent to Peru by his brother Edward with a modest budget and instructions to focus on the unglamorous work of measuring the color and brightness of southern hemisphere stars. By 1893, however, he had devoted most of his labors to the Moon and Mars, and his constant requests for more money had led his reluctant brother to send him over a million dollars (in present-day currency). His observations had given him "a colossal reputation," Edward acknowledged in a telegram to William, but they had provided nothing of interest for the computers of the Harvard College Observatory. When Edward finally recalled his brother from Peru, William boarded a ship back to Boston with his assistant, Andrew Ellicott Douglass.[13]

Shortly after disembarking in Boston, Pickering and Douglass were introduced to Percival Lowell. They soon learned that Lowell, one of Boston's wealthiest men, had a keen interest in building an observatory that could eclipse the capability of the one they had left behind in Peru. Lowell, for his part, quickly decided that Pickering and Douglass could help him unravel what—or rather who—had created the canals of Mars.[14]

Creating the Mars World System

Contemporaries framed Lowell's Martian ambitions as a radical departure from his previous interests, but they were actually a homecoming. As a child he had observed the shifting appearance of Mars through a small telescope on the rooftop of his family home. At Harvard he had given a commencement address on the nebular hypothesis, which would later underpin his understanding of

Martian history. After graduation he had traveled to Asia, where he wrote Orientalist travel narratives that amused a colonial audience by combining astronomers' theories with the social Darwinism he had absorbed in Boston. In Japan he declared with typical hubris, "We behold . . . in the case of man the same spectacle that we see in the case of the moon, the spectacle of a world that has died of old age." When his travels began to bore him, he considered a new project, to explain how life inevitably evolves on every planet. He was abroad when he heard that Schiaparelli's eyesight had started to fade. Now Lowell saw an opportunity for someone to, like a conquistador, fully realize and exploit the discovery of a new world. Using his fortune, Lowell decided to build the ideal telescope for observing Mars. Armed with this instrument, he would not only map the canals in unprecedented detail, he would also explain their purpose and origin.[15]

By 1893, breakthroughs in glassmaking and optical design had permitted the construction of colossal refractors that, it seemed, would reveal the surface of Mars with new clarity by gathering more light and enabling greater magnifications. It soon became clear, however, that these spectacular instruments did not provide the great advance many astronomers had anticipated. Newspapers eagerly covered the construction of the world's largest refractor at the US Naval Observatory in Washington, DC, for example, yet the telescope showed less detail on Mars than Schiaparelli's much smaller refractor.[16]

Details visible in small instruments were limited by the physics of diffraction, but they were less sensitive to the physics of the atmosphere than larger instruments were. By using small telescopes under a steady sky—in "good seeing," as astronomers call it—observers could wait for moments in which the light gathered by their instruments passed through a single atmospheric convection cell. For fractions of a second, planetary details could suddenly snap into exquisite focus. Light entering large instruments passed through a wider slice of Earth's atmosphere, and therefore through more roiling convection cells. Moments of clarity were much harder to come by, but they revealed unparalleled detail. These atmospheric complexities provoked bitter debates between professional and amateur astronomers over whether enormous, expensive new refractors really provided superior views of the planets, especially Mars.[17]

In 1887, before Edward dispatched William to Peru to observe stars, a mechanical engineer named Uriah Boyden had bequeathed over $8 million (in 2024 USD) to the Harvard Observatory so Edward and William Pickering could install a telescope at a location free from "impediments . . . owing to atmospheric influences." After setting a bulky reflector at various locations in Colorado, culminating with the summit of Pike Peak, the brothers realized that air could be unsteady even at high elevations, where there was less atmosphere to look through. To maximize the performance of big telescopes, it was important to find the right location, not just elevation. It was this insight that led Edward to send William to Peru.[18]

Ironically, it was on a rare night when atmospheric turbulence precluded detailed observation of Mars that Pickering made his most important breakthrough at Arequipa. While trying to explain the cause of the night's poor seeing, he developed a novel method for systematically measuring atmospheric turbulence on a ten-point scale. If seeing could be measured according to a common scale, Pickering later told Lowell in Boston, it could also be rigorously compared from place to place, so that locations could be found where seeing would consistently be ideal. Pickering, Lowell now realized, had "brought into cooperation a practically new instrument, the air itself," which turned out to be "the most vital one of all." If Pickering's scale were used to measure nightly seeing, a large telescope could, in theory, be installed at a location where the atmosphere was sufficiently stable for it to combine the clarity of a smaller instrument with the light-gathering power of a much larger telescope. With "monasteries in the wilds," Lowell predicted, "dedicated to astronomy as in the past to faith," the mysteries of Mars would be unveiled.[19]

Pickering and Douglass received leaves of absence from Harvard to help Lowell establish an observatory somewhere in those "wilds." It would have to be built before the opposition of October 1894, Lowell decided, and because it was harder to coordinate construction overseas, that meant that it would need to be located in the United States. It could not function east of the Mississippi, where "smoke from multiplying factories," Lowell wrote, joined "electric lighting to help put out the stars." Because "water in the air is the great unsettler," it would need to be in the arid Southwest, ideally high up where the air was thin, but not on a mountain where air currents rushing up or down would degrade visibility. Promoters of frontier settlement, named pioneer boosters, advertised the superior climate of Arizona. In the spring of 1894 Lowell therefore sent Douglass to Arizona to survey high-altitude sites for an observatory. Douglass's travels were eagerly followed by local newspapers and sparked competition between municipal governments and businesses for the money and prestige the new observatory—or, some believed, university—would bring.[20]

Lowell asked Douglass to gather, at every stop, the longest records of rainfall measurements that newly established weather stations could provide. Lowell graphed these measurements, seeing them as proxies for humidity and cloudiness. While Lowell reconstructed the climate of each potential site for his observatory, Douglass used a portable telescope to test the stability of the local atmosphere. In theory, climate data allowed Lowell to identify the atmospheric mechanisms that explained the seeing conditions Douglass observed with his telescope. Yet rainfall reconstructions did not offer a complete picture of regional climatic conditions, and Douglass could only measure seeing for a few nights in any one location. Lowell's conclusions about the suitability of possible sites for his observatory therefore depended on extrapolations derived from limited observations and a lot of wishful thinking. There was no way for him to know whether these sites were, on average, climatically ideal for astronomy, or merely ideal while Douglass visited them. Still, when Douglass reported excellent seeing, including observing two new comets, at Flagstaff, which at that time was a fledgling ranching and lumber settlement on the Coconino Plateau, Lowell made his choice. With the enthusiastic support of the Flagstaff community, in 1894 the Lowell Observatory went up on a mesa thereafter known as Mars Hill.[21]

That summer, as Mars wheeled toward its biannual opposition, Lowell arrived in Flagstaff and began observing the planet with refractors borrowed from Harvard. He knew that, according to the nebular hypothesis, Mars had formed before Earth, out of a ring that spun free from the rotating Sun (Chapter 6). Nineteenth-century scientists had refined the hypothesis by adding a process of planetary evolution whereby worlds cooled, dried, and shrank as they aged. Now, atop the Mars-colored sandstone of his mesa, breathing in the thin, dry air, Lowell concluded that Mars, being older than Earth, was also drier and less hospitable for life. The foundation for his subsequent observations was a more realistic view of the planet than had prevailed earlier in the century.[22]

After a few weeks on Mars Hill, Lowell found that individual canals would suddenly appear, as obvious as steel engravings, in flashes of good seeing. Already he had fully developed the theory that would occupy the rest of his life. He knew that Flammarion, riding in a hot air balloon thousands of feet above the Rhine and Loire Rivers, had found that terrestrial rivers were "reduced to a mere thread" by distance but "admirably traced" by the meadows that border them. Lowell concluded that a similar system must prevail on Mars. Lines of vegetation sprouted around a vast irrigation system that funneled water from the Martian poles to triangular relay stations and ultimately circular oases that the likes of Pickering had confused with lakes.[23]

When Douglass observed canals cutting through the supposed seas of Mars, the last piece of the puzzle seemed to snap into place. The dark areas of Mars were not blue oceans but cropland: vast fields of vegetation fertilized by the oases that faded from green to ochre as summer ended in each hemisphere. Lowell decided that behind this "elegant and elaborate pattern" was an intelligence older than and therefore superior to humanity, capable of moving mountains in the low Martian gravity. To Lowell, Mars seemed almost perfectly flat—ideal terrain for a world-spanning network of canals. With grueling diligence, Lowell and his team eventually charted more than four hundred canals. Of these, they decided that more than fifty showed signs of "germination," or doubling. Lowell and Douglass also mapped nearly two hundred oases, including one that seemed to connect no fewer than seventeen canals. No point on Mars appeared to extend more than three hundred miles from a canal.[24]

Lowell introduced his Martian world system with a flurry of popular articles, a series of wildly successful lectures, and finally a book, *Mars,* that sealed his celebrity. He presented himself as an explorer, following in the footsteps of Schiaparelli and making discoveries that dwarfed those made by the agents of European empires on Earth. Despite, or perhaps because of, the rapturous reception by the press and public, some astronomers were skeptical, though few expressed objections that, on their own logic, were especially convincing then or now. Schiaparelli, who had worked out a similar explanation for the canals at around the same time, supported Lowell's vision of Mars—although he, unlike Lowell, would keep an open mind to alternative explanations for the planet's strange appearance.[25]

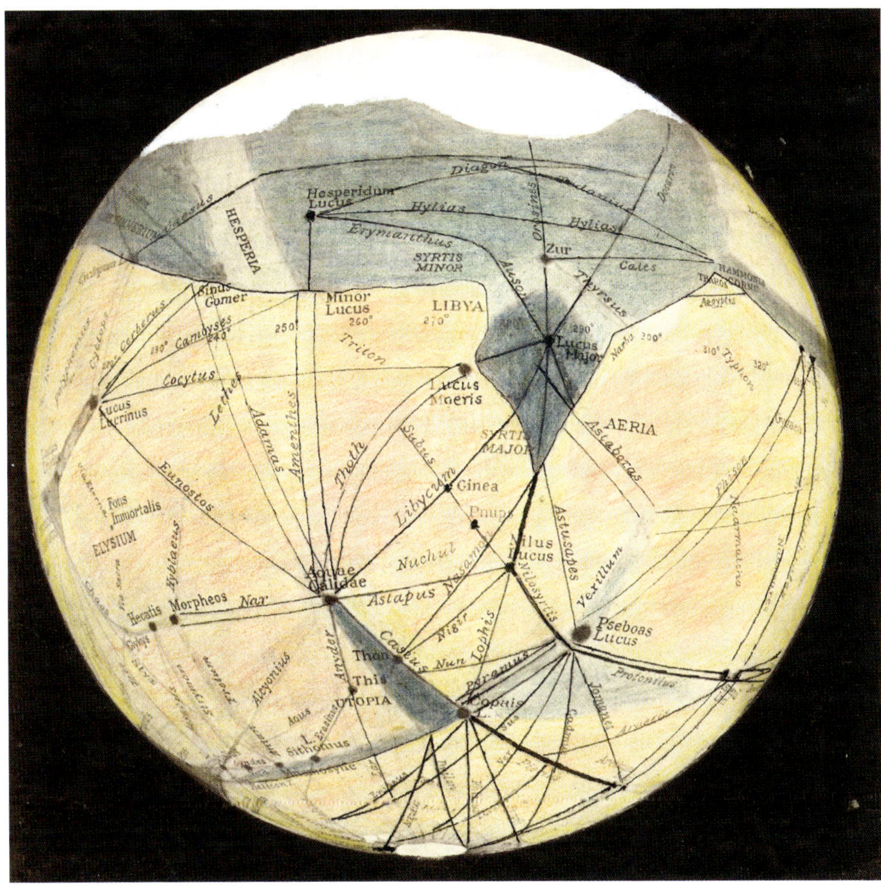

Canals, oases, and meltwater lakes on a map of Mars published by Percival Lowell in 1905.

Mars became front-page news at every opposition, and thousands of observers scoured the planet for canals. Even those who initially doubted their existence eventually saw them. One such observer was the respected amateur Percy Molesworth. The soft shadings in his Mars sketches from the early 1890s suggest that he was initially a canal skeptic, but he was hunting for canals by the turn of the century. His notebooks hint that he did not quite believe what he persuaded his eye to see. In 1900 he wrote that one canal was "very faint but darkening," and a scratched-out note reported that it was "only seen late in the [opposition]." Notes that Molesworth did not cross out reveal that he could not see another supposedly prominent canal "till late in the [opposition]." Even then it was "faint and narrow." Yet another canal was "well-defined," except on December 31, when it mysteriously disappeared. Other canals seemed "diffuse," "faint at first," or "faint and narrow." For most observers, in fact, the canals were forever just on the threshold of perception. "During some very clear, still night," the *Tacoma Times* assured its readers in 1909, "a glimpse may be had of a canal or two if a four-inch telescope is used." It was a common refrain in newspapers written at the time.[26]

Explaining a Vast Delusion

Of course, there are no canals on Mars. For over a century, academics of all kinds have struggled to explain why so many saw them anyway, including astronomers whose eyesight, telescopes, and judgment were widely celebrated in their time. The best-known but least convincing explanations focus narrowly on the flaws and foibles of Lowell and his closest allies. One holds that amateur astronomers like Lowell were more willing than sober-minded professionals to speculate on the basis of flimsy evidence. According to this view, amateurs popularized the canal theory over the objections of serious scientists. In fact, there was no rigid distinction between amateur and professional in late nineteenth-century astronomy. Lowell's observations of the planet were, for a while, unsurpassed in their rigor, and in any case influential professionals were key proponents of the canal hypothesis.[27]

Another explanation posits that Lowell and his supporters erred by seeking, and thus inevitably seeing, only what they had expected to see. This idea has some merit. A similar tendency to extrapolate from inadequate evidence had led Lowell to select Flagstaff for his observatory. Yet it is unlikely that Schiaparelli intended to detect canals. And later many of the foremost canal proponents used painstaking means to remove bias from their observations. Lowell, for example, feared "the brain" could fool him into seeing canals that were not really there. To ensure that his observations were accurate, he took steps like rigorously quantifying the environmental changes he saw on Mars. Working with his assistants, he not only measured apparent fluctuations in the visibility of the canals but also calculated the volume of water they seemed capable of transporting and the degree of solar radiation that would have reached different canals at different times of the year.[28]

A slightly more persuasive explanation suggests that prominent canal supporters suffered from compromised eyesight or used flawed optics. These same observers, however, excelled at distinguishing between, or "splitting," orbiting "double stars," a notoriously challenging optical feat. Some have proposed that Lowell suffered from astigmatism, a visual condition that could have led him to mistake veins in his eyes for a network of Martian canals. Yet no observer ever claimed to see the entire canal network at once, for "no air is so steady," Lowell wrote, "as to permit continuous view of the finer detail." Like Schiaparelli, Lowell and his assistants glimpsed only a handful of canals at a time, in flashes of good seeing, and then drafted quick sketches that could be combined to form a global map of the canal network. Moreover, a large, diverse group of astronomers reported having glimpsed the canals.[29]

More compelling explanations consider the cultural, economic, and political contexts in which the canal theory became a popular sensation. The historian of science K. Maria D. Lane first drew attention to how the science of geography gave credibility to canal-covered maps, and prestige to the observers who published them. Joshua Nall, another historian of science, more recently suggested that the combination of the penny press and the use of the telegram distorted the discourse of astronomy, so that observers ended up giving undue legitimacy to canals. William Sheehan offered a multicausal, psychological explanation: Lowell's dominance over the staff in his observatory, who saw on Mars what he wanted them to see; the tendency of the eye to link discontinuous markings in flashes of good seeing; and the tendency of the mind to arrange complexity into geometric shapes before it transforms stimuli into stable perception.[30]

All of these explanations seem relevant to both the origin and the popularity of the canal theory. Yet a critical piece of the puzzle is missing: the way in which observers perceived and interpreted environmental changes on Mars, in Earth's atmosphere, and across Earth's surface. We have already seen that dust storms originally led astronomers to conclude that Mars resembled Earth and therefore that liquid water must flow across its surface. Without that assumption, Schiaparelli and other canal proponents might have been more hesitant to conclude that linear features on Mars were canals. Ironically, the changes in the dark regions of Mars that were caused by dust storms also convinced Lowell that those regions could not be seas, but instead must be cropland harvested by Martians.[31]

Once the canal theory attained a level of scientific credibility, change on the Martian surface became even more important. To Schiaparelli, the true purpose of the canals was revealed when spring came to a Martian hemisphere and a blue belt appeared around the edge of a melting ice cap. As Lowell put it, the polar bands "developed the peculiar property of retreating with the cap as the latter shrank," before making inroads into and across the cap itself. These changes convinced both Lowell and Schiaparelli that the bands had to be meltwater seas. Schiaparelli wrote that when the apparent seas appeared, "the canals of the surrounding region become blacker and wider." Then, wrote Lowell, a "wave of darkening" spread south from the poles as water flowed along the canals to the fields where Martians grew their crops. These changes inspired Schiaparelli and especially Lowell to transform a puzzling observation of linear features on Mars into a theory for a hydraulic system that, according to Lowell, could only be the creation of intelligent builders.[32]

The darkening wave and fluctuating blue bands helped canal proponents defend their theory from skeptics who could not see canals when Mars was closest to Earth. Because the advance of the alleged Martian vegetation depended on the supposed seasonal appearance of polar meltwater, which had little relation to the position of Mars relative to Earth, it made sense that canals could be hard to see at some oppositions. "As the planet goes away, the canals come out," Lowell wrote. Skeptics had simply looked "at the wrong time." Only keen observers like Lowell could glimpse the "skeletons" of canals after their leafy outlines withered.[33]

In a sense, canal proponents were right. The Martian ice caps are partly composed of water. But that water sublimates directly into the Martian atmosphere as the ice caps retreat; that is, it turns directly into gas, without going through a liquid stage. In each Martian hemisphere, dusky regions do appear around the ice caps as they sublimate in the spring. Eventually the caps are crisscrossed with dark rifts. The true color of these dark features, however, is closer to brown than blue, because they are dunes exposed by the retreating ice. At times, winds swirling around the ice caps compound the contrast between the bright ice and

Dark dunes surround the northern polar cap of Mars in a 2016 photograph by the Indian Space Research Organisation's Mars Orbiter Mission, also known as Mangalyaan.

the dark dunes by clearing bright surface dust from around the cap. Other times, white clouds at lower latitudes accentuate the contrast by giving the dunes the appearance of a belt between bright regions. Moreover, darkness does spread toward the equator in the spring of each Martian hemisphere. The cause is not, however, vegetation responding to the flow of water and the coming of summer. Instead, the sublimation of gas from the ice caps creates seasonal winds that blow bright dust off of ever more southerly land, which then looks darker. "Sufficiently like us to seem in part decipherable," Lowell cautioned, "Mars is yet sufficiently unlike to baffle the very conjecture it starts." This was truer than he could imagine.[34]

Changes of a different sort convinced many that the canal builders were not only real but also interested in Earth. Beginning in the 1870s, Schiaparelli and other observers reported that they had seen lights on Mars, near the terminator, the border line that divides Martian night from day. By 1895 a front-page article in the *Cook County Herald,* a Minnesota paper, announced, "It is no longer to be denied that some very strange and mysterious things are going on on Mars." Journalists, Mars enthusiasts, even some seasoned observers interpreted the "projections" as evidence that Martians were attempting to contact humanity.[35]

After observing many lights in 1894, Douglass concluded that they must be sunlight reflected off high-altitude clouds. When Lowell saw "two brilliant star points suddenly flash out" from a Martian ice cap, he assumed he had noticed sunlight glinting off ice. Such sightings, Lowell wrote, held as much interest "for workers on Mars" as "a new comet possesses for astronomy," and each sparked a wave of speculative headlines in the penny press. In 1895, prominent newspapers even reported that astronomers had found features on Mars spelling out "The Almighty" in Hebrew.[36]

Letters to the Lowell Observatory soon followed. "I observe from the papers that . . . you observed a projection of some kind on the planet Mars," Daniel Parks, a lawyer, wrote to Douglass in late 1900. Could the Martians have constructed a monument big enough to see from Earth, he wondered? Then in 1903 astonished astronomers at Lowell Observatory observed an entire "islet of light" roving across the otherwise unilluminated hemisphere of Mars. Lowell accurately insisted that the yellow color of the projection revealed it to be a dust cloud, and dismissed the possibility, breathlessly reported in the popular press, that it signified an attempt by Martians to message Earth. Nevertheless, once Schiaparelli and Lowell had articulated their canal theory, any anomalous lights or projections on Mars seemed like further evidence for Martian civilization.[37]

Environmental changes on Earth also inspired and entrenched the canal hypothesis. In the equatorial expanses of the Pacific Ocean, water regularly warms in El Niño events, then cools during a La Niña. The pattern of warming and cooling usually endures for no longer than a year or two, but it involves so much water that it profoundly rearranges worldwide weather patterns. An El Niño, for example, weakens the Indian and West African monsoons, and can thereby cause severe drought. A La Niña has the opposite effect, strengthening the monsoons to the point where floods may cause widespread devastation. Yet not all El Niño and La Niña events are alike. During four crucial periods in the time of the canal controversy, powerful El Niño events occurred year after year, bringing drought to French and British colonies across West Africa and the Indian subcontinent.[38]

First, a strong El Niño coincided with Schiaparelli's "discovery" of the canals in 1877. Then alternating strong El Niño and La Niña events recurred between 1888 and 1894, the period when the Mars boom captured public attention. These events culminated in an extreme La Niña that dried out Arizona just as Lowell established his observatory there. Next a series of unusually strong El Niño events closed out the nineteenth century and continued into the twentieth century, precisely when the canals became a popular and scientific sensation. Finally, four very strong El Niño events occurred in the years 1912 to 1915, the final years of Lowell's life, when the canal theory retained much of its influence and popularity.[39]

Colonialism brought the planetary scale of these droughts into focus, and ensured that drought and its frightening consequences preoccupied scholars, government officials, and journalists, at just the time when astronomers developed the canal theory. Colonial authorities did much more than chronicle the spread of drought and famine. Many forced local communities to abandon traditional agricultural and pastoral practices, along with granaries, wells, irrigation systems, and other safeguards against drought. Authorities compelled indigenous farmers to grow cash crops, such as tea or cotton, or participate in destructive mining or logging operations. The flow of resources extracted from the colonies enriched imperial metropoles, helped them to industrialize, and allowed them to forge a global economy increasingly centered on London. Many rural communities on the imperial periphery were impoverished, left vulnerable to fluctuations in commodity prices, and imperiled by pollution. Before the onset of the British Raj, many communities in India had survived monsoon failures. After colonization, the monsoon failures created destructive droughts that provoked staggering harvest failures, and contaminated meager stores of potable water with human or animal waste. Famine, malnourishment, and consequent outbreaks of epidemic disease soon followed. These eventually led to widespread revolt against colonial authorities.[40]

After helping to cause famines, colonial governments worsened them, or at least did little to alleviate them. Prevailing laissez-faire economic ideology discouraged governments from intervening in the market to, for example, halt grain exports or provide aid for the needy. Administrators also repeatedly exploited famine to expand their control over colonized peoples. Collectively, the combination of colonial mismanagement and drought in the closing decades of the nineteenth century contributed to the deaths of tens of millions across South Asia and Africa. Amid such widespread devastation on Earth, the idea that Martians had engineered their planet to survive the ultimate drought naturally captured widespread attention and, for a while, convinced many scientists.[41]

Canal enthusiasts also found inspiration in the relatively recent discovery of geological deep time. Emerging theories blamed glacial periods on changes in the shape of Earth's elliptical orbit. Such theories appeared to explain how the climate of Mars could be so extreme that vast quantities of meltwater could drain from the poles toward the equator when winter on the planet yielded to spring. In the United States, westward expansion also led to the discovery of marine fossils across dry land. Canal proponents came to believe that Earth had lost much of its water, meaning that it was following Mars on the path to planetary desiccation. As white settlers began to colonize arid land, canal proponents leaned on the nebular hypothesis (which held that the planets were formed one after another, and dried as they aged) to assume that deserts were new to Earth. They were like the "first gray hairs in [a] man," Lowell wrote, and it seemed they would spread until Earth fully resembled Mars. The canal theory both strengthened and was strengthened by this idea. Some scientists responded to the perceived desiccation of Africa, Australia, and Brazil by drafting schemes for extreme hydraulic interventions that would have impressed even a Martian, and won widespread support from settlers eager to perpetuate white supremacy.[42]

Meanwhile, on a previously unimagined scale, hydraulic engineers and laborers across the colonial world dammed valleys, straightened rivers, walled off the sea, and drained bogs. Canals were their most spectacular undertakings. In the United States alone, between 1780 and 1860, laborers and engineers constructed some sixty-eight hundred kilometers of navigable canals, a figure almost exactly equivalent to the diameter of Mars. No project compared to the Suez Canal, which by shortening the distance for waterborne travel between East and West promised nothing less to its proponents than the harmonious union of the commerce and culture of Eurasia. For ten years, work unfolded across more than a hundred miles of scorching desert, as a vast hierarchy of European bureaucrats and engineers harnessed the grueling labor of some four hundred thousand men, many of whom died during their labor, until at last the waters of the Red Sea mingled with those of the Mediterranean.[43]

The growing mastery of colonial powers over water directly influenced canal proponents and indirectly enhanced the legitimacy of their canal theory. Schiaparelli, for example, had studied hydraulic engineering at the University of Turin when the Suez Canal was first conceived, and he became an astronomer in Milan soon after the groundbreaking in Egypt. His interests may have attuned him to see linear features on Mars, and his training certainly led him to draw those features as though he were plotting artificial waterways on Earth. More broadly, the unprecedented scale at which canal building unfolded, the revolutionary power of the technologies it required, the new kind of workplace—the large, industrialized construction site—that it created, and the vast, hierarchized labor force it demanded: all popularized the idea that canals, like railroads, were milestones of modernity. If God had seeded the planets with life, as many still believed, it stood to reason that an aging, drying world would be littered with the canals of a more advanced civilization.[44]

Once the canal theory had been established, atmospheric change on Earth—or its relative absence—helped entrench its influence. By the first decade of the twentieth century, the largest telescopes in the world outperformed even the gigantic refractor that Lowell eventually installed at his observatory, and some astronomers who used the new instruments claimed they were "too powerful for canals," meaning they were capable of resolving the purported canals into subtler and less linear features. The canal theory began to depend on the idea that Lowell and his assistants could see with a smaller instrument what astronomers armed with bigger telescopes could not. Lowell insisted on the superiority of his eyesight, experience, methods, and, above all, the seeing at his observatory, which he insisted had no equal anywhere on Earth. To the inhabitants of

Flagstaff, however, he complained that seeing on Mars Hill was "exceedingly disappointing" during winter, which Douglass could not have known after scouting the site in spring 1894. The atmosphere was so unsteady that in the next year, Lowell corresponded with Mexican president Porfirio Díaz to facilitate a temporary relocation of his observatory to Tacubaya, Mexico. The move proved exhausting, leading Lowell to write that he had suffered "a complete breakdown of the body." Still, more than any other observer, Lowell had the stamina, money, and desire to observe Mars under the steadiest atmosphere possible, and that gave legitimacy to the canal theory into the twentieth century.[45]

The canal theory was an idea partly inspired and popularized by the confluence of colonialism and perceived environmental change, on both Mars and Earth. "Underlying all the particular phenomena presented to us by the planet's disk," Lowell wrote, "is the fundamental one of change," because it was change that revealed Mars to be "not a dead but a living world." It was a world that did not seem to have long to live, though, and to many the implications for humanity were disturbing.[46]

15

Meeting the Martians

The humorist, inventor, and poet Charles Cros was in some respects Lowell's polar opposite. Poverty denied him the easy confidence that elites like Lowell tend to have. And a careless streak deprived Cros of patents for color photography or the phonograph that might have secured his financial success. According to the historian Howard Sutton, Cros had "remarkable eyes that were by turns dreary, quizzical, and tragic," and a nervous, jittery personality. "Words tumbled out of his mouth" in confusing bursts. Yet they flowed easily on the page, and there, at least, he resembled Lowell.[1]

He was also fascinated by Mars. Even before Schiaparelli sketched his canals, Cros concluded that lights on the planet were the work of Martians who were trying to communicate with Earth. In 1869 he proposed a means of sending a response. A signal would need to travel fast—only light seemed fast enough—and it would have to be repeated, for it was regular recurrence that would distinguish it from the random noise of the universe. The trouble was that light diminished with distance. A solution, Cros concluded, lay in focusing intense light—"electric light is best"—with gigantic, parabolic mirrors, preferably at night (and perhaps in the Arctic, where it was night for the entire winter). "The inhabitants of the planet Venus or those of Mars," Cros wrote, could then "see, on the dark edge of the Earth's disk, a luminous point." However, they might mistake it for "an active volcano or some other unexplained optical effect; in a word, a natural phenomenon." The key to interplanetary communication would be to ensure that the signal, the light, underwent "modifications such that its intended origin and its purpose [did] not remain in doubt." For Cros, those modifications could transmit a numerical code that could then be used to efficiently send further messages.[2]

It was the clearest articulation to date of the principles of communication between two worlds: principles that today inform the well-funded effort to listen for extraterrestrial signals. Some ridiculed the idea, but Cros had a response. "Beware of the poet," he wrote, "Who, being hungry, / Might, in the end, / Put some bullets in your head." Cros died young, perhaps from a combination of alcoholism and tuberculosis, yet his ideas assumed new influence with the apparent discovery of Martian canals. The canal builders seemed to reveal that humans were latecomers to an interplanetary conversation, and they hinted that humanity's future would be defined by environmental transformation and decline. By the turn of the century, environmental changes on Mars had elicited widespread fears of interplanetary invasion. In subsequent decades, popular fiction, in various media, exploited and perpetuated those fears, occasionally to critique imperialism on Earth. By inspiring and then strengthening the canal theory, the fluctuating atmospheres of Earth and Mars had helped create new understandings of the global risks and opportunities shared by all humans.[3]

Imagining a Martian Anthropocene

Scholars and science fiction authors alike have long speculated that the genuine discovery of extraterrestrial life will provoke disruptive cultural and social responses on Earth, culminating in a profound rethinking of what it means to be human.[4] This was doubly true for the apparent discovery of life on Mars, since the Martians, as an aged species, seemed to presage humanity's future. "O'er all there comes a shadow and a fear," the president of the Royal Astronomical Society of Canada told an audience in 1897, misquoting the English poet Thomas Hood; the works of "Martian intelligence" seemed both wondrous and terrifying.[5]

They also appeared to offer a surprising lesson for human observers. According to Lowell's assistant Vesto Slipher, the magnitude of public interest in Mars, which seemed to eclipse the public's interest in nearly every scientific subject, reflected the capacity for the canals to inspire a reimagining of humanity's past, present, and future.[6] At the height of his theory's popularity, Lowell wrote that the canals revealed how "the true history of man has consisted not in his squabbles with his kind but in his steady conquest of all earth's animals except himself." The Martians had seemingly achieved complete dominance over their dying world. Likewise, "man . . . has enslaved all that he could; he is busy in exterminating the rest. From this he has gone on to turn the very forces of nature to his own ends." According to Lowell, "Subjugation carries its telltale in its train; for it alters the face of its habitat to its own ends. Already man has begun to leave his mark on this his globe in deforestation, in canalization, in communication. So far his towns and his tillage are more partial than complete. But the time is coming when the earth will bear his imprint, and his alone. What he chooses, will survive; what he pleases, will lapse, and the landscape itself will become the carved object of his handiwork."[7]

This was one of the first full articulations of a concept similar to the Anthropocene (Chapter 12), a term now used by scholars and activists to express the magnitude of human interventions into the chemical cycles and energy flows that constitute the Earth system. The Anthropocene idea is popular today partly because it helps contextualize and communicate the human erosion of the conditions that make Earth habitable, most importantly the planet's biodiversity. At the turn of the twentieth century, though, the canals were interpreted as revealing that an intelligent species could prolong its existence by destroying the biodiversity of its planet. "As the brain develops," Lowell assumed, it had no choice but to "take possession of its world," in part because worlds inevitably grew uninhabitable. To survive, a big-brained species had to construct its own water cycle by reshaping the landscape of its planet, and to eliminate any life that did not serve its own ends. Indeed, Lowell's use of evolutionary theory provides one more explanation for the popularity of his Martian world system. Scientists, journalists, and laypeople were only too ready to accept an explanation for the canals that combined cutting-edge theories of planetary, Darwinian, and social Darwinian evolution.[8]

Leading canal proponents agreed that the Martians had seized control of their planet, but they differed sharply over the social and political significance of that feat. Lowell admired Darwin, and saw in Darwin's career a model of his own struggle to convince the ignorant of a grand truth in nature. He believed that any policy that helped those whom nature had ostensibly made weaker threatened the survival of civilization. He had therefore long scorned both the cooperative spirit he identified in the culture of East Asia and the growing movements for women's suffrage, trade unionism, immigration, socialism, and communism across much of the West. To him, imagination most easily arose from rugged individualism—and imaginative leaps allowed civilizations to survive in the competitive world described by social Darwinists (Chapter 6). Martian achievements, Lowell stressed, revealed the superiority of a ruthlessly hierarchical society, led by imaginative elites who had found a way to survive on a dying world.[9]

By contrast, Schiaparelli understood the canals as a triumph of "collective socialism," a communal response that preserved a "socialist paradise" in the face of environmental catastrophe. When others suggested that the canals revealed that the destiny of intelligent life was to die with its planet, in 1898 the US astronomer Garrett Serviss imagined how the Martians could mine an asteroid to prosper in spite of the environmental decline of their planet. It was the first articulation of asteroid mining, a proposed solution to environmental scarcity that would gain popularity in the late twentieth century (Chapter 18). The relatively optimistic impressions of the canals advanced by Schiparelli and Serviss would inspire two of the most influential science fiction stories about Martian life—out of the roughly one hundred written before the First World War.[10]

In Germany the author Kurd Lasswitz, in his science fiction novel *On Two Planets,* imagined an attempt to colonize Earth by the Nume, a Martian civilization. The Nume had established a socialist utopia on Mars by, among other things, supplanting fossil fuels with solar power. Lasswitz's novel, which was banned by the Nazis, offered a vision of social, technological, and environmental progress that later inspired the German engineers who helped launch the Space Age.[11]

In Russia, Alexander Bogdanov, a onetime associate of Vladimir Lenin, took a different approach. In his science fiction novel *Red Star,* Mars's harsh environment, by requiring the construction of canals and therefore collective labor on a vast scale, enabled a smoother transition to communism than was possible on Earth. The price was an accelerating exploitation of the Martian environment that eventually destabilized the planet's climate. Bogdanov's Martians debated whether to exterminate humans and thereby occupy Earth, but they decided to settle Venus instead, believing that Earthlings could be elevated to communist brotherhood. Far from being a mere flight of fancy, Bogdanov's *Red Star* contributed to a utopian milieu that would soon provide fertile soil for revolution in Russia.[12]

Part of the appeal of the canal theory was that it could be used to support whatever political ideology its proponents espoused. It did inspire new ways of imagining humanity's past and future, but it did not lead to a truly widespread rejection of existing ideas and ideals. Since at least 1960, when the Brookings Institution, a Washington, DC, think tank, advised NASA on the implications of space travel, theorists have warned that the discovery of an alien civilization could result in social upheaval by undermining the anthropocentric foundations of human ideologies. The apparent discovery of the Martians suggests that such fears might be overblown.[13]

Sending a Message to Mars

Early in the nineteenth century, as the idea of Mars as a second Earth was just beginning to take shape, the German mathematician Carl Friedrich Gauss proposed contacting the Martians by carving enormous geometric shapes across the Siberian forest. The Austrian astronomer Joseph Johann von Littrow suggested burning shapes into the countryside. Efforts to communicate with Mars assumed greater urgency when reports of unexplained lights on the red planet seemed to suggest that the Martians were trying to contact Earth. Other plans imagined focusing onto Mars the light of the setting Sun, using mirrors arrayed on mountains or perhaps the Eiffel Tower. These systems would cost money, but in 1891 Flammarion announced that a French widow would bequeath her fortune to anyone who could, within ten years, communicate with another world. Now even the famed inventor Alexander Graham Bell planned "to devise a means of interplanetary communication that would transfer information by a 'shaft of light.'" Some proposed using the powers of the mind; indeed, Flammarion knew that mystics had long claimed to use "psychical" means to communicate with other worlds. Élise-Catherine Müller, a medium known as Hélène Smith, claimed to have taken an astral journey to Mars, where she glimpsed people indistinguishable from humans "save that both sexes wore the same costume." She did not inherit the fortune.[14]

Without a common language, it seemed that a message from Earth to Mars could at best merely communicate the existence of another civilization. In a fictional account of what he considered a likely future, Flammarion wrote that it would take centuries before scientists could decipher the "hieroglyphic symbols, addressed in vain for several thousand years by the inhabitants of the planet Mars to the earth." In 1896, however, Sir Francis Galton argued that Martians could easily modulate the length of signals transmitted to Earth using a heliograph (a mirror that focuses sunlight) and thereby communicate the mutually intelligible basics of geometry, or the structure of the solar system. With that worked out, a language could be deciphered—and the same method could be used to signal civilizations in other star systems.[15]

Serbian American inventor Nikola Tesla soon became the leading advocate for interplanetary communication. Tesla had developed a series of technologies that harnessed the power of alternating current, which provided electricity more efficiently and over longer distances than competing direct current. As Tesla grew rich and famous in the 1890s, he started imagining new applications for powerful electrical devices, including the transmission and reception of messages from Mars or Venus. To build these devices, he built and moved to an isolated laboratory in Colorado Springs, where inexpensive hydroelectric power allowed him to channel millions of volts of electricity.[16]

In July 1899 he made improvements to the sensitivity of his receivers to devise new ways of detecting the distant approach of thunderstorms, which fascinated him. On the night of July 28, Tesla was alone in his laboratory when his instruments registered a weak and mysterious signal that seemed to repeat a simple sequence: one, two, three. At first he may well have thought that the signal originated from the storm he had been tracking. Yet even then, he later reflected, "I felt as though I were present at the birth of a new knowledge or the revelation of a great truth." As the days passed, it dawned on him that he might have been "the first to hear the greeting of one planet to another." By tracking an environmental change on Earth, it seemed he had detected a message from the alien environment of another world.[17]

Soon after his apparent discovery, Tesla considered how humans could harness electricity to shape Earth as the Martians had transformed their world. In December he confided to the journalist Julian Hawthorne that the "art of transmitting electrical energy through the natural media" could well make it possible for humanity to effect "wonderful changes and transformation" to the surface of Earth, changes that "are, to all evidence, now being wrought by intelligent beings on a neighboring planet." Tesla also started thinking about how he could harness electricity to respond to the signals he had heard. In 1900 he published a design for an electrical oscillator that could generate nearly limitless energies. Tesla argued that his invention made it "perfectly practicable" to produce an "electrical movement of such magnitude that, without the slightest doubt, its effect [would] be perceptible on some of our nearer planets, as Venus and Mars." The vast energies of electricity, rather than reflected light, would catch the attention of the Martians (or Venusians). "That we can send a message to a planet is certain," Tesla concluded, adding, "that we can get an answer is probable: man is not the only being in the Infinite gifted with a mind."[18]

Tesla had not yet revealed his apparent discovery of an alien message. Finally, in December, he was one of several prominent figures to receive an invitation from the American Red Cross to predict humanity's greatest possible achievement in the coming century. He responded by revealing that he had detected "electrical actions, which have appeared inexplicable," that had convinced him that someday humanity would "have a message from another world."

In February 1901 the editors of *Collier's Weekly,* an influential American magazine, asked Tesla to comment on series of discoveries he made in Colorado that promised to revolutionize science. The "most gratifying," Tesla wrote in response, concerned "certain feeble electrical disturbances" that had to be "of planetary origin." It was time, Tesla concluded, "for the electrician to join the astronomer in the exploration of our neighboring worlds." When *Collier's* published Tesla's description of his discovery, an editor's note added that "[prominent] scientists . . . are disposed to agree with Mr. Tesla in his startling deductions." It was an achievement to "amaze the entire universe."[19]

The *New York Sun* also announced with great fanfare that Tesla had detected a signal from Mars. Tesla quickly wrote a letter to the editor that left open the possibility that the message had come from Venus. Six years later, however, he concluded that it had indeed come from Mars, which had been close to Earth in July 1899. The signal had lasted only as long as Mars stayed in the night sky

An artist's conception of a volcanic eruption on Io, with Jupiter in the background.

before setting, and "there existed no wireless plant other than mine that could produce a disturbance perceptible in a radius of more than a few miles."[20]

Tesla's sensitive receivers may indeed have detected electromagnetic radiation with a "planetary origin," but one more distant and even stranger than Tesla could have imagined. Deep within Jupiter, a vast sea of electrically conducting, metallic hydrogen powers a magnetic field nearly twenty thousand times stronger than Earth's. Charged particles of sulfur and oxygen gush from volcanoes across Jupiter's innermost moon, Io, and accelerate to nearly the speed of light in the Jovian magnetic field. The energy created by this crackling electricity creates auroras on both Io and Jupiter, and it inflates the largest structure in the solar system not formed by the Sun: the Jovian magnetic field.[21]

Electrons rushing through Jupiter's magnetic field accelerate and release radio waves at frequencies between 10 and 40 megahertz, precisely where Tesla's receivers were particularly sensitive. When Io and its auroral footprint on Jupiter are visible from Earth, Jupiter's radio signals surge in intensity. Sometimes they click in ways that make them sound like static from a thunderstorm on Earth; occasionally their clicks come in triplets that resemble an alien signal. In late July 1899, Jupiter and Io were in the sky over Colorado, broadcasting the radio waves created by their interactions, while Mars was visible and the Sun, Moon, and Venus were not. Environmental entanglements around the fifth planet from the Sun may have convinced Tesla that he had received an artificial message from the fourth.[22]

As Tesla grew older, poorer, and more eccentric, he insisted with increasing conviction that he had detected a message from Mars. Print journalists now identified him first and foremost as an inventor who had received signals from Mars. They fanned public interest in some of his most ambitious schemes—such as a massive wireless transmission tower he left partly constructed in Shoreham, New York—by portraying them as efforts to contact Mars. Indeed, Tesla's dream of detecting and perhaps communicating with the beings who had apparently messaged him in 1899 inspired some of his most visionary, and at the time fanciful, ideas for new technologies. During a 1923 interview, for example, he announced that he had designed a mechanical eye: a kind of Palantir to spy on Earthlings and Martians alike.[23]

By then Tesla had been eclipsed as an interplanetary communicator by the Italian radio pioneer Guglielmo Marconi, who repeatedly inspired headlines by announcing he had detected "very queer sounds . . . which might come from somewhere outside the Earth"—likely Mars. The US inventor Charles Steinmetz doubted these claims but argued that if the United States was as serious about communicating with Mars as it had been about fighting the First World War—if it, for example, concentrated the entire electrical output of the country into a single sending station, at the cost of, say, $17 billion (in 2024 USD)—then "it is not at all improbable that the plan would succeed." Astronomer David Todd meanwhile suggested various schemes to launch wireless receivers in balloons, where they could listen for Martian signals beyond Earth's obstructive atmosphere.[24]

Most proposals, however, imagined enormous environmental interventions on Earth that Martians would be able to make out across space. These plans included a mirror fifty miles wide, a sign crafted with enormous strips of black cloth, and a vast network of lights that could be turned on and off. Mars enthusiasts argued that building them would be worth the exorbitant price. "Some day we may tell the Martians all about our great war," an article in *Popular Science* suggested in 1919; "perhaps we will learn from an older and wiser planet how we ought to run the Earth." When Mars moved toward a close opposition in 1924, Swiss astronomers sent a message using a mountaintop heliograph. Todd persuaded radio operators across Europe and the Western Hemisphere to observe a "National Radio Silence Day" on August 21–23 by temporarily suspending transmissions and carefully listening for possible signals from Mars. At 7:12 a.m. on August 21, operators at Point Grey Wireless Station in Vancouver indeed reported hearing a "mysterious" radio pattern consisting of four groups of four pulses each. Newspapers around the world briefly announced that Mars had contacted Earth, but scientists soon concluded that the signals had natural origins, perhaps a thunderstorm. Meanwhile, the US Navy had launched an airship from Washington, DC, that would relay signals to a cryptographer on the ground. To interpret those signals, Todd had worked with the inventor Charles Francis Jenkins to develop a "radio camera," a device that transformed radio signals into optical flashes that were recorded on photographic paper. Todd was astonished to find that the signals appeared to resemble a face when they were translated to paper, and newspapers once again breathlessly reported a signal from Mars. Jenkins himself, however, rejected the possibility, so the results remained frustratingly inconclusive. Interest in interplanetary communication faded as the prospects for Martian life dimmed in subsequent years.[25]

By 1960, advances in radio astronomy, initially spurred by breakthroughs in military radar, sparked renewed interest in what the astronomer Frank Drake termed the "search for extraterrestrial intelligence," or SETI. Powerful new radio telescopes enabled scientists to listen not just for signals from Mars or Venus, now widely considered devoid of intelligent life, but also from worlds orbiting distant stars. Scientists established organizations, notably the SETI Institute, dedicated to undertaking SETI efforts and harnessed the enthusiasm of thousands of amateur observers. Today SETI research is rapidly expanding in both scope and technological sophistication, benefiting from increased funding and closer integration within mainstream science. Although SETI research has yet to identify a definitive extraterrestrial signal, it continues to captivate public imagination, amplified through popular books such as Carl Sagan's *Contact* and Liu Cixin's *The Three-Body Problem.*[26]

Methods to contact Mars, popularized by animator-producer Max Fleischer. Top: an enormous mirror to dazzle the Martians. Bottom: motors control strips of black fabric in the Sahara Desert that can be turned on or off to wink at Mars.

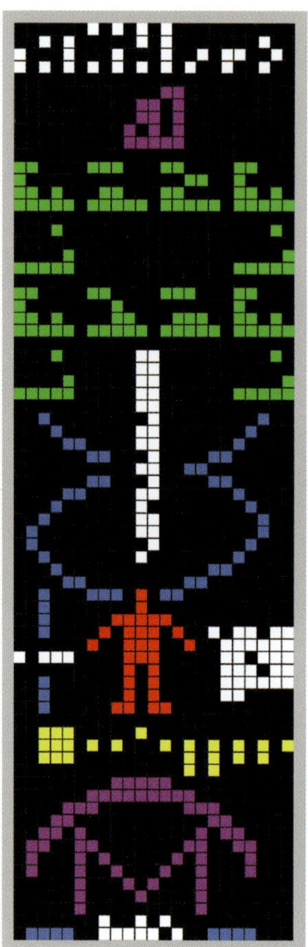

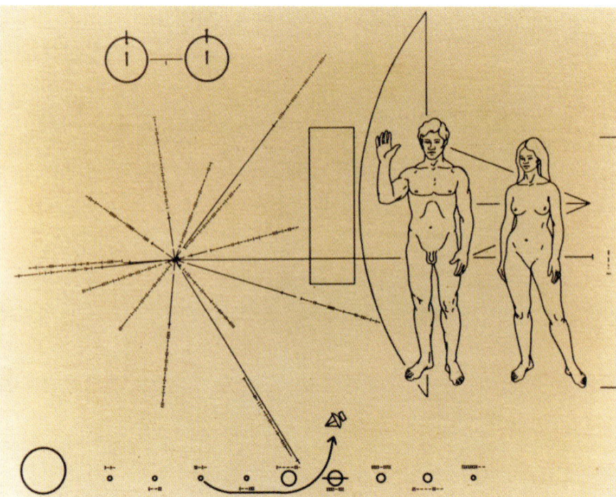

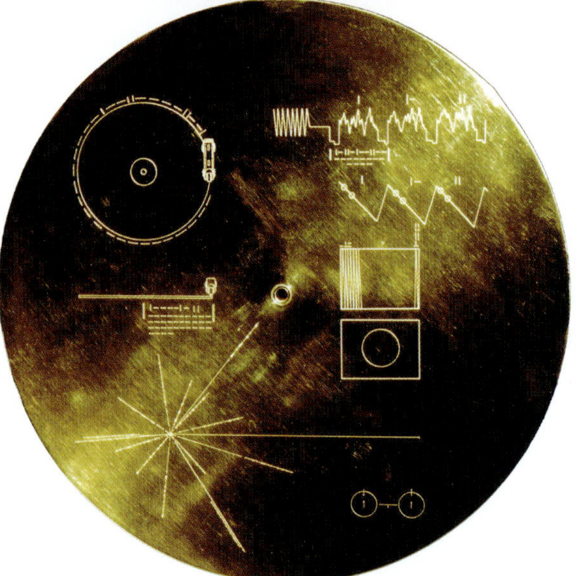

Attempts at METI. Left: a message beamed from the Arecibo radar telescope to the Hercules Globular Cluster, crafted in 1973 by Frank Drake, Carl Sagan, and other scientists, uses numbers, shapes, and a map of the solar system. Top right: The plaque on the now-interstellar *Pioneer 10* spacecraft includes a map of the solar system and the Sun's position in our galaxy. Bottom right: The golden record on the *Voyager* probes, the first spacecraft to leave the solar system, includes a map of the solar system relative to the position of fourteen pulsars (rapidly rotating neutron stars), as well as diagrams that represent the scientific knowledge of humanity.

Many of the most important ideas in today's SETI—especially the subset of the search known as active SETI, or messaging to extraterrestrial intelligence (METI)—emerged out of efforts to communicate with Mars. Like modern SETI advocates, pluralists considered what means an alien civilization would use to send a message to other planets—light? electricity? radio?—and debated how best to reply. The reasons they proposed for sending interplanetary messages anticipated reasons later offered by SETI scientists, including the benefits of learning from aliens, who they assumed would undoubtedly be older and thus wiser than Earthlings. Pluralists debated how to communicate and decided to use the languages of mapping and mathematics, which many SETI practitioners still believe are likely to be common across technological civilizations. They also confronted the challenge of distinguishing between artificial and natural signals, which has long bedeviled SETI efforts.[27]

In the pages of the *New York Times*, proposals to listen for a signal from Mars even inspired some of the first concerns that knowledge from another world "for which we are unprepared" might destabilize affairs on Earth. This worry may have resonated with particular force as many countries struggled to rebuild after the First World War and a 1918 influenza pandemic that may have killed over 50 million people. Plans to reply, however, did not inspire public conversations about the existential risk involved in alerting alien civilizations to humanity's presence on Earth. Concerns about these risks today render METI controversial. Nineteenth- and twentieth-century contact proponents appear to have assumed that Martians were simply too close to hide from.[28]

Imagining a Martian Invasion

In the nineteenth century, intellectuals increasingly confronted the limits of humanity. As we have seen, scholars in different disciplines had come to fear for the sustainability of civilizations, owing in part to a fluctuating Sun and climate. Meanwhile, a new class of scientists, the physicists, had developed the laws of thermodynamics and come to suspect that the universe would end through the inevitable accumulation of entropy. Geographers had mapped most of the inhabitable areas of Earth, and former frontiers were increasingly controlled by colonial governments. By the turn of the twentieth century, scholars, journalists, and authors frightened and fascinated the public with visions of the end of the world. In that context, the canals of Mars provided a sobering vision of the distant but inescapable future of Earth. Some speculated that they also revealed a more urgent threat. The canal theory implied that Martians were both technologically superior to humanity and desperate for a change in scenery. They seemed to have plenty of motivation to invade Earth. Did they have the capability?[29]

At the height of Mars mania, none explored these questions more eloquently than H. G. Wells. Educated as a scientist and a convert to socialism, Wells was an outspoken critic of British imperialism and a proponent of women's rights. A Darwinist, he believed Martians on their dying world would have evolved to be radically different from humans, "beyond the most bizarre imaginings of nightmare." Yet in his 1897 novel *The War of the Worlds,* Wells imagined Martians with disturbing similarities to terrestrial colonizers: as the literary scholar Robert Markley put it, they had "the technologies to overspread planets, exhaust available resources, reshape what is left of the environment to their own ends, and then seek new territories, new worlds, to invade." Although it is easy to read colonizers' fear of the colonized in that period's warnings of invasion from Mars, Wells presented an even more frightening prospect: What if the Martians did to Europeans what Europeans had done to others?[30]

"Men like Schiaparelli watched the red planet," Wells wrote, "but failed to interpret the fluctuating appearances of the markings they mapped so well." In Wells's narrative, the lights of 1894 were each "a mass of flaming gas . . . moving with an enormous velocity towards this earth," carrying Martian invaders who, with intellects "vast and cool and unsympathetic," occupied and despoiled southern England. Even for readers accustomed to invasion literature—and there was a lot of it in the years before the First World War, as geopolitical tensions surged across Europe—the brutal description of colonialism in the heart of the world's preeminent empire came as a shock. So did the vivid descriptions in *The War of the Worlds* of bodily violation, social disintegration, and a hollowed-out London. They offered an eerie foreshadowing of the wars that would soon befall Europe, and more importantly, critiqued the common assumption that colonialism pulled all peoples up a universal ladder of progress. In the end, the Martians imagined by Wells succumb to the same ignoble fate that had long thwarted British colonizers in tropical environments: diseases deadly for interlopers who came from cooler, drier climates.[31]

After the publication of *The War of the Worlds,* newspapers exploited every observation of apparent change on the Martian surface to stoke public fears and thereby sell more papers. During the planetary opposition of 1922, for example, the *Washington Times* warned that "the ice and snow cap around the poles of Mars have spread further than ever," and "the Martians must soon find means to move to some better world or be wiped out of existence." The stage seemed set for "a descent upon the earth by the inhabitants of Mars."[32]

In comic strips, novels, and short stories, science fiction that imagined an alien invasion also took off following the success of *The War of the Worlds.* The genre reached new heights when, in the 1930s, radio emerged as the primary medium for escapist entertainment in the United States—largely because, during the Great Depression, even impoverished families could afford a simple radio set. The obscure *Mercury Theatre on the Air* radio show, with the actors and producers Orson Welles, John Houseman, and Paul Stewart, famously agreed to adapt *The War of the Worlds* for their 1938 Halloween broadcast. Inspired by breaking news bulletins that chronicled the descent to war across Eurasia, they planned a parody news broadcast that placed the Martian invasion near Princeton, New Jersey. They scrapped the dramatic scenes and references to the passage of time that were common in contemporary radio shows, and their actors studied actual news broadcasts to report the Martian invasion with realistic alarm. At the last minute, Welles altered the pacing of the broadcast so that it started slow and gathered momentum toward its intermission—which came unusually late in the show.[33]

When *The War of the Worlds* aired on October 30, 1938, all of these changes fooled at least some listeners into believing that an invasion was really underway. On the following day, newspapers across the United States reported "a wave of mass hysteria," as the *New York Times* put it, that "disrupted households, interrupted religious services, created traffic jams and clogged communications systems." The thousands of letters sent to the Federal Communications Commission and *Mercury Theatre on the Air* confirm that *The War of the Worlds* provoked widespread confusion and fear. They also reveal that popular responses ran the gamut from panic to delight. Above all, the broadcast inspired a surge of communication along networks that bound listeners to one another and to different kinds of news and entertainment media. For an earlier generation the Moon hoax had revealed the power and peril of a new form of mass media, the penny daily. Now the *War of the Worlds* broadcast had done the same for radio.[34]

The most important outcome of the *War of the Worlds* radio broadcast may be the schisms it revealed and encouraged in the changing media landscape of the United States. Print journalists played up popular panic to discredit radio, which had an immediacy that seemed to threaten the slower newspaper business. Spurred by newspaper coverage of the broadcast that characterized radio as a threat to public order, politicians in Washington, DC, raised the possibility of radio censorship. Provided an opening by the newspaper accounts, Adolf Hitler even described the broadcast as "evidence of the decadence and corrupt condition of democracy." Certainly it did not help that Welles, Houseman, and Stewart had deliberately exploited popular trust in authority, at one point by ensuring listeners that, because "radio has a responsibility to serve in the public interest at all times," the broadcast would be turned over to the "state militia at Trenton." The irony, of course, was that print journalism had encouraged sensational fears about Mars that Welles, Houseman, and Stewart merely exploited.[35]

The *War of the Worlds* broadcast inspired similar radio programs during the 1940s in Brazil, Chile, and Ecuador, where panicked crowds rioted upon discovering the hoax. By then the environmental changes that helped inspire the canal theory had reverberated through popular culture for more than seventy-five years. They had spurred efforts to reconceptualize humanity's history and future, to invent the technology and language of interplanetary communication, and to imagine alien invasion. Above all, they had shaped and strengthened an emerging discourse, among authors, inventors, journalists, scholars, and some politicians, that considered the existential risks and opportunities common to humanity. The most important expressions of that discourse would soon involve climate change, on Earth and on Mars.

16

Reshaping Planetary Climates

For more than a decade Andrew Douglass dutifully served William Pickering and then Percival Lowell, the two men most responsible for transforming widespread but mostly unheralded observations of Mars into a popular explosion of interest in Martian aliens. Douglass was an exacting young scientist, and his rigor made him Lowell's most respected advocate among astronomers. When Lowell claimed to discover features suspiciously similar to canals on Venus and then the moons of Jupiter, though, Douglass started to question his mentor's methods (Chapter 6). In 1901 Douglass privately excoriated Lowell's tendency to find "facts in support of some speculation." When Lowell discovered Douglass's betrayal that summer, he unceremoniously sacked his longtime assistant.[1]

Douglass was devastated. Working for a while as a probate judge and a Spanish teacher, he struggled to find employment as an astronomer. Eventually he returned to an idea that may have first occurred to him in 1894, when Lowell directed him to gather rainfall measurements across Arizona. Lowell had used these measurements not only to find an ideal site for his observatory, but also to seek correlations between solar activity and terrestrial weather. When Douglass arrived in Flagstaff, the work may have inspired him to count the growth rings of a pine log on Mars Hill. He likely knew that eighteenth-century scholars, inspired by the severe winters and droughts of the Little Ice Age, had established that trees add a ring in every growing season, that the width of each ring could correlate to temperature or precipitation during a growing season, and that the thickness of rings in different trees could be compared to reconstruct trends in weather. Aware of rising interest in connections between solar activity and terrestrial climate, in 1901 Douglass wondered whether tree rings could reveal centuries of change in the Sun's influence on Earth (Chapter 2). The great achievement of his career would be to develop a method by which scientists could use tree rings to test the basic idea behind Lowell's canal theory: that a planet's climate could change, forcing civilizations to adapt.[2]

In 1905 Douglass accepted a position at the newly established University of Arizona. Even before he arrived at the university's Tucson campus, he had developed the principles of dendrochronology, the systematic study of human and environmental change as recorded in tree rings. Beginning at a log yard in Flagstaff, he collected tree ring samples from long-lived giant sequoias and ponderosa pines across the American Southwest. Then, he identified shared changes in the width of growth rings extracted from many different trees. The new technique of cross-dating revealed that these changes were correlated to shifts in precipitation that had been recorded by meteorological instruments. Now that he knew that tree rings could serve as a proxy for weather, he graphed how their width had changed over the lifespan of his trees. When he determined that precipitation patterns across American Southwest had repeatedly varied in past centuries, he had, for the first time, firmly established what astronomers before him could only suspect: that climate changed on human timescales (Chapter 2).[3]

Douglass's collaborator, Ellsworth Huntington, argued that those changes explained the course of human history. A prominent eugenicist, Huntington believed that the climate of western Europe and eastern North America was uniquely suitable for the advance of civilization, an idea that had once been used to justify European imperialism. The idea suggested not only that Europeans were superior to other peoples, and thus justified in ruling them, but also that a shift in climate could easily bring about civilizational collapse.[4]

Most geographers, anthropologists, sociologists, and historians rejected Huntington's claims that climate change had caused the rise and fall of empires. A few were uneasy with the racist implications of Huntington's ideas, but most simply would not accept that environments could change so profoundly that civilizations would have to respond. Still, the work of Douglass and Huntington would establish a foundation, however imperfect, for the later emergence of fields of study that identified genuine connections between human and climatic histories.[5]

Were it not for the Martian environmental changes that helped inspire the canal theory, Douglass and Huntington might not have launched their study of the history of climate change on Earth. Scholars who study that history today contribute to the development of climate adaptation policies and help shape public discourse around reducing greenhouse gas emissions. In a sense, then, nineteenth-century changes in Martian environments are helping to shape the present-day course of global warming. Indeed, the most important influence on history of Martian dust storms has been to reveal the vulnerability of Earth's climate to human industry and warfare. Growing awareness of that vulnerability helped inspire schemes to transform the Martian climate so it has no more dust storms, and can sustain human life in the event of the ultimate catastrophe on Earth.

Unraveling the Canals of Mars

By the first decade of the twentieth century, Lowell Observatory was the nerve center of an international media operation that converted astronomers' observations of environmental changes on Mars into a steady stream of urgent newspaper updates on the Martian canal builders. Lowell and his assistants published their observations in the *Annals of the Lowell Observatory,* a widely disseminated publication, and Lowell himself wrote newspaper editors to inform them about Martian changes and their significance. The editors would ask Lowell to help shape their coverage of these changes, occasionally paying him to comment on newspaper articles that had already been published by rival publications. They also asked Lowell to comment on changes on Mars that had been observed by other astronomers, or to respond to astronomers' criticisms of his observations. Lowell used and was used by the press, a symbiosis that amplified the hold of the canal theory on the popular imagination.[6]

Among astronomers, though, Lowell's claim that canals also covered Venus began to erode faith in his vision of Mars. Walter Maunder codirected an experiment that asked schoolboys to sketch far away, featureless disks. The boys drew disks that were covered with lines, suggesting that Lowell's canals were also illusions. With Lowell's hesitant support, Douglass had already undertaken similar, private tests that partly confirmed Maunder's results. The "small boy theory," as Lowell called it, cast further doubt on the existence of the canals.[7]

Lowell realized that a photograph could convince his skeptics, yet the long exposure times of contemporary cameras made them unable to capture those moments in which planetary detail suddenly appeared through Earth's turbulent atmosphere. He hired Carl Otto Lampland, an American astronomer and photographer who had developed a new camera capable of taking rapid sequential exposures. Now Lowell was able to publish photographs of Mars made possible, he said, by the use of new technology in the superior seeing that prevailed over Flagstaff. The groundbreaking photographs contained hints of linear features that led some to accept them as proof of the canal theory. Indeed, during the opposition of 1909, the French astronomer Eugène Michel Antoniadi used a refractor of unprecedented size to apparently confirm the existence of the canals charted by Schiaparelli. But Antoniadi argued that when the atmosphere was steady, most of the canals mapped by Lowell and his assistants resolved into discontinuous, irregular markings. Lowell, who viewed Antoniadi more as a student than a rival, insisted on the superiority of his seeing, experience, and techniques. Still, support for his canal theory slumped in the face of Antoniadi's observations.[8]

When Lowell died in 1916, Vesto Slipher, the new director of his observatory, memorialized him as an "ideal of a man of science," a conqueror of nature who had "found a great truth trodden underfoot" that was doubted only by "bigots." In fact, few astronomers now subscribed wholesale to Lowell's canal theory, and fewer still believed that another round of sketches, photographs, or spectroscopic measurements could definitively prove or disprove it. Then, during the extremely close opposition of 1924, two teams of astronomers equipped their telescopes with vacuum thermocouples, devices that convert thermal radiation into electrical signals. Using the thermocouples, the astronomers estimated that temperatures across the dark regions of Mars could reach 20 degrees Celsius, much higher than temperatures in lighter regions. Canal proponents had argued that the dark regions of Mars were irrigated farmland, and now it seemed that those regions were warm enough to support plant life. Even when new spectroscopic analyses failed to find evidence for water or oxygen on Mars, the changing appearance of the Martian surface continued to convince some scientists that these analyses were wrong, or that vegetation had found a way to survive on Mars without oxygen. Meanwhile, amateur astronomers searched for and sketched canals. Later, the map used by NASA to plan its first robotic mission to Mars would include them.[9]

By the 1960s, Mars still seemed the planet most likely to harbor life. In the context of the Cold War, the discovery of even microorganisms on Mars promised to enhance the prestige of the nation responsible for it. As a result, only the Moon eclipsed Mars as a destination for robotic spacecraft. The first Soviet attempt to reach Mars, launched soon after the 1962 Cuban Missile Crisis, failed when the rocket's upper stage broke apart. All of its fragments traveled along trajectories that took them over Canada and the United States, and US radar operators briefly thought a nuclear attack was underway. The second attempt had to be postponed due to power shortages in the wake of a Soviet effort to trigger an electromagnetic pulse with an atmospheric nuclear explosion (Chapter 4). After launch, the probe made it into interplanetary space but fell silent before it could reach Mars. The first spacecraft to fly by Mars ended up being NASA's *Mariner 4*, in 1965. Its blurry photographs showed a cratered and seemingly lifeless surface, more similar to the Moon than to Earth.[10]

Mariner 4 seemed to reveal that Mars was a less exciting destination than previously imagined, and worse, that discoveries made there would have little value in signaling national strength. Political support for NASA's planetary exploration program declined. The probe had photographed only slivers of the Martian surface, though, and exobiologists argued that it had not ruled out the presence of microbial life. Both NASA and the Soviet space program prepared more robotic visitors and confronted the same opposition rhythm that astronomers using telescopes had experienced for centuries. NASA waited for the opposition of 1969 to launch its next Mars-bound spacecraft, *Mariner 6* and *Mariner 7*. Again, both probes passed over heavily cratered terrain that resembled the barren lunar landscape. Yet they detected water and possibly—briefly, tentatively—evidence for methane, a potential byproduct of microscopic life. Anticipation mounted for 1971, when Mars reached perihelic opposition and would be relatively easy to reach from Earth. For the first time, NASA prepared to send orbiters, *Mariner 8* and *Mariner 9,* that would map the Martian surface, and possibly chart environments capable of harboring life.[11]

Soviet scientists and engineers, meanwhile, planned to launch three sophisticated spacecraft to Mars. The first, *M-71,* never left orbit, but the second and third—*Mars 2* and *Mars 3*—successfully departed on their own mapping missions. As they traveled, astronomers at an observatory in South Africa monitored the Martian atmosphere. On September 22, 1971, they noticed a "bright spot" in the southern hemisphere of Mars, a dust storm that quickly spread to cover two-thirds of the planet's circumference. A second storm swirled out from the sprawling Valles Marineris canyon system, then merged with the first. Within weeks the storm shrouded the entire Martian surface.[12]

As *Mars 2* and *Mars 3* approached the red planet in late November, their mapping mission looked ever less promising. Both spacecraft carried landers that housed tiny rovers on skis. A computer error, however, caused the *Mars 2* lander to collide with Mars before it could open its parachute. The *Mars 3* lander made it safely to the surface, but lasted only seconds before it fell silent. A Soviet engineer later speculated that dust from the storm had overwhelmed its transmitters. Meanwhile, the orbiters set about their programmed task of photographing Mars, which was still enveloped by its planet-encircling dust storm. Both orbiters returned valuable data on the gravity, magnetic field, and atmosphere of Mars, but they mapped a featureless globe.[13]

By contrast, the computers aboard NASA's *Mariner* spacecraft could be reprogrammed from Earth. That was of little use to *Mariner 8,* which crashed into the Atlantic Ocean. *Mariner 9,* however, arrived at Mars on November 14, 1971, becoming the first spacecraft to orbit a planet other than Earth. NASA engineers delayed its mapping mission until January, when the dust storm started to abate and the largest volcanoes began to emerge from the haze. Now *Mariner 9* slipped into a closer orbit, and compiled detailed maps of the Martian surface that transformed how scientists understood Mars. The maps definitively established that there were no canals on the planet's surface. Yet they also revealed the telltale signs of dried-up rivers and oceans. Mars had cooled and dried, but that process was further along than even Lowell had imagined.[14]

Dust Storms and Nuclear Winter

The canal controversy had contradictory consequences for astronomy. On the one hand, as early twentieth-century astronomers became more skeptical about lunar or Martian canals, they became convinced that little could be established about the nature of the environments of other planets. Until the advent of the Space Age, most turned their attention instead to the shape and structure of the cosmos. Ironically, measurements made at Lowell Observatory encouraged that turn by providing the first evidence for the expansion of the universe, and by establishing that telescopes in areas with a consistently stable atmosphere could enable astronomers to peer farther into the cosmos. Lowell's insistence that the atmosphere over Flagstaff provided superior seeing helped inspire the construction of more than a hundred mammoth observatories atop arid mountains in Arizona, Hawaii, and Chile alone, where the air is both thin and stable.[15]

On the other hand, fictional depictions of the canals of Mars excited a generation of children, some of whom would mature as scientists when the Space Age revolutionized the study of planetary environments. One such scientist was Carl Sagan. Just before beginning his doctorate in 1956, Sagan had planned to observe Mars with his mentor, Gerard Kuiper, as the planet reached opposition. At McDonald Observatory in Fort Davis, Texas, they were foiled by two dust storms: one on Mars and the other in Texas. With nothing to do, they discussed the likelihood of vegetation and canals on the Martian surface. For Sagan, it was a rekindling of a childhood fascination inspired by the books of American science fiction author Edgar Rice Burroughs, which described swashbuckling adventures on a canal-covered Mars. After joining Cornell University in 1968, Sagan devoted much of his attention to the study of Mars—and decorated his office door with a map of the Martian surface imagined by Burroughs.[16]

In 1971, history was repeating itself. Sagan and his former student, the American astrophysicist James Pollack, were on the *Mariner 9* imaging team. Pollack also led a separate group at NASA's Ames Research Center that focused on the study of planetary atmospheres. No imaging could be done during the dust storm, but it dawned on Sagan and Pollack that *Mariner 9* had detected higher-than-expected temperatures in the planet's upper atmosphere, and lower-than-expected temperatures at its surface. Now they passed the time by developing a method for calculating how suspended dust affected temperatures at different altitudes. The similar atmospheric pressures at the Martian surface and Earth's stratosphere, combined with the decidedly Earthlike landscapes eventually charted by the *Mariner 9* mapping project, in time encouraged Sagan, Pollack, and another of Sagan's former students, the American physicist Owen Brian Toon, to apply their method to a study of terrestrial volcanoes. After receiving some unexpected help from a series of well-timed eruptions, including the spectacular explosion of Mount Saint Helens in 1980, they confirmed that eruptions could cool Earth's climate (Chapter 1).[17]

Sagan, Pollack, and Toon discussed whether the consequences of a nuclear war could mirror those of Martian dust storms and terrestrial volcanoes. Since the 1950s the Department of Defense had sponsored studies on the climatic effects of nuclear war, some of which suggested that a war might create effects similar to those of a volcanic eruption. Science fiction authors had long speculated that nuclear war would cool Earth's climate, and by the 1970s public studies started to suggest that, indeed, nuclear weapons could profoundly affect Earth's atmosphere. Then in 1980 an interdisciplinary team announced that an asteroid impact had likely ended the reign of the non-avian dinosaurs (Chapter 19). The team argued that the impact had created the sorts of changes to Earth's climate that would have been caused by a supersized volcanic eruption, and that these climatic changes were what killed some three-quarters of the world's plant and animal life.[18]

By using their method for studying Martian dust alongside a simple radiative-convective model (a computer simulation of terrestrial temperatures), Pollack and Toon confirmed that the asteroid impact would have halted photosynthesis by shrouding the Earth in dust. The result was disturbing, and not only because it suggested that cosmic impacts posed an existential threat to Earth (Chapter 19). Cold War tensions were reaching levels unseen since the Cuban Missile Crisis, and the Department of Defense was planning a major shift in how it used its nuclear arsenal. Instead of targeting Soviet cities by detonating nuclear bombs far above ground, it would target nuclear missile fields by exploding bombs on the surface. The trouble was that "ground bursts" would lift tremendous quantities of dust into the atmosphere. The Defense Nuclear Agency therefore asked Pollack's group to explore how nuclear war could alter Earth's climate.[19]

Thomas Ackerman, a member of Pollack's team at Ames, suspected that because even all-out nuclear war could never lift nearly as much dust into the atmosphere as an asteroid impact, it could not meaningfully change Earth's climate. Yet Pollack and his colleagues sought other analogies, studying, for example, the cooling effect of haze left in the stable Arctic atmosphere by Soviet coal mining. In 1982 they stumbled across a key breakthrough. That year, Paul Crutzen, a Dutch meteorologist who a decade earlier had helped reveal the vulnerability of the ozone layer to human disruption, and the American chemist John Birks published a study that compared the atmospheric effects of nuclear war to those of forest fires. Like a gigantic forest fire, Crutzen and Birks argued, a nuclear holocaust would loft vast quantities of soot high into the stratosphere and thereby cool the lower atmosphere. Pollack and his colleagues realized that in order to determine the climatic effects of nuclear war, they would need to model the cooling effects of both dust and smoke.[20]

Luckily, their study of the Martian dust storm and volcanic eruptions had enabled them to develop a computer model for simulating aerosols in the atmosphere. By using this model in concert with their radiative-convective model, they arrived at a bleak result. A major nuclear war would indeed cool Earth like a catastrophic asteroid impact and, by halting photosynthesis, rule out agriculture for survivors across the Northern Hemisphere. Sagan added that Martian dust studies showed that "nuclear winter" would quickly spill into the Southern Hemisphere, so that no place on Earth would be safe.[21]

Fresh off filming the documentary *Cosmos* that cemented his celebrity, Sagan organized a major publicity campaign to communicate the nuclear winter hypothesis. Department of Defense projects had assumed there would be a quick recovery in the wake of nuclear war. Instead it now seemed that any such war would so devastate the Earth's atmosphere and biosphere that it could lead to human extinction. For the first time, a majority of the US public believed that nuclear war would lead to the total destruction of the United States.[22]

Public opinion shifted just as the Reagan administration undertook a massive arms buildup that to millions now seemed potentially suicidal. In response to popular and political pressure, administration officials at first used the possibility of nuclear winter to promote costly plans for a space-based shield against incoming missiles (Chapter 19). The Soviet government tried to exploit the idea of nuclear winter to criticize Western militarism, while dissidents and peace activists either explicitly or implicitly used it to criticize Soviet policies. US military analysts eventually estimated that a nuclear winter would not be too cold for agriculture; they preferred the term "nuclear autumn." Reagan, however, had learned about research that connected the eruption of Mount Tambora, in 1815, to a wave of global cooling (Chapter 3). The climatic consequences of the far greater destruction unleashed by all-out nuclear war, he feared, could indeed "wipe out the earth as we know it."[23]

Ultimately, the discovery of "a doomsday machine hiding in the arsenals of the United States and the Soviet Union," as Sagan put it, helped persuade both Reagan and Soviet first secretary Mikhail Gorbachev to reduce nuclear tensions. Across the West, it also created a lasting alliance between advocates for arms control and environmental regulation. Once again, changing environments on Mars had strengthened fears, in this case richly justified, of an apocalyptic future on Earth.[24]

Building an Earthlike Mars

Mariner 9 not only encouraged new understandings of a threat to life on Earth, it also renewed the quest for life on Mars. The red planet had turned out to be a much more dynamic world than the first Mariner missions had suggested, and there was strong evidence that it had once closely resembled Earth. There were even hints that the Martian surface harbored microbes that had evolved to survive its desiccation. In 1975 NASA's ambitious *Viking* spacecraft therefore included landers with instruments meant to detect carbon-based life. When they reached Mars, cameras aboard the probes revealed that landing sites painstakingly selected on the basis of *Mariner 9* photographs would actually pose unacceptable risks. Dust had remained in the atmosphere for the entire *Mariner* mission, obscuring boulders that could have destroyed the landers. Scientists chose new landing sites that were safer but were far from the craters, volcanoes, and channels that were most interesting to geologists and biologists. Amid much public scrutiny, both landers descended to the Martian surface and, with robotic arms, scooped up samples of its regolith. Subjecting the samples to water and heat, they tested for the metabolic byproducts of microbial life. To the frustration of many scientists, the results were inconclusive. Part of the reason may be that scientists had once again assumed too great a similarity between Earth and Mars. The study of extremophiles—microorganisms that survive in extreme conditions—in the driest deserts of Earth has suggested that microorganisms adapted to the alien aridity and temperatures of Mars would have died when exposed to the heat and moisture of the *Viking* tests.[25]

In the United States, popular and scientific interest in Mars waned after the Viking missions offered no definitive evidence of Martian life. Yet in the 1990s the discovery of the ALH84001 meteorite, a wave of innovative robotic missions to Mars, and popular science fiction set on a realistic Mars, from *Total Recall* and *Mars Attacks* to Kim Stanley Robinson's *Red Mars,* all converged to create renewed enthusiasm for Martian exploration. Since then, a growing armada of orbiters and rovers have revealed that Mars was, and may remain, a habitable world. These machines have tracked remarkable changes in environments across the Martian surface, from the trickle of water down slopes to the swirl of dust devils across dunes. The images and videos they beamed to Earth have popularized a vision of Mars as a dynamic world whose landscapes are navigated, exploited, even altered by human action.[26]

The world's largest space agencies and companies are now developing the technologies to land astronauts on Mars and eventually permanently settle the planet. The astrobiologist Jacob Haqq-Misra has identified three broad reasons. Exploration is, perhaps, the most obvious: the allure of scientific discovery, the desire for new resources. Then there is transformation: the possibility of developing new economic, political, or social arrangements on another world, far from Earth, which may teach us how to live more harmoniously and sustainably back home. Mitigation may be especially influential for proponents of settlement: mitigation of existential risks back home. A colony on Mars seems to offer the prospect of a plan—or planet—B, a fallback in the event that something like a nuclear winter exterminates or forever diminishes humanity on Earth.[27]

Still, the allure of Martian settlement cannot be reduced to cold logic. It owes much to the perpetuation in science fiction of a Lowellian Mars, an alien, romantic, and unforgiving desert world such as Arrakis in *Dune,* or Tatooine in *Star Wars.* Depictions of such planets combine Orientalist fantasies of life in the Middle East with nostalgia for the settler heroism of the American Wild

West. Landers and rovers have returned countless pictures of the real Mars, but it is impossible to see them without recalling the harsh but habitable desert worlds of popular science fiction.[28]

In reality the Martian surface is profoundly hostile to human life. It is frigid, irradiated, and toxic, and the atmosphere is so thin that it might as well be outer space. It may be that the planet can only be settled by terrestrial microorganisms and machines. Indeed, even before *Mariner 4* reached Mars in 1965, scientists suggested terraforming the planet to make it more habitable for humans. As orbiters and rovers began to reveal that Mars once resembled Earth, and still has a substantial store of the chemicals necessary to sustain human life, proposals proliferated for restoring the planet to its primordial state, and thereby creating a genuine second home for humanity.[29]

In 1984 the chemist James Lovelock, for example, imagined a future in which disarmament and ozone hole alarmism together inspire a plan to crash hundreds of defunct nuclear missiles, each loaded with suddenly useless chlorofluorocarbon (CFC) gas, into the Martian surface (Chapter 8). A potent greenhouse gas, the CFC would warm the planet until astronauts could introduce algae imported from similarly cold and dry Antarctica. The algae would supercharge the warming and transform the atmosphere until a Martian version of Gaia established itself. Other scientists and authors have subsequently imagined more realistic geoengineering schemes to heat up the planet and increase its stock of volatile chemicals—for example, by steering an icy asteroid into its atmosphere, saturating its surface with bacteria, spreading genetically engineered lichens, focusing sunlight onto its ice caps with orbiting mirrors, detonating nuclear weapons on its ice, spraying dark material across the ice, releasing vast quantities of greenhouse gases into its atmosphere, or all of the above. A new study finds that the first pulse of warming could be easily achieved by manufacturing sunlight-absorbing particles from Martian ingredients, which would then readily blow into the atmosphere.[30]

Some of these ideas would send off-world the gravest threats to terrestrial environments in order to resurrect the ancient environment of Mars. Most would take centuries to complete. All would annihilate the present-day environment of Mars, and, quite possibly, the microorganisms it might sustain. Some scientists have advocated a thorough search for Martian life before the first steps of any terraforming program are taken. Carl Sagan argued that if human robots find Martian microbes, "Mars should be left to the Martians." It seems unlikely today that the rights of microbes will outweigh the desires of sapient life on Mars, any more than they do on Earth. SpaceX, for example, has explicitly made terraforming the ultimate goal of its nascent Mars colonization program. Unlike Venus, Mars is too small to ever truly be Earth's twin, but it is also easier than Venus to return to its primordial state (Chapter 8).[31]

In the twentieth century, the dust storms of Mars helped reveal existential threats to the climate of Earth and the survival of humanity. It is conceivable that, in this century, entrepreneurs and scientists will begin to transform Mars into a refuge for humanity in the case of a climate apocalypse on Earth. The creation of the Earthlike world imagined by nineteenth-century astronomers would, in time, destroy the alien climate that today permits planet-encircling dust storms. If so, those storms will have helped bring about their own annihilations by influencing the culture of a species on a neighboring world.

CONCLUSION

A Planet for Dreamers

In the early hours of October 6, 2020, I found a shadowy nook along the flanks of a local church, pulled off my pandemic mask, and unpacked a gleaming white refractor. Only rabbits rustling in the bushes nearby, and the occasional bat fluttering overhead, broke the predawn silence. Mars was just days away from a particularly close opposition, and it seemed almost impossibly brilliant in the southern sky. When I peered through my eyepiece, there it was: shimmering and salmon-colored, adorned with dusky streaks around a brilliant ice cap.

I imagined myself leaping through time, to those thrilling decades when Mars seemed to resemble Earth and the Martians offered a vision of our future. Then, for the briefest instant, the atmosphere aligned above me. To my astonishment, a latticework of straight lines snapped into focus across the southern hemisphere of Mars, looking for all the world like a network of canals. An instant later, the lines disappeared amid renewed turbulence. I sat back for a minute, stunned. Canal enthusiasts, I realized, had seen something real: a genuine connection between dynamic environments on two worlds, across sixty million kilometers of space.

That connection once convinced astronomers that Mars was a second, smaller Earth, and then that it was a dying Earth, transformed into a vast machine to sustain an alien civilization. Efforts at interplanetary communication and fears of invasion, expressed and exploited in new genres of fiction, accompanied the apparent discovery of extraterrestrial life. So did a new understanding of humanity's history and future, one that recognized how populations had changed, and would be changed by, Earth's environment. After the robotic visitors of the Space Age revealed Mars for what it really was, the planet's environmental changes suggested that human annihilation could be even closer at hand. Plans for transforming Mars into the Earthlike planet once imagined by astronomers—a planet capable of providing insurance in the case of catastrophe on Earth—inspired the most ambitious schemes for Mars settlement, which may be realized this century by the world's wealthiest men.

The strange history of Martian observation may also offer a glimpse of another future. In May 2016 scientists announced the discovery of a telltale wobble in the rotation of the star nearest our own: Proxima Centauri, a red dwarf just over four light years away. A planet not much bigger than our own tugs on that star, ever so slightly, and it receives about as much stellar radiation as Earth does. At the time of this writing, even the best telescopes cannot distinguish its reflected glow from the glare of its star. Soon, though, scientists will directly discern a pale red dot: the planet, warmed by its little star. What will that show us?[1]

The first images will be unimpressive, but they will grow sharper with time. If astronomers spot the telltale spectroscopic signatures of life in the atmosphere of the planet—an unlikely prospect, given periodic eruptions from its star—both scientific and public speculation will reach a frenzied pitch. Recently an apparent signal from the planet's star system made global headlines; that signal, it turned out, actually came from human technology. Yet the pale red dot around Proxima Centauri may soon create an echo of the controversy and excitement that once swirled around Mars.[2]

PART V

Comets and Asteroids

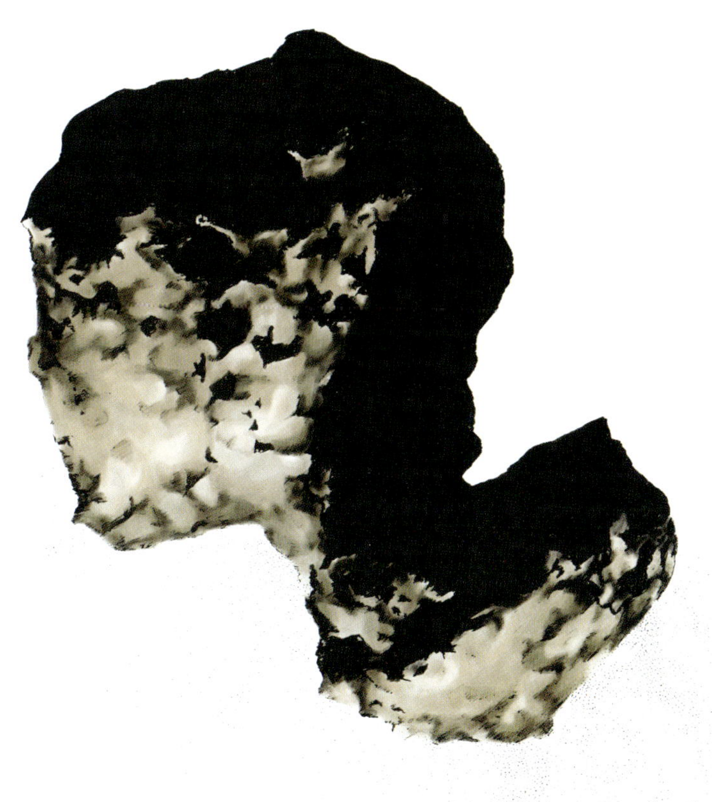

Comet 67P/Churyumov–Gerasimenko.

"Comets signifie corruption of the ayre. They are signes of Earthquakes, of wars, changing of kingdomes, great dearth of Corne, yea a common death of man and beast."
—Leonard Digges, *A Prognostication Everlasting* (1596)

"It would be stupid for humanity to let itself be eliminated by a rock."
—Robert Nichols, "Will We Be Ready . . . When Worlds Collide?" (1992)

"Consciously, I came into this world with a comet," Percival Lowell mused shortly before his death, remembering when, at three years of age, he gawked as Donati's Comet loomed over the horizon. It was the brightest comet of the nineteenth century, and it introduced Lowell to a variable universe. "I can see yet a small boy half way up a turning staircase," he recalled, "gazing with all his soul into the evening sky where the stranger stood." For centuries, others had been similarly inspired. In 1577 a healer-herbalist named Katharina Guldenmann walked her six-year-old son, Johannes Kepler, to a prospect where both admired the Great Comet of 1577. Camille Flammarion's first memory was of being in a crowd of people, gawking at a comet suspended above the sunset like a "luminous feather." The memory, he reflected in his memoirs, undoubtedly "shaped my future interest in astronomy." A young Isaac Newton loaded paper lanterns with candles and sent them aloft in kites, hoping to terrify countryfolk who, he thought, would confuse them with comets.[1]

I still remember the fragrant smell and gentle breeze on an evening in late spring, 1996, as I stood beside my childhood vegetable garden, gawking, open-mouthed, at Comet Hyakutake—a ghostly white sword slashing across the sky. And I recall shivering beside a dirt road the following spring, at the first blush of dawn, peering over a budding maple tree at the brilliant, blue and white tails of Comet Hale-Bopp. Few childhood memories left so strong an impression.

Comets are memorable because they reveal with unusual clarity that the cosmos around Earth is alive with change and that it moves in ways utterly alien to human perceptions of time and space. Hyakutake and Hale-Bopp, for example, will not return for millennia. They are bound for the outermost reaches of our solar system: the Oort Cloud, a sphere teeming with a trillion comets that extends to where the last tendrils of the Sun's gravity fade into interstellar space. How exactly the Oort Cloud formed remains a mystery. After a young Jupiter ballooned in size by absorbing the light elements and ices that accumulated in the outer solar system, it may have moved closer to the Sun and destabilized the orbits of the other giant planets. That, in turn, could have launched countless little worlds toward the fringes of the solar system, and given birth to the Oort Cloud. Alternatively, the Sun may have captured today's comets in close encounters with the nascent stars that swirled nearby as discs of primordial matter. Jostling between passing stars still dislodges comets today, sending them tumbling toward the Sun—and us.[2]

A smaller reservoir of comets, the Kuiper Belt, extends from Neptune to about fifty times farther from the Sun than Earth's orbit. These comets are true remnants from the earliest days of the solar system: icy worlds that were always where they are today. Yet owing to collisions or the gravitational pull of giant planets, they too may lose energy and slip into narrower orbits. The deep gravity wells of the giants, especially Jupiter, may then entrap them, or wrench them toward us.[3]

Inbound comets pass swarms of other little worlds as they tumble into the inner solar system. These worlds—asteroids—often orbit nearer the Sun than comets, and many lack the ices that, on comets, melt to form tails in the inner solar system. Yet asteroids come in a bewildering variety of orbits, shapes, sizes, and compositions, which can make it hard to distinguish them from comets. The largest conglomeration of asteroids, the belt just beyond the orbit of Mars, is a vast collection of planetary embryos that never coalesced, owing perhaps to the perturbing influence of migrating Jupiter. Gaps in this belt still reveal the present-day reach of Jupiter's gravity, which combines with the far weaker gravity of little Mars to tug and nudge asteroids out of their orbits—an echo of similar processes in the Kuiper Belt. An asteroid trickle continually flows into the inner solar system and gathers into another swarm, this one rushing around and through Earth's orbit around the Sun. Pushed and prodded by the planets, near-Earth asteroids (NEAs) join a small collection of inert comets; together they are called near-Earth objects (NEOs). Some of these slingshot around planets and escape back to the outer solar system; others meet a more violent end.[4]

Comets have been associated with disaster since ancient times. But only in the seventeenth century did scholars popularize the idea that comets could directly devastate Earth's environment. Thereafter, popular fears of a cosmic collision or near-collision swelled with the appearance of every bright comet. Following the nineteenth-century discovery of asteroids, close approaches of asteroids to Earth also raised popular concerns. By the dawn of the Space Age, scientists, engineers, journalists, and science fiction authors had together created an enduring, Janus-faced characterization of asteroids and comets. On the one hand, they posed existential dangers to Earth's environment and the survival of human civilization; asteroids even seemed capable of hindering the free movement of human spacecraft through the solar system. On the other hand, as remnants of the primordial solar system, their environments had unparalleled scientific value. Their abundant resources also seemed capable of unlocking a future of limitless abundance on Earth and in space, a prospect that seemed particularly inviting in the 1970s, when there were widespread fears of commodity and energy shortages, and environmental degradation.

Asteroids collided with Earth throughout the twentieth century and exploded with devastating force in its atmosphere. Motivated in part by these collisions, and the prospect of bigger impacts in the future, in the 1970s a small group of largely American scientists launched the first, humble efforts to catalogue NEOs. Not long after, a scientific consensus slowly emerged that impacts, by changing Earth's climate, had fundamentally altered the ancient trajectory of life's evolution. Beyond encouraging research into the seemingly urgent threat of a nuclear winter, in the 1980s this scientific insight helped motivate the first concerted efforts to imagine how to mitigate and manage the risk of impacts. Even though these efforts captured widespread media attention, they were not taken seriously by policymakers and did not prompt a response from governments.

Everything changed in the years 1989 to 1998. The steady growth of asteroid and comet surveys had enabled scientists to detect and predict a series of impacts and near-impacts across the solar system, including a series of spectacular collisions between Comet Shoemaker-Levy 9 and Jupiter. These discoveries resonated in a fast-changing political, economic, social, and cultural landscape that reflected the end of the Cold War and on the whole encouraged seeing the impact threat as a serious natural hazard. Owing in part to congressional pressure, NASA firmly incorporated asteroids and comets in mission planning, and the US Air Force provided resources to expand existing survey programs.

These programs revealed more close passes of comets by Earth, and suggested (wrongly, it turned out) that catastrophic impacts were imminent. The impact threat therefore continued to motivate public and congressional pressure in the United States. Survey programs grew more ambitious; planning for a possible impact became more sophisticated and comprehensible to the general public; and robotic probes were designed to visit comets and asteroids. Yet only when the Obama administration prioritized space commercialization and set an NEO as a primary target for human spaceflight did NASA belatedly prioritize what by then had come to be known as "planetary defense."

Spacecraft have since sampled, altered, and redirected asteroids and comets. The first private efforts to mine these worlds have failed, but grander ambitions by a novel collection of companies and space agencies may soon become reality. The utopian promise of bringing asteroids and comets firmly within the scope of human action beckons more strongly than ever while the cataclysmic potential of a collision seems less likely than before.

17

Warnings in the Heavens

At the end of the nineteenth century, the apocalypse was in the air. In newspaper reports, science fiction, public lectures, and scientific articles, environmental changes on Mars and the Sun helped fuel a popular discourse on the possibility of human extinction. When astronomers used spectroscopes to detect poisonous gases in comets, Camille Flammarion saw an opportunity to contribute cutting-edge science to the fin de siècle milieu. In 1894 he published a popular science fiction book in which a twenty-fifth-century woman computes that a comet with a toxic, carbonic-oxide tail is headed for Earth. The daily papers, which in Flammarion's narrative had long since abandoned any pretense of truthful reporting, "broadcast this alarming news . . . embellished with sinister comments and numberless interviews in which the most astonishing statements were attributed to scientists." The public of the future, accustomed to such sensationalism, paid little heed until it was clear that the comet would saturate the atmosphere with carbon dioxide. Ultimately, half of humanity suffocated in the poisoned air.[1]

It was always likely that comets would join the apocalyptic discourse of the late nineteenth century. Across Europe, Mesopotamia, East Asia, India, and Mesoamerica, diverse peoples had long associated the appearance of a comet with calamity. Cometary "apparitions" provided warnings of war, drought, disease, or the untimely demise of political leaders. Yet between the sixteenth and nineteenth centuries, scholars' gradual realization that comets traveled far beyond Earth, and carried toxic gas, inspired a slow, uneven, but ultimately widespread reimagining of the connection between comets and disaster. Around the turn of the twentieth century, comets really did transform environments on Earth and across the solar system. Scientists did not recognize these changes for what they were. Instead, astronomers captivated publics across Europe and the United States with the spurious idea that a comet's tail could imminently poison Earth's atmosphere. By the beginning of the twentieth century, comets were widely seen as potential sources of environmental degradation, and perhaps even existential risk. Yet it was unclear whether they really had caused a planetwide catastrophe, or whether they were likely to in the imminent future.

From Earthbound Portents to Cosmic Visitors

It was twilight on November 13, 1577. Tycho Brahe, the keenest observer of the heavens in the pre-telescopic era, gripped his fishing pole and scanned the water, waiting for the nibble of his evening meal. He could not keep himself from glancing at the sky. There, to his astonishment, he noticed the telltale smudge of a comet, emerging in the darkening west. Five years earlier, the sudden appearance of a new star, a supernova, had moved him to doubt that the heavens were really the perfect and immutable realm imagined in Aristotle's cosmology. Now he retreated to his island observatory, Uraniborg, in the Øresund Strait. Uraniborg may have been the largest and most sophisticated observatory built before the invention of the telescope. Using its unrivaled instruments and the measurements of other astronomers across Europe, Brahe determined that the comet's position relative to background stars scarcely changed when it was observed from different locations. By contrast, the Moon's position, relative to those stars, did shift when viewed from different positions.[2]

For Brahe, the implications were startling. To understand why, look at something near you. View it with one eye closed, then with the other eye closed instead. Notice that the object seems to shift its position. Now look at something far away, again while closing first one eye and then the other. Its position should not change nearly as much. The distance between your eyes has allowed you to use the parallax method, the apparent shift of an object when viewed from two locations. The farther away the locations are from one another, or the closer the object is to the observer, the more it seems to change its position. Because the comet of 1577 changed its position less than the Moon when viewed from different locations, Brahe concluded that it was farther from Earth than the Moon was. That, in turn, seemed to confirm that the heavens could change, and that Aristotle had erred in believing comets to be exhalations in Earth's atmosphere.[3]

Historians once argued that for learned Europeans, the Great Comet of 1577 revealed that truth could not be obtained through ancient wisdom. In this simple narrative, the inevitable consequence was a mass adoption of Copernican heliocentrism, and a newfound acceptance of empirical observation that later inspired the Scientific Revolution. In fact, educated Europeans sampled the ideas of Aristotle's ancient rivals, notably the Stoics, who imagined an interconnected and dynamic cosmos, to accommodate new observations within their own, idiosyncratic versions of Aristotle's worldview. Measurements of the comet's distance did not overturn Aristotelian thought, but instead helped to diversify it. While the comet seemed to challenge the distinction in Aristotelian cosmology between terrestrial and superlunary realms, it did not refute the idea that Earth lay at the heart of the universe.[4]

It was Brahe's student, Johannes Kepler, whose observations of comets would begin to challenge that idea. Kepler argued that the movement of brilliant comets across the sky in 1607 and 1618 could only be explained if Earth itself was moving, which of course would mean it was not the stationary center of the universe. Telescopic observations soon suggested that comets appeared and brightened as they approached Earth, then faded and disappeared as they retreated back into deep space. By the middle of the seventeenth century, many European scholars agreed that comets were celestial objects with distant origins.[5]

Yet what those origins were, exactly, remained a matter of heated debate. Like many of his contemporaries, Kepler believed that whales and other sea monsters spontaneously generated in the open ocean. He suggested that comets might be formed in the same way, from impurities in the aether that he believed permeated the solar system. Others proposed that comets were created by exhalations from the Sun or the atmospheres of giant planets. When, in 1644, René Descartes attempted to develop an alternative to Aristotelian physics by formulating principles

Citizens in Rotterdam gather to gawk at comet C/1680 V1. Lieven Verschuier, *Staartster (komeet) boven Rotterdam,* Museum Rotterdam, 1680.

for the motion of matter, he argued that comets were planets that had spun free from the vortices surrounding other stars. In 1651 and 1664, the appearance of bright comets with trajectories that plausibly originated in Sirius, the brightest star in the night sky, seemed to support these views. Once again the telescope offered a partial answer. Observations by respected scholars, including Christiaan Huygens and Robert Hooke, seemed to indicate that a comet's tail formed from a solid and slowly dissolving nucleus. Comets were not merely wisps of gas that had escaped from stars or planets.[6]

In late 1680 a spectacular morning comet slipped behind the Sun and then re-emerged with a tail that filled the evening sky. Observers determined that the morning and evening comets were the same object, and their measurements suggested that it moved in a parabolic orbit. Isaac Newton eventually agreed.

His new theory of gravity enabled him to develop a mathematical model that explained how and why comets follow parabolic paths around the Sun.[7]

Using Newton's model, Edmond Halley determined that another comet, in 1682, had an orbital period of approximately seventy-six years. Echoing scholars who scoured archives to find recurring patterns in climate, Halley uncovered records of similarly bright comets in 1531 and 1607. Convinced that these were appearances of the same celestial object, Halley predicted its return in 1758. It was a prediction he would not live to see fulfilled, but it came true. A little-known German astronomer named Johann Georg Palitzsch spotted the comet reappearing on schedule. The return of Halley's Comet sensationally vindicated Newtonian physics and for the moment established the eminence of celestial mechanics over other forms of astronomy. After nearly two centuries of observation, ancient beliefs that comets were creations of sin or warnings from God had given way, among scholars, to a new consensus that comets were natural objects from the distant reaches of the solar system that traveled in elongated but precisely measurable orbits.[8]

Comets and Existential Risk

Embedded within this new consensus was a startling idea. If comets were tangible objects that traveled with great speed in orbits that intersected those of the planets, then it seemed that in any given year a comet could strike Earth, with potentially cataclysmic results. Given the immensity of the solar system and the relatively small size of cometary nuclei visible in telescopes, the likelihood of a direct impact seemed low. Yet the tails that flowed from those nuclei were now known to be immense, and full of gas. Contemporary climatology depended in part on humoral theory, which suggested that the quality of the air differed from place to place, making some regions healthier than others. Beginning with Kepler, scholars warned that if the tail of a comet grazed Earth's atmosphere, its poisonous gases could leave much, if not all, of the planet inhospitable to human life.[9]

By the end of the seventeenth century, it seemed that Earth's environments could retain traces of cataclysmic encounters with comets in the distant past. As we have seen, new interpretations of the archives of nature were beginning to suggest that, as Halley put it, "the Earth seems as if it were new made out of the Ruins of an old World." While many scholars attempted to use wayward boulders or fossils to confirm that the biblical deluge had broken the old world, Halley tried to imagine how God had created the flood in the first place. He proposed that the collision or near-collision of a comet nucleus with Earth had unleashed immense tidal waves and altered the axis and speed of Earth's rotation. It might have created new ice sheets, he speculated, that cooled northern Canada until it was much chillier than Europe at the same latitude.[10]

Halley had proposed one of the first natural explanations for ancient climate change. Two years later the English theologian and mathematician William Whiston refined these ideas. He proposed that, on November 28, 2349 BCE, the close approach of a comet had fractured Earth's crust, releasing subterranean water that caused Noah's Flood. The new consensus that comets traveled in parabolic orbits suggested that another close pass loomed in Earth's future. The comet would be back, Whiston predicted, and this time it would bring about a universal conflagration by pitching Earth into the Sun. When Halley wrongly calculated that the comet of 1680 would return in 575 years, Whiston decided that its next apparition would be responsible for the world's end. His ideas were well received, widely reprinted, and influential in shaping a catastrophist vision of Earth's history that prevailed among scholars until the late eighteenth century.[11]

But was the threat posed by comets really limited to the distant past and future? In 1705 Halley calculated that the great comets of the seventeenth century could have passed exceptionally near Earth, had they come but a little sooner or later. A "Shock of the Celestial Bodies," Halley warned, seemed possible in any given year. Then in 1770 a comet passed a little more than two million kilometers from Earth, about five times the average distance to the Moon from Earth. It was a close shave by astronomical standards. By now a confluence of natural catastrophes—including, most importantly, a calamitous 1755 earthquake in Lisbon—and the development of probabilistic theories that mathematized chance had led some European scholars to develop new ways of identifying and interpreting risk. Meanwhile, a popular revolution in numeracy led some to experiment with communicating risk through quantification.[12]

One of the pioneers was the prominent French astronomer Jérôme Lalande. In 1773 he undertook a statistical analysis confirming that a comet could strike

Earth, with devastating effect. Lalande had hoped to clarify what he considered an unlikely but potentially catastrophic risk and prepared a lecture on the subject. Yet on April 21 the session at the Académie des Sciences in Paris where he was to speak ran out of time, and Lalande's lecture was canceled. Rumors spread across Paris that the Académie had censored Lalande's prediction of an imminent collision of a comet with Earth. Widespread panic spread across France and then to other countries, despite attempts by the *Gazette de France,* the official newspaper of the monarchy, to assure its readers that Lalande had not predicted a specific impact. Beginning in the Académie, the furor prompted perhaps the first open discussion about how best to communicate low-likelihood, high-consequence risks to the public. According to the historian of science Ampollini Ilaria, some focused on Lalande's use of the phrase "not impossible," which in popular discourse meant "entirely probable." Others claimed that statistical analysis inherently gave a veneer of credibility to popular concerns and that the potential consequences of a risk did more to inspire fear than its probability of occurring. Voltaire, for his part, ridiculed Lalande by writing that scholars could not be sure when comets would strike Earth, or what the consequences of such a collision would be. Doomsday predictions, he pointed out, had always served to benefit the prophets who made them.[13]

An article in the journal *Affiches de Province* stressed that it was the relatively recent reimagining of comets as astronomical objects that had fanned anxieties over a potential impact, rather than the ancient interpretation of comets as portents from heaven. In fact, scientific interpretations of the risk posed by comets had not replaced but instead coexisted with popular associations between comets and divine displeasure. In the right context, the interactions between these comet perceptions could provoke social upheaval.[14]

Across the newly formed United States, for example, congregations that adhered to traditional, evangelical Protestantism had dwindled, owing both to the fearsome toll of the Revolution and the steady tide of westward migration. Protestant ministers feared the obvious disinterest of the new federal government in promoting Protestantism, and perceived a steady decline in its prospects. They responded with the Second Great Awakening, a wave of impassioned preaching and conversion that, ironically, fractured Protestant congregations and provided fertile ground for unorthodox beliefs.[15]

With the Awakening in full swing, in 1832 the Baptist preacher William Miller caused a sensation when he declared that the Second Coming of Christ would occur on October 22, 1844. The timing could not have been better. The 1832 apparition of Biela's Comet sparked widespread predictions of Earth's destruction when calculations by German astronomer Heinrich Olbers suggested that Earth would pass through the comet's coma, the gaseous envelope that surrounds the nucleus. In early 1843 the shocking apparition of a comet bright enough to be visible during the day seemed to vindicate Miller's prediction that the world would end in the following year. The comet's appearance encouraged a wave of conversion to apocalyptic millenarianism. According to a later observer, the panic did not reach many Black communities in the United States, for the oppressed always "hoped to hear Gabriel's trump[et]." When the comet faded and Christ did not make his scheduled return, the "Great Disappointment" that followed would lead a new denomination, the Adventists, to proclaim that a process had been set in motion that would still culminate in the Second Coming.[16]

Warnings and Close Calls

Ancient interpretations of comets as portents survived into the twentieth century, but evidence was mounting that instead of heralding a global catastrophe, comets could cause one. At 7:17 a.m. on June 30, 1908, something detonated over the Podkamennaya Tunguska River, in eastern Siberia, with the force of a thousand Hiroshima bombs. The blast flattened some eighty million trees across nearly twenty-two hundred square kilometers. Of the region's Evenki population, at least three lost their lives, hundreds lost their homes, and thousands recounted their trauma in later interviews with Soviet researchers.[17]

The remoteness of the Podkamennaya Tunguska River from Russian population centers and the disruptions caused by war and revolution meant that nearly two decades passed before Russian scientists arrived to investigate the disaster. Scientists proposed diverse causes for the explosion, and conspiracy theories spread throughout the Soviet Union. By the 1970s, however, scientists on both sides of the Iron Curtain agreed that either a small comet or an asteroid had detonated in the atmosphere, or else that a large comet had grazed the atmosphere, triggering an explosion. It was now clear that had the comet

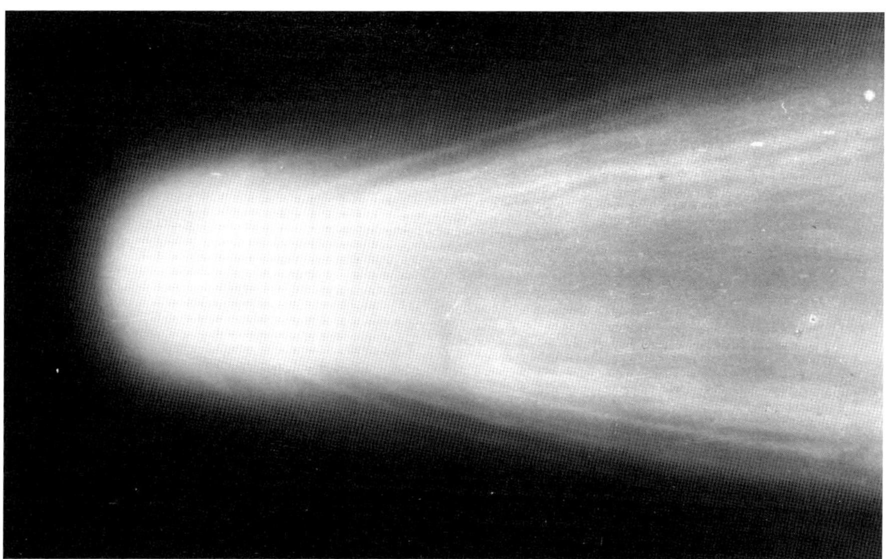

Halley's Comet. Photographed using the Hale Telescope at the Palomar Observatory on May 8, 1910.

or asteroid arrived just a few minutes earlier, Earth would have been in a slightly different point in its orbit. The impact could then have occurred in a densely populated region, killing millions and altering the course of history.[18]

Rather than considering the magnitude of the disaster that had been narrowly avoided, public attention in Europe and North America focused on the expected return of Halley's Comet. By chance, Flammarion was with the German astronomer Max Wolf when the latter, peering through his telescope, noticed the telltale smudge of the comet as it reemerged from the outer solar system. It did not take long before observatories around the world began tracking the comet to determine its precise path. The American astronomer Frank Seagrave predicted that, for a few hours on the night of May 18, 1909, Earth would spin into the comet's tail. "Nothing will happen," he stressed, "the end of the world will not come." Nevertheless, the announcement inspired sensational headlines, all the more after the Jesuit astronomer Charles Charroppin predicted that the comet would collide with Earth. Public concern reached new heights in January 1910, when crowds gathered to gawk at an entirely unexpected comet bright enough to be visible during the day.[19]

Then, in February, the American astronomer Edwin Frost and his assistant at Yerkes Observatory in Wisconsin used a spectroscope to identify cyanogen in the tail of Halley's Comet. "Cyanogen is a very deadly poison," the *New York Times* reported. Astronomers debated the "probable effect on the earth should it pass through the comet's tail." Some believed there was no reason for concern, but the *Times* explained that "Prof. Flammarion is of the opinion that the cyanogen gas would impregnate the atmosphere and possibly snuff out all life on the planet." Had the catastrophe in Flammarion's apocalyptic fiction come five centuries early? Certainly his cynical characterization of the news media seemed prescient. On both shores of the Atlantic, in local and national newspapers alike, dramatic headlines broadcast the possible end of the world. Many astronomers, Flammarion included, struggled to reassure the public, advising them to put aside fears that were only "worthy of barbaric ages." Meanwhile, charlatans sold comet pills, inhalers, and even umbrellas that would ostensibly protect the wise from celestial cyanogen. As the tail swept over Earth, women in Chicago allegedly sealed doors and windows; others reported smelling sulfur. Newspapers reported that to escape the horror of suffocation, some committed suicide. In one sad case, a man apparently murdered his family to save them from the agony of asphyxiation. Rising geopolitical tensions may have set the stage for widespread panic, and it would not be long before many Europeans, fighting in the First World War, perished in trenches filled with poison gas. The fear sparked by Halley's Comet foreshadowed the terror of chemical warfare and conveyed a warning that a sixteenth-century astrologer might have recognized—a warning that went unheeded.[20]

Meanwhile, another world had sent a different kind of warning to Earth. In the late 1870s the expansion of the Great Red Spot on Jupiter, which may have emerged just a few decades earlier, amazed astronomers. A decade earlier, the controversy regarding the apparent disappearance of the lunar crater Linné had created a large corps of amateur astronomers. Now they joined professionals to keep Jupiter under constant observation. It did not take them long to realize that mysterious dark markings suddenly appeared in the Jovian clouds. On October 28, 1880, for example, Henry Chamberlain Russell, an Australian astronomer and meteorologist, was "astonished to see a large spot of decidedly blue colour" around Jupiter's South Temperate Zone, round and "more dense in the centre than at the edges." On February 19, 1885, another Australian astronomer,

William Ernest Cooke, noticed five "striking" markings that looked like nothing else on Jupiter.[21]

By the final decade of the nineteenth century, the "evanescent nature" of some especially dark spots on Jupiter had been well established among astronomers. It soon grew clear that these spots could fundamentally transform the appearance of Jupiter. In the early morning hours of August 13, 1928, the British astronomer Bertrand Meigh Peek suddenly noticed a large dark spot in Jupiter's south equatorial belt, then another, and another, until a vast belt of "irregular dark areas and spots" encircled the planet. The Great Red Spot churned through but did not entirely destroy these dark regions. As it did, its own appearance changed profoundly. Just over a decade later, Earl Slipher at Lowell Observatory reported sudden dark spots that, to him, were even more striking than those of 1928. Observers wondered whether they had glimpsed a dark Jovian surface or unimaginably large volcanic eruptions.[22] Only later did the real cause for the spots become clear: impacts of asteroids and comets that could have doomed human civilization had they struck Earth instead.

By the early twentieth century, comets were widely seen as potential sources of existential risk. Scientists, though, had ignored or misinterpreted the environmental upheavals created by comets on Earth and across the solar system, so the nature of that risk was difficult for them to define. It was not clear whether cosmic objects had actually struck Earth, or whether they were likely to in the future. In the coming decades, close encounters with asteroids, a kind of cosmic object unimagined before the nineteenth century, would begin to clarify the impact risk and suggest ways for governments to respond.

18

Encountering Near-Earth Objects

For forty-two years, Daniel Moreau Barringer enjoyed the typical life of a wealthy white American. He had swept through the halls of the country's elite universities, then made his fortune by purchasing gold and silver mines in Arizona. Yet one hot October night in 1902, his life of wealth and comfort took a detour from which it never recovered.[1]

While smoking on the porch of an old hotel in Tucson, Samuel Holsinger, an agent of the US Department of the Interior, told Barringer about a giant crater just five hundred kilometers north of his mines. Barringer had studied geology at Harvard and was "naturally incredulous" when Holsinger told him of a theory, "held by some of the people living in the neighborhood," that "this great hole in the earth's surface had been produced by the impact of an iron body falling out of space." Yet the idea nagged at him. In 1903 he purchased a mining claim to the crater and established a company, Standard Iron, in the hopes of extracting its riches.[2]

It was a bold gamble that promised to elevate Barringer up from the ranks of wealthy but unremarkable men. A mine at "Meteor Crater" could have been the metallic equivalent of the oil wells that, in the United States, had helped create the gilded age. Yet after nearly three decades of grueling, financially ruinous effort, Barringer still had not found his motherlode of extraterrestrial iron. In 1929 he hired Forest Ray Moulton, an astronomer, to appease his financial backers by confirming that the object responsible for Meteor Crater had buried itself beneath the Arizona desert. Moulton soon found that Barringer had made a fatal miscalculation. At the speed with which it had collided with Earth, the object responsible for Meteor Crater had simply exploded. The overwhelming majority of its metal had ended up in the atmosphere, not the desert. Barringer refused to accept that he had squandered so much of his life and fortune. Within a week after receiving Moulton's report, he died of a massive heart attack.[3]

Barringer had been right about one thing: the force responsible for Meteor Crater was not of this Earth. Scientists now know that an asteroid forged the crater by smashing into Earth some fifty thousand years ago. Almost exactly one century before Barringer resolved to mine the crater, astronomers discovered the first asteroids. By the time Barringer died, it was clear that asteroids swarmed around Earth in disturbing numbers. It was war that began to reveal the magnitude of the risk they posed. The similarity between craters hollowed out by newly powerful bombs and natural craters on Earth and other worlds convinced scientists that asteroids periodically collided with planets. At the height of the Cold War, engineers started to imagine how newly potent weapons could be used to protect the entire planet from that threat. Responding to fears of resource depletion on Earth, they also drafted the first concrete plans to realize Barringer's dream of mining the wealth within asteroids—in outer space, rather than the desert of Arizona.

Discovering Asteroids

Efforts to intercept and exploit asteroids originated from a desire to restore order to the solar system. In 1766 German astronomer Johann Titius spelled out what astronomers had long known: the distance between each planet and the Sun approximately doubled from Mercury to Saturn. The one exception was a gap where a planet should reside between Mars and Jupiter. "Should the Lord Architect have left that space empty?" Titius asked. "Certainly not" answered fellow German Johann Bode in 1772. Just as it seemed inconceivable that planets would be left uselessly devoid of life, it was unimaginable that a divinely structured solar system could be that messy. On the first day of 1801 an Italian priest and astronomer named Giuseppe Piazzi found what seemed like the missing world between Mars and Jupiter. Yet it was so small that it looked like a point of light through his eyepiece. Astronomers eventually called it an "asteroid"—meaning starlike.[4]

By the middle of the nineteenth century it was clear that a swarm of these strange little objects filled the gap between Mars and Jupiter, a space now called the asteroid belt. Astronomers initially believed they were debris left over from the destruction of a full-fledged planet, a calamity that cast doubts on Earth's long-term survival. Yet it soon became clear that asteroid orbits did not cluster around a single origin, as they might have if they originated from the debris of a ruined planet. Many astronomers came to believe, correctly, that asteroids were actually planetary building blocks that never had a chance to assemble. As this explanation gained favor, the development of astrophotography enabled astronomers to discover a new asteroid, Eros, which, to their surprise, did not move within the asteroid belt. Instead its orbit brought it perilously close to Earth.[5]

In the eighteenth century, the eruptions of the volcanoes Vesuvius and Etna, combined with reports of eruptions on the Moon, had led some scholars to argue that meteorites originated in lunar volcanoes. Yet others pointed out that such an origin did not square with the trajectory of incoming meteorites, many of which were fragments of asteroids behaving as tiny independent bodies orbiting the Sun. Astronomers eventually concluded that the solar system teemed with countless little rocks that occasionally struck Earth. By the final decade of the nineteenth century, it was clear that the smallest asteroids and asteroid fragments routinely collided with Earth as meteorites, and that even larger asteroids followed orbits that could also allow them to impact Earth.[6]

In 1891 Grove Karl Gilbert, the chief geologist of the US Geological Survey (USGS), heard that fragments of meteoric iron had been found at Meteor Crater, then known as Coon Butte, in the Arizona desert. Gilbert already suspected that the Moon's craters had been forged through impacts. "The thought of examining the scar produced on the earth by the collision of a star," he later recalled, "was so attractive that I desired to visit the crater." He was delayed by work in Washington, DC, so he sent an assistant, whose description of the crater only whetted his appetite to observe it himself. Finally, in October 1891, Gilbert and a team of geologists boarded a train for Flagstaff, Arizona, and from there traveled in a horse-drawn carriage to the nearby crater. For seventeen days they scoured the area for magnetic anomalies and surveyed its topography. When they uncovered no evidence for a buried metallic asteroid,

Gilbert concluded that a steam explosion, similar to the one that had recently blown apart the Krakatoa volcano, had hollowed out Coon Butte.[7]

The visit to Arizona reinforced Gilbert's interest in the process responsible for cratering. Upon returning to Washington he turned his attention from the Meteor Crater to craters on the Moon. Observing the lunar surface from the Naval Observatory, he concluded that the size of the largest lunar craters, the rays that surround them, and the low elevation of the flat "floors" within their rims all revealed that they were caused by impacts, rather than "the ordinary volcanoes of the earth." Selenographers dismissed Gilbert's theory, for good reason. If asteroids and comets had indeed collided with the Moon, they should have smashed into the lunar surface from all angles. In that case, many lunar craters would be ovals. Yet nearly all of them were circles. In response, Gilbert argued that the Moon had coalesced gradually out of a cloud of smaller bodies, the last of which plummeted vertically into the emerging lunar surface to create circular craters.[8]

Close study of bomb craters in the First World War soon led a growing group of scientists to develop a different explanation. Explosions, they realized, always created circular craters, and asteroids traveled fast enough to explode when they collided with a planet. The craters of the Moon now seemed more likely to have been caused by explosive impacts. Moulton's study of Meteor Crater suggested that such impacts had also occurred on Earth.

In 1949 the American physicist Ralph Baldwin, who had worked on fuses during the Second World War, published a book, *The Face of the Moon,* that identified similarities between bomb craters and craters on the Moon. The idea carried with it a "disturbing factor," Baldwin wrote, for "there is no assurance that these meteoric impacts have all been restricted to the past."[9]

The new technology of the Cold War encouraged a broad reappraisal of the significance of comet and asteroid impacts on Earth and across the solar system. Geology would take the lead, owing largely to the efforts of a young graduate student named Eugene "Gene" Shoemaker. While working at the USGS in 1953, Shoemaker joined the Megaton Ice-Contained Explosion (MICE) project. Led by the Atomic Energy Commission, MICE aimed to generate plutonium by enveloping nuclear explosions in ice or salt. Shoemaker soon realized that nuclear bomb craters had features absent in volcanic craters, such curled rims with rock layers realigned in reverse order.[10]

A striking resemblance. Top: Meteor Crater in Arizona. Bottom: Sedan Crater at the Nevada Test Site.

Shoemaker could now theorize about the mechanisms by which explosions that abruptly release immense energies forged craters with distinct features. Nuclear bombs caused such explosions; so did comet and asteroid impacts, but on a grander scale. Shoemaker therefore convinced the Barringer family to let him apply his theory to a careful study of Meteor Crater. His PhD dissertation in 1959 showed how impact shockwaves had hollowed out the crater, incinerated the asteroid responsible for it, reversed the layering of nearby rocks, and left behind a telltale quartz signature. Motivated by the growing importance of the Moon in the early Space Race, Shoemaker identified the telltale features created by impact explosions in nearly every feature visible on the lunar surface. He had played a key role in forging a new field, planetary geology, that reinterpreted impacts as a fundamental force that had shaped every rocky world, including Earth.[11]

Diverting Asteroids and Protecting Earth

The devastation wrought on Earth by newly powerful bombs—especially nuclear bombs—had helped unveil the threat posed by asteroid and comet impacts. But could bombs exacerbate that threat? Mere months before the Cuban Missile Crisis, the engineer Dandridge Cole made headlines by suggesting that the Soviets could soon divert an asteroid to Earth using nuclear weapons. "The energy release at impact on the earth would be on the order of one thousand times as great as the energy of a multimegaton bomb," Cole warned. And worse, the impact could masquerade "as a natural catastrophe," so "the attacker could hope to escape blame and retribution." In a world of mutually assured destruction, it was a disturbing possibility, though it was soon eclipsed by more likely threats.[12]

Nuclear weapons also seemed to provide a means of defense against the impact threat. In 1949, at California's Mount Palomar Observatory, the German astronomer Walter Baade discovered a large, Earth-crossing asteroid with an eccentric orbit that brought it nearer to the Sun than Mercury. Astronomers named it Icarus. After plotting its orbit, they realized that in June 1968 it would pass just more than six million kilometers from Earth. By now astronomers had tracked many asteroids as they passed much closer to Earth. Yet influenced no doubt by the era's climate of nuclear paranoia, the American astronomer Robert Richardson wrote that any deviation in the asteroid's orbit could redirect it to Earth. In that case, he concluded, it might be necessary to blow it up with a nuclear weapon. Just over a year later, Australian nuclear physicist Stuart Thomas Butler warned that Icarus could smash into Earth with the power of a thousand hydrogen bombs. Newspapers falsely announced that the United States had partnered with Britain and the Soviet Union to explore whether a nuclear-powered rocket could divert the asteroid.[13]

These reports seem to have fueled widespread, persistent public fears. Both US and Soviet space planners had proposed detonating nuclear devices on or near the Moon, and both sides had exploded nuclear weapons in space. Another nuclear program, this one to divert an asteroid, seemed entirely plausible, and the popularity of Velikovsky's theories about close encounters between Earth, Mars, and Venus (Chapter 7) gave cosmic catastrophes a new veneer of plausibility. Walter Sullivan, a reporter at the *New York Times,* described "nervous telephone calls to this and other newspapers" in the wake of Butler's claims. Even when Icarus passed near Mercury in early 1968 and newspaper headlines announced that the planet's gravity had not, in fact, set the asteroid on a collision course with Earth, some readers continued to fear "catastrophic fires, winds, and far-flung destruction" as the asteroid made its closest approach.[14]

The most significant response to the Icarus scare would come out of MIT. Just before classes started in early 1967, Paul Sandorff in MIT's Department of Aeronautics and Astronautics posted an announcement on bulletin boards across campus: that spring, his graduate class would develop a plan to prevent Icarus from hitting Earth. Sandorff understood that Icarus did not actually pose a threat, but he hoped his students could determine whether the United States could deal with an asteroid that did seem likely to collide with Earth. It was with a "skeptical, and almost cynical, attitude" that twenty-one students registered for the class, and, they later wrote, the "first glimpse of . . . Sandorff as he entered the classroom did little to change our attitudes."[15]

Yet the magnitude of public concern focused their attention. As the months passed, Sandorff's students organized themselves into specialized groups, learned to work together, listened to visiting experts, and toured Air Force and NASA facilities. They quickly realized that the orbit of Icarus would be easiest to change

Paul Sandorff teaches his class at MIT.

while the asteroid was moving slowly at aphelion, its farthest point from the Sun, yet their compressed timetable meant they would have to deflect it while it raced toward Earth at over one hundred thousand kilometers per hour. It seemed that only a nuclear bomb could generate enough energy to redirect the asteroid while it traveled at that speed. The students believed that a hydrogen bomb with a yield of one thousand megatons, nearly seventy thousand times more powerful than the bomb that had devastated Hiroshima, could entirely destroy Icarus if detonated on the asteroid's surface. Yet they doubted that it could be designed and manufactured in time. Smaller weapons, they worried, would only fragment Icarus, and the leftover pieces could then cause just as much destruction as the intact asteroid.[16]

With that in mind, the students proposed an effort to deflect Icarus by detonating a series of hundred-megaton hydrogen bombs a hundred feet above the asteroid's surface. They determined that six Saturn V rockets, which would soon launch astronauts to the Moon, could instead be loaded with the bombs, and they worked out when and how each could be guided by military radar to reach Icarus. If the first bombs managed to redirect Icarus, the next in the sequence could target and destroy any fragments that had separated from the asteroid. The total effort, the students concluded, would require about 1 percent of US GDP—roughly the same share devoted to the contemporary Apollo program—and have a roughly 71 percent chance of deflecting the asteroid. On May 22 they presented their plan to a packed house at MIT's Kresge Auditorium. With "tremendous enthusiasm," they later wrote, they followed the subsequent "appearance of frightening articles about our project in at least 30 newspapers from coast to coast."[17]

Of course, scientists understood that Icarus had no chance of colliding with Earth, and the recommendations drafted by MIT students therefore had no immediate impact on NASA or the Department of Defense. Yet by taking public concerns seriously, Project Icarus developed principles that would guide future efforts to address the impact threat. Cosmic impacts were natural hazards that could be mitigated through detailed planning and government intervention. Furthermore, the threat posed by asteroids and comets needed to be confronted in space, rather than prepared for on the ground. It was important to have advance warning of a potential impact, because it was far easier to intercept a dangerous asteroid when it was farthest from the Sun. Deflection was preferable to destruction, and only nuclear devices detonated well above the surface of an incoming body could accomplish deflection at short notice. Finally, fragmentation was a risk that required redundancy—in the case of Project Icarus, multiple warheads—in order to mitigate.

Still, the project depended on the assumption that Icarus would survive a nearby nuclear explosion without shattering into a cloud of debris. To confirm whether this was possible, the MIT team stressed that efforts to characterize the chemical and physical properties of asteroid environments had to be expanded. More importantly, there was an urgent need for a program that could survey the sky to identify unknown asteroids or comets and plot their orbits so that any likely to collide with Earth could be intercepted while they were farthest from the Sun. On June 14, 1968, Icarus finally passed by Earth, enabling scientists for the first time to use radar to precisely determine the size, shape, and reflectiveness of an asteroid. However, because the signal returned by radar diminishes rapidly with distance, characterization efforts depended on sending spacecraft to asteroids and comets.[18]

NASA engineers thought of asteroids and comets first and foremost as opportunities for exploration, rather than threats to humanity. This perspective motivated serious planning for characterization missions. The primordial environments of asteroids and comets fascinated some scientists because they seemed essentially unchanged since the origins of the solar system. Soon after the first Apollo missions landed astronauts on the Moon, two engineers at the University of California San Diego, Hannes Alfven and Gustav Arrhenius, submitted a memo to NASA that proposed sending a robotic or crewed spacecraft to an asteroid. NASA had by then also drafted tentative ideas for robotic missions to asteroids or comets, but the Alfven and Arrhenius proposal catalyzed years of more detailed planning, and even suggestions that a joint expedition could be undertaken with the Soviet space program.[19]

Plans for asteroid and comet missions were motivated by more than science. In 1961 Su-Shu Huang, a physicist at the Goddard Space Flight Center, asked what it would take for the United States to "conquer nature" by "reshaping" the solar system. He found that near-Earth objects could be diverted to orbit Earth, perhaps by launching the asteroid's own material into space as propulsion. At the following year's World's Fair in Chicago, Vice President Lyndon Johnson predicted that an asteroid would someday be steered "close to Earth to provide a vast source of mineral wealth to our factories."[20]

In theory, the Saturn V would have permitted crewed expeditions to Earth-crossing asteroids. Yet the end of the Apollo program also spelled the end of the giant rocket. Human space flight was now limited to low Earth orbit. In the same year that the last Apollo mission reached the Moon, the Club of Rome—a worldwide network of scholars, industrialists, and government officials assembled to evaluate global environmental and social trajectories—published a sensational report warning that without concerted action, demographic and economic growth would soon breach unavoidable planetary limits. Playing on environmental and technological anxieties, a wave of literature, headlined by Paul and Anne Ehrlich's best-selling *Population Bomb,* had already made a similar point (Chapter 8). Soaring oil prices, economic stagflation, and the increasingly obvious effects of pesticides and unrestrained pollution all suggested that the crisis was closer than even the Club of Rome had predicted. It seemed to some that even basic commodities such as copper, nickel, and iron would quickly grow prohibitively difficult to access, and a sharp decline in living standards would follow.[21]

A prevailing sense of narrowing horizons and looming ecological collapse motivated contradictory calls for conservation, regulation, deregulation, and expanded mining and drilling. "Where on Earth can new supplies be found?" an MIT analysis asked in 1976; "perhaps," it concluded, "the answer is not on Earth at all." The world's third-largest impact crater—Ontario's Sudbury Basin, where NASA had trained Apollo astronauts to recognize rocks created by impacts—abounded with reserves of nickel and other metals that seemed to suggest that asteroids were rich reservoirs of the commodities Earth would soon need. There were "significant economic, environmental and political incentives for the utilization of extraterrestrial resources," the MIT report emphasized, adding that "no insurmountable technical problems exist to prevent such utilization." What was urgently needed, however, was a survey program to detect Earth-crossing asteroids. If one could be intercepted while passing near Earth, it might be deflected into Earth's orbit, or at least its metals could be retrieved without expending too much energy.[22]

With meager funds and inadequate instruments, a small group of astronomers and geologists had in fact already launched the first survey programs. In 1972, for example, American geologist Eleanor Helin collaborated with Shoemaker to launch the first systematic NEO survey. Shoemaker feared that an unknown asteroid could impact Earth. Working with him enabled Helin to become the first woman to regularly use the instruments at Palomar Observatory. On most clear nights she photographed a different part of the sky, keeping the shutter of her camera open for twenty- and ten-minute exposures. Then, using a binocular microscope, she would study the photographs, hoping to detect the telltale streak of an asteroid moving across the background stars. It was grueling, unglamorous work. "Astronomers, your colleagues, people you'd see in the cafeteria," Helin later reflected, believed that anyone who did it must be "of questionable mentality." Maybe they had a point. It took six months before Helin and Shoemaker identified their first NEO, and their progress improved only a little over the course of the decade.[23]

Nascent survey programs attracted little attention, but the idea that comets and asteroids could imperil human survival while, paradoxically, supplying enough resources to sustain a limitless human future in space, fascinated the public. Realistic, "hard" science fiction had come of age by the 1970s. And as we have seen the popularity of Velikovsky's ideas had spurred public debates over the possibility of cosmic collisions (Chapter 7). Stories by the era's most popular science fiction authors—including Isaac Asimov, Arthur C. Clarke, Larry Niven, and Jerry Pournelle—imagined the consequences of an impact with unprecedented detail. Niven and Pournelle's *Lucifer's Hammer,* which chronicles efforts by survivors to recover from an Earth-altering comet impact, resonated with particular force at a time of new fears for the survival of civilization. Newspapers also reported on close encounters with asteroids, and the new survey programs meant that there were ever more to announce. Project Icarus even inspired a popular movie, *Meteor,* in which a comet shatters an asteroid, sending its pieces careening toward Earth. The Soviet Union and United States combine forces to vaporize a particularly menacing piece, but a smaller fragment destroys the World Trade Center.[24]

Meteor flopped, and so did the decade's efforts to exploit and develop defenses against asteroids and comets. By the end of the 1970s, NASA officials had decided that asteroids and comets were prohibitively difficult to reach, that mining schemes were premature, and that their dwindling funds should be diverted to worthier programs, such as the Space Shuttle. In California a handful of scientists continued poorly funded survey programs that detected asteroids at a pitifully slow rate. The dawn of the Space Age had created a widespread conception of asteroids and comets as potentially dangerous but, in the long run, plausibly lucrative. Still, for much of the 1970s only speculation over the distant and mysterious Tunguska Event provided any hint that impacts had ever shaped the course of life on Earth. That would soon change.

19

Giggles and Close Calls

On the first day of September 1969, a young couple woke to the sound of gunfire in Tripoli, Libya. "The streets were empty," Milly Alvarez later recalled, "except for military tanks and troops." The morning passed in uncertainty. An hour past noon, Milly and her husband, Walter Alvarez, an American petroleum geologist, learned what had happened: a coup, led by Colonel Muammar Gaddafi. It would be three months before they could leave the country. As they passed the time exploring the ruins of Roman cities in Libya, Walter decided to apply for a North Atlantic Treaty Organization (NATO) postdoctoral fellowship to work in volcanic terrain north of Rome. When he received the fellowship, Milly and Walter boarded a ferry for Italy.[1]

Not long after, they joined another American geologist, William Lowrie, on a limestone outcrop in Gubbio, a picturesque Italian town. By measuring the magnetic mineral grains in limestone, the team hoped to determine whether the Adriatic microplate had rotated during the formation of the Apennine Mountains. It would have been a minor elaboration of the theory of plate tectonics, which held that Earth's crust consists of pieces, called plates, that float on the hot, liquid mantle. To their surprise, the team uncovered something more interesting: shifts in the orientation of magnetic minerals in successive rock layers. Scientists already knew that Earth's magnetic field had abruptly and repeatedly reversed, but Walter and William realized that microscopic fossils in the limestone rock layers could be used to date these magnetic field reversals. While studying the fossils, they found something even more intriguing.

A thin, clay deposit seemed to separate the fossil-rich layer of the late Cretaceous geological epoch, sixty-six million years ago, from the barren and relatively lifeless layer of the early Paleocene that followed it.[2]

Walter wondered whether the clay had settled suddenly or gradually. He consulted his father, Luis Alvarez, a prominent nuclear physicist. Luis suggested measuring the quantity of iridium within the clay, because iridium is regularly deposited on Earth's surface by meteors and meteorites. If the rate of meteor and meteorite bombardment had remained steady over time, then the amount of iridium in the clay would indicate how quickly it had formed. To everyone's surprise, the clay turned out to be saturated with iridium, more than could be explained even if it had accumulated very gradually. After evaluating other explanations, father and son determined that, at the end of the Cretaceous, an asteroid rich in iridium had smashed into Earth with the energy of a hundred trillion tons of TNT, ten thousand times the combined megatonnage of the Cold War nuclear arsenals at their height. By using the explosion of Krakatoa as an example, they determined that the impact had converted sixty times the asteroid's mass into dust, which then enveloped Earth. The result was an apocalyptic winter that drove the dinosaurs to extinction, along with about three-quarters of all species. Here was a mechanism that clearly linked the theoretical threat posed by impacts to an emerging danger, climate change, that was beginning to seem both tangible and urgent (Chapter 8).[3]

For more than three centuries, scholars had speculated that cosmic impacts could devastate life on Earth. By the 1970s they had established that relatively small impacts had occurred, yet they had found no concrete evidence for an impact large enough to imperil the biosphere. It was therefore impossible to gauge whether impacts actually posed an existential risk to humanity, let alone to determine the seriousness of that risk, the amount of money that should be spent to mitigate it, and how it could best be mitigated. The possibility of an impact had inspired campus thought experiments, campy movies, sensationalistic newspaper articles, and small-scale asteroid survey programs, yet it had not motivated serious concern.[4]

The Alvarez hypothesis changed the equation. In the 1980s and early 1990s, a growing group of mostly American scientists, weapons experts, and policy advisors came to believe that impacts posed an existential risk to life on Earth.

They attempted to determine the magnitude of the risk and how to effectively mobilize public and political support for efforts to reduce it. A remarkable series of asteroid hits and near-misses focused public attention on their efforts and led politicians to demand a plan for mitigating the impact risk. Asteroid scientists believed that such a plan could focus entirely on detecting large near-Earth objects. As NEOs were found and their orbits plotted, the mathematical risk of an impact would decline, provided none were on a collision course with Earth. If a detection program found an asteroid that was on course for Earth, a collision would likely happen decades or even centuries in the future. There would be ample time to design a system to intercept the asteroid and change its orbit so it would not strike Earth. Scientists and engineers who worked on weapons systems, however, argued that risk mitigation urgently needed to address how NEOs, both large and small, could be deflected or destroyed. They insisted that a detection program could never identify every asteroid capable of incinerating a city before it neared Earth. Therefore, interception systems had to be deployed as soon as possible and kept on hair-trigger alert. The sensational nuclear-armed interception plans proposed by weapons specialists, along with the public criticism of these ideas by outspoken asteroid scientists, derailed efforts to confront the risk posed by comet and asteroid impacts.

The Snowmass Workshop and the Conceptualization of Existential Risk

Days after the Alvarez team published its results, the possibility that humanity could follow the dinosaurs to extinction led a NASA advisory council to recommend a new approach to detecting and, if necessary, intercepting and deflecting, potentially threatening NEOs. Spurred by the call to action, in 1981 NASA sponsored a workshop in Snowmass, Colorado, to consider every dimension of the impact threat. The Snowmass workshop was chaired by Gene Shoemaker and convened about thirty participants from NASA, the USGS, government laboratories, and American universities.[5]

Despite knowing little about the characteristics of NEOs, the rate of impacts on Earth, and the resilience of modern civilization, participants sought to quantify and thereby clarify the risk posed by asteroids and comets. To that end they consulted the influential Rasmussen Reactor Safety Study, which in 1972 had assessed the risks of nuclear energy generation and compared them to other kinds of hazards. Climate models and the Alvarez hypothesis convinced participants that impacts by large asteroids, a kilometer across and bigger, would imperil human survival by transforming the atmosphere. Rough estimates of the number of undiscovered NEOs, and analyses of the rate of crater-causing impacts on Earth and other worlds, convinced them that an asteroid large enough to cause catastrophic climate change could hit Earth as often as once every three thousand years. Now the math was simple. Workshop participants took the total population of Earth and the aggregate size of the global economy, assumed a complete loss in the event of a large impact, and then divided the total cost by an estimate of the average number of years between impacts. Using this formula, they found that infrequent collisions by large asteroids posed a greater risk to humanity than more common impacts by smaller bodies incapable of altering the global climate. In fact, the average person's risk of dying from a large impact seemed comparable to that of dying in a plane crash. "Our tabulated statistics should be cause for great alarm," a draft of the Snowmass workshop report stressed, for impacts turned out to be "at least as important as many other kinds of disasters with which societies are concerned enough to protect themselves at some expense."[6]

Besides serving as an effective communication tool, the new formula enabled participants to offer a seemingly objective value for how much should be spent on mitigating impact risk. The annual amount, they argued, should be equal to the average annual cost of an impact, which now seemed to run in the millions of dollars. Given the existential threat posed by large impacts, the draft report recommended that some of these funds be allocated to an observatory, run by asteroid scientist Tom Gehrels at Kitt Peak, Arizona, that would focus on detecting large NEOs. A Spacewatch Camera at the observatory would use a charge-coupled detector (CCD), a light-sensitive imaging sensor recently invented at the Jet Propulsion Laboratory, to automate the search. Within a few decades it should have been able to identify most big asteroids capable of striking Earth. The report also urged a new emphasis on asteroid characterization in NASA planning. By designing robotic missions to asteroids, the report explained, engineers would learn how to design spacecraft for both asteroid deflection and mining efforts. Rarely had scientists made such an explicit link between the threats posed by asteroids and the opportunities asteroids seemed to provide.[7]

Although their math highlighted the threat posed by large bodies, Snowmass workshop participants agreed that smaller asteroids could still pose existential risks in an era of heightened nuclear competition. If a Tunguska-sized asteroid or comet exploded above a major city, they worried, military detection systems could mistake it for a nuclear attack, and a retaliatory strike could bring about Armageddon as surely as a larger impact.[8] This emphasis on the potentially destabilizing consequences of smaller impacts was new, and there was more than Cold War nuclear tensions to justify it. Evidence was mounting that small asteroids smashed into Earth routinely. Scientists knew that meteors had sparked wildfires across the Amazon in August 1930 and that a meteorite had hollowed out some 122 craters in Siberia in February 1947. A small asteroid had exploded with the force of an atomic bomb near Revelstoke, British Columbia, in March 1965. In August 1972, eyewitnesses watched as a larger asteroid streaked into the atmosphere over Utah before bouncing back into space. Had its course been just a little different, it would have excavated a crater some five hundred meters across. In 1981 it seemed only a matter of time before such an impact destroyed a city.[9] The Snowmass report therefore called on NASA to alert "appropriate U.S. Government officers" that an impact could be confused with a nuclear strike.

By admitting that small impacts could imperil humanity, the report undermined its recommendation to prioritize detection and characterization efforts that would focus on large asteroids and benefit asteroid scientists. Small asteroids capable of destroying cities were more abundant than large ones, and were much more likely to strike Earth. They were so dim, however, that surveys would be unlikely to detect them until they were very close to Earth. The same problem applied to unknown comets, which were usually too far from the Sun to be visible. While less likely to impact Earth than asteroids, comets were also impossible to detect until they were just months, perhaps weeks, away. As a result, a comprehensive asteroid and comet defense plan seemed to require both a detection program and nuclear-armed interceptors that could strike at short notice.[10]

With that in mind, Snowmass participants debated whether their report should mention interception at all. Some asteroid scientists suspected that the media would focus on sensational plans to destroy asteroids, inviting public ridicule. Others feared that if they succeeded in alerting policymakers and the public to the risk posed by impacts, and if interception were then prioritized in resulting efforts to mitigate the risk, they would trigger a campaign to militarize space that would have destabilizing effects on Earth. In the Cold War, efforts to reduce one existential risk could end up increasing another. Shoemaker decided that the Snowmass report would omit mention of nuclear weapons or interception systems, but briefly acknowledge that an explosive would be necessary to disintegrate an object that seemed likely to collide within two years.[11]

Owing perhaps to Shoemaker's many other commitments, the Snowmass report was never published. Yet news of the workshop leaked to the press. Newspapers focused on the apparent attempt to conceal the necessity of using nuclear weapons for interception and speculated that rival powers might someday use bombs to aim asteroids at nuclear missile silos. In all the coverage, detection and characterization programs received little attention.[12]

The Snowmass workshop set the terms for subsequent discussions of the impact threat. But it did not lead to new funding for detection, characterization, or deflection programs. Many within the small but growing community of asteroid scientists lamented what they called a "giggle factor" among US politicians and the public. Close encounters between Earth and other celestial bodies had long inspired widespread but fleeting public fears. Yet efforts to mitigate the threat ran into an extreme version of the skepticism that had either slowed or derailed the twentieth century's other attempts at reducing low-likelihood, high-consequence risks, such as solar storms, global warming, and microbial back contamination. It was, perhaps, the availability heuristic—the cognitive bias that leads people to judge the likelihood of events based on how readily they can recall examples of those events occurring, rather than their actual statistical probability—that most weakened public support for urgent action to reduce the risk of catastrophic impacts. Not even the risk calculus developed at Snowmass, which linked the impact risk to more tangible threats, enabled asteroid scientists to make much headway. There was simply little incentive for policymakers to prioritize a long-term threat to the entire globe when more familiar problems concerned their constituents on political timescales.[13]

Still, the Alvarez hypothesis had moved scientists in other disciplines to seriously consider the impact possibility. During the 1980s a wave of studies sought to identify new impacts in the geological record. Other publications sought to detect patterns in the frequency of past impacts, which could then be tied to celestial causes and used to predict future hazards. Media outlets covered this new science, reported the occasional NEO flyby, and published visionary articles on the prospects for asteroid mining. Yet with the Snowmass workshop increasingly becoming old news, news outlets scarcely reported on asteroid detection programs or interception schemes. By the late 1980s the vast majority of large NEOs had still not been observed, and their composition remained unknown. The impact threat still seemed more like science fiction than science fact. But that would soon change.[14]

Close Calls and New Opportunities

On March 23, 1989, just ten days after a geomagnetic superstorm raised awareness of the risk posed by solar storms (Chapter 4), an asteroid passed within seven hundred thousand kilometers of Earth. This was roughly twice the average distance between Earth and the Moon and the closest any asteroid had come in more than fifty years. It approached from the direction of the Sun; lost in the glare, it went entirely undetected as it hurtled past Earth. Eight days passed before the American astronomer Henry Holt noticed it while comparing exposures at Palomar Observatory. When NASA finally announced the flyby, the news made headlines around the world. Newspapers, including the *Washington Times,* stressed that humanity could have succumbed to a "years-long nuclear winter," had the asteroid approached Earth just six hours earlier.[15] While many reports focused on the possibility of using nuclear bombs to deflect asteroids, scientists were able to focus unprecedented attention on the dismal state of detection programs. If a threatening object could not be detected well in advance, Holt told the *New York Times,* "we'd all have to head for our basements and hope for the best."[16]

The asteroid, temporarily designated 1989 FC, frightened many Americans and spurred them to write their congressional representatives, who in subsequent weeks asked NASA's administrators for legislative affairs what the agency could do to address the impact risk. The responses they received all followed the same template, and none were reassuring. An August 1989 letter to Senator Bob Graham of Florida, for example, concluded that NASA officials "simply do not have enough information to understand how great the danger of such a collision is and what we could do about it."[17]

At the American Institute of Aeronautics and Astronautics (AIAA), a professional society for aerospace engineers, the physicist Edward Tagliaferri turned his attention to the impact risk. After consulting asteroid scientists, Tagliaferri grew convinced that there was a genuine risk of a catastrophic impact, and that efforts to detect asteroids were woefully inadequate. On behalf of the AIAA, he drafted a white paper calling for a new program to detect threatening asteroids and the development of systems that could deflect or destroy these asteroids. After his white paper was narrowly approved by the AIAA's board of directors, it was distributed to every member of the US House of Representatives and Senate.[18]

The House Space Subcommittee on Science and Technology now directed NASA to sponsor two workshops on the impact threat: one on detection and another on interception. The first would develop a plan for sharply increasing the detection rate of asteroids that approached Earth, and the second would explore how these asteroids could be intercepted and destroyed. Meanwhile, interest in the impact risk surged beyond Congress. At the AIAA's national meeting in 1990, Vice President Dan Quayle called for an effort to detect menacing asteroids and, if necessary, do something about them. In January 1991 a bus-sized asteroid sparked new fears when it passed between the Moon and Earth. Newspaper reports emphasized that it would have released the energy of five Hiroshima bombs had it collided with Earth. Two months later, scientists identified the Chicxulub Crater, a vast structure beneath the Yucatán Peninsula, as the impact site of the asteroid that caused the dinosaurs' extinction.[19]

Amid rising public concern over impact risk, in July 1991 David Morrison, an American planetary scientist who had attended the Snowmass workshop, chaired the detection workshop that Congress had directed NASA to sponsor. Just months before NASA announced the detection of 1989 FC, Morrison had coauthored a popular book with another planetary scientist, Clark Chapman, that revealed the full conclusions of the Snowmass workshop. The book opened with a terrifying scenario: a small asteroid, like the Tunguska meteor, detonates over a military base, setting off a civilization-ending nuclear war. The participants at Morrison's detection workshop briefly considered that possibility. But they concluded that the rare impact of asteroids larger than one kilometer in diameter posed the gravest risk to humanity. The risk calculus introduced at Snowmass provided a seemingly objective reason for that assessment. Only collisions by large asteroids could set off a global winter capable of imperiling human survival, which meant that their cost per year would be greater than that of smaller impacts.[20]

This simple math obscured the reality that scientists did not actually know how big an asteroid would have to be in order for its impact to have global climatic effects. Much depended on the speed of the asteroid as it approached Earth and where it landed. Scientists could not even precisely determine the size of an asteroid with the optical telescopes astronomers used for survey programs. A dim asteroid could be small, or it could be large but dark. Most importantly, scientists had no way to determine how likely it was for a small impact to trigger nuclear war. If that scenario was indeed plausible, as Morrison and Chapman had suggested, then small impacts could pose a greater existential risk than larger ones.[21]

After his workshop Morrison drafted a report on it for Congress, calling for a Spaceguard Survey—terminology borrowed, to Arthur C. Clarke's delight, from the asteroid defense system in Clarke's 1973 book, *Rendezvous with Rama.* Spaceguard would require six new telescopes and a center for program coordination, all of which would cost $115 million to build and $23 million annually to operate (in 2024 USD). Computer models indicated that the network would discover some five hundred Earth-crossing asteroids per month, and that within twenty-five years it would detect 90 percent of all potentially threatening asteroids larger than one kilometer in diameter. If none of these asteroids were on a collision course with Earth, Spaceguard would bring about a 75 percent reduction in the risk of a surprise impact large enough to threaten civilization.[22]

The second workshop, on interception, was slated to convene in January 1992 at Los Alamos National Laboratory. It would be chaired by John Rather, a weapons specialist who was the assistant director for space technology at NASA headquarters. Ninety-two participants had signed up. Asteroid scientists would describe the threat facing Earth, while engineers, military officers, and physicists would devise means of dealing with it. Many among the latter group were affiliated with the Strategic Defense Initiative (SDI), a missile defense system proposed by President Ronald Reagan in 1983 that was widely known as Star Wars. Reagan had envisioned using lasers and other directed energy weapons in space, but by the end of the Cold War, the SDI Organization (SDIO) called instead for an orbiting network of small interceptors, labeled the Brilliant Pebbles defense system, that would swarm rising ballistic missiles in the event of a nuclear war. After the 1991 collapse of the Soviet Union, the soaring cost and dwindling need for Brilliant Pebbles meant that the future of such systems had grown uncertain. Weapons specialists had an urgent need to justify continued funding for Brilliant Pebbles and other expensive defense systems.[23]

The impact threat seemed to be an almost miraculous answer to their problems. Soon after the asteroid 1989 FC made global headlines, Lowell Wood, the physicist who had proposed Brilliant Pebbles, established that the system could detect and destroy not only rising missiles, but also small descending asteroids. If the asteroid risk was taken seriously by politicians and the public, the future of Brilliant Pebbles seemed secure. However, the Spaceguard report that Morrison had drafted after the detection workshop presented a new obstacle. If policymakers believed that the impact risk could be mitigated merely by funding a detection effort, there would be no perceived need for a vastly more expensive Asteroid Defense Initiative. After all, the odds were vanishingly small that the Spaceguard Survey would identify a large asteroid that had a good chance of hitting Earth on a timescale that required an interception and deflection effort.[24]

With this threat in mind, in September 1991 Rather wrote a letter to the asteroid scientist Faith Vilas lambasting the "orthodox" approach of the detection workshop, with its focus on big asteroids and their assumption that Congress would fund nothing more expensive than a modest telescope network. The workshop, Rather argued, had dismissed the idea of using "unconventional platforms" because astronomers would not be able to control or benefit from them. Rather found it baffling that small impacts had been ignored in the Spaceguard proposal. "Let one hit a densely populated area in the United States," he wrote; the media would be merciless. The correct approach to the impact risk was a "thorough systems study to find first the best and most complete way to deal with the threat, and then to seek ways to minimize the cost."[25]

In October, Rather wrote to Morrison. "Congress," he stressed, "wants to be informed about the whole range of threats and options." He asked Morrison to add to his report at least a paragraph calling for acquisition and guidance systems capable of creating a "sphere of detectability surrounding Earth." Rather's interception workshop would then build on these ideas to propose a "robust" program for destroying unexpected asteroids, many of which would be small.[26]

By now it was plain that different assessments of the risks posed by asteroids of different sizes, and of comets versus asteroids, would shape which systems appeared best suited for defending Earth, and in turn whether asteroid scientists or weapons specialists stood to gain most from the expected influx of federal funding. In the coming months these different assessments would undermine efforts to convince Congress, and in turn the US public, that the impact threat was serious enough to require well-funded mitigation efforts. It is tempting to conclude that leading asteroid scientists and weapons specialists developed distinct risk interpretations only to cynically benefit their own careers and communities. Certainly, neither group made an effort to consult specialists in other fields, such as the humanities and social sciences, who might have helped them understand how societies were likely to respond to smaller impacts that did not threaten global extinction.[27]

Yet although asteroid scientists' and weapons specialists' differing interpretations of risk may have originated partly from self-interest, they also emerged organically from the priorities, methods, and experiences of the scientists in each group. Planetary scientists and astronomers who focused on asteroids or comets had scraped by for years with inadequate funding to confront a problem that most Americans did not take seriously. Gehrels, for instance, advertised his Spacewatch effort in the *Wall Street Journal* and solicited small donations from 230 individual donors. According to the astrophysicist Gerrit Verschuur, one donor, a "gentleman in Hiroshima," sent a "$500 check twice a year" that he did not "want his wife to know about." Asteroid scientists had refined a low-cost technology, the CCD camera, that enhanced the capabilities of traditional optical telescopes. Even when upgraded with CCDs, however, these telescopes could for the most part only detect large asteroids. It was therefore natural for asteroid scientists to focus on identifying these sizable asteroids with the tools they had developed. It was equally logical for them to quantify risk in a way that prioritized the global climatic disruptions caused by the impact of large asteroids. With this risk analysis in hand, asteroid scientists could attempt to convince policymakers and the public that detection programs directed at large asteroids should receive modest federal funding.[28]

Weapons specialists, by contrast, had long benefited from exorbitant federal funding. After all, their purpose had been to mitigate the existential threat that seemed most realistic to politicians in Washington, DC. The Reagan administration had encouraged them to experiment with speculative or impractical technology so that they could develop a novel and complete answer to a problem—an overwhelming nuclear assault—that brooked no partial solutions. It made perfect sense for them to design a system that would provide an impenetrable shield against the cosmic "enemy."[29]

With the interception workshop just weeks away, Morrison responded to Rather, accusing him of violating good scientific practice. "I am personally confused by your style of pulling secret or ill-defined ideas out of a hat," Morrison berated Rather, "and expecting anyone to take them seriously." Rather, Morrison wrote, risked undermining the "credibility of the entire effort" to mitigate the impact hazard with his "irresponsible" focus on asteroids that were both less threatening and harder to deal with.[30]

Infighting, the Media, and a Missed Chance to Reduce Existential Risk

By January the stage was set for an explosive meeting at Los Alamos, and Rather closed the meeting to the press. The interception workshop included many of the asteroid scientists who had contributed to the Spaceguard report. Brian Marsden, director of the Minor Planet Center at the International Astronomical Union, articulated a widespread opinion when he wrote, "I was initially rather apprehensive about meeting with 'SDI types,'" adding that all the same "it seemed to me that we ended up with a rather united view" on the impact hazard. Indeed, Rather drafted a report summarizing the workshop that recommended prioritizing efforts to detect large asteroids.[31]

The trouble was that weapons specialists like Rather viewed a detection program as only a first step in a broader effort to build a comprehensive asteroid

A 1991 painting imagines a last-ditch effort to intercept an incoming asteroid.

and comet defense system. To that end, they argued that nuclear weapons should not be decommissioned with the collapse of the Soviet Union. Instead they should be repurposed for use against asteroids. The weapons specialists proposed using former intercontinental ballistic missiles (ICBMs) for asteroid deflection experiments, and they explored how the most exotic SDI technologies could be incorporated into an interception program.[32]

The workshop's keynote speaker, the hydrogen bomb pioneer Edward Teller, laid out a step-by-step plan of attack. A detection program would need to begin first, he argued, but it should be followed by efforts to visit asteroids and finally deflect them using nuclear weapons. In private, however, Teller stressed that deflection experiments should begin as soon as possible, using rockets now available at a discount from the former Soviet Union. The end of the Cold War, he believed, provided an opportunity for former enemies to unite around a shared threat, using weapons they had once reserved to use against each another.[33]

These proposals provoked a range of reactions among the workshop's asteroid scientists, all of whom had come of age in an era of intense nuclear anxiety. Scientists such as Gehrels, whose Spacewatch program had benefited from a trickle of Department of Defense funding in the 1980s, had few objections. Others feared that an interception program would siphon away funds for a detection program or, worse, perpetuate the deployment of nuclear weapons on hair-trigger alert just as the threat of nuclear war seemed to be waning. The American astronomer Steven Ostro raised an even more worrisome prospect. An interception program would allow militaries to redirect asteroids not only away from Earth but also toward a hostile nation. Such a program, he argued, could "pose a far greater risk to humanity than 'natural' asteroid impacts!"[34]

After Rather had prepared a draft report summarizing the interception workshop, Morrison wrote letters to his colleagues in which he lambasted "dumb" proposals for future technologies that targeted the ostensibly "negligible" hazard posed by small impacts. Chapman was less generous. In his own letter to workshop participants, he wrote that the draft report proposed "an unjustified, profligate, expensive, and unwarranted program that is potentially dangerous to humanity and is thus immoral." Rather, he argued, had mischaracterized the opinions of workshop participants to foolishly prioritize asteroids smaller than five hundred meters in diameter that posed no risk to civilization.

He had done so to justify an armada of nuclear interceptors that, Chapman wrote, posed "a danger of misuse and / or accidental war—which we thought had diminished due to the end of the Cold War—that is enormous compared with the NEO hazard." Chapman demanded that the report be rewritten. If it was not, he threatened to rally other asteroid scientists in a campaign to discredit the report among scientific organizations, policymakers, and the media.[35]

Rather, evidently outraged, scribbled "blackmail!" in the margin of his copy of Chapman's letter. Even some asteroid scientists seem to have disagreed with Chapman's approach. Helin faxed Rather a copy of an anonymously authored parody letter, ostensibly addressed to Vice President Quayle, that mocked Chapman's criticisms and their tone. Rather assured the American physicist Duncan Steel that "the Chapman / Morrison diatribe has been highly visible, but has had little real impact." He promised not to heed what "two prima donnas would like to force me to say."[36]

Shoemaker tried to resolve the dispute by suggesting that the final draft of the report openly acknowledge criticisms by asteroid scientists. That mollified Morrison but not Chapman, who made good on his threat by approaching the press. The response was explosive, and ultimately destructive to everyone at the detection and interception workshops. In exposing the schism between the asteroid and weapons communities, journalists conflated detection and interception as equally cynical schemes to exploit taxpayers. Criticism came from both ends of the political spectrum. "By what twisted priorities would a nation lead a charge against space invaders," an article in the *Los Angeles Times* concluded, "while bucking world pressure to combat global warming?" An article in the *Washington Times* had a solution: "If there's anything in outer space we ought to blast out of the skies, its [*sic*] not dumb, harmless asteroids but the lawmakers and bureaucrats who spend their time dreaming up government programs that more properly belong in *Amazing Stories*."[37]

Chapman's account of the interception workshop resonated in the popular press partly because it tapped into long-standing trends in US political opinion, some of which accelerated with the end of the Cold War. Beginning in the 1970s, painful recessions and high inflation helped bring about a rightward turn in US politics that weakened public support for large government programs. The legacy of the Vietnam War, a series of scandals involving wasteful military spending, and the collapse of an existentially threatening foreign power at the end of the Cold War meant that, by the early 1990s, the military establishment and federally funded weapons scientists were both viewed with deep cynicism. Politicians, including President George H. W. Bush, argued that cuts to defense spending amounted to a "peace dividend" that would provide broad economic benefits. NASA, meanwhile, had only begun to rebound from a nadir in its prestige and popularity that followed from a series of costly equipment failures—none more traumatic than the disastrous explosion of the Space Shuttle *Challenger*—and the erosion of its utility in signaling US superiority over the Soviet Union. The Clinton administration and the new Republican Congress soon slashed NASA's annual budget by well over $2 billion (in 2024 USD), bringing it to below 1 percent of annual US GDP for the first time since 1961. A wave of layoffs and resignations accompanied the cancellation of programs that included a robotic mission to a comet. It was an inauspicious time to propose an exorbitantly expensive asteroid interception system that would be designed by federal scientists and managed by the military.[38]

Leaked letters between asteroid scientists, many of whom were associated with NASA, and weapons specialists, most of whom were connected to the Department of Defense, fueled accusations that the SDIO was attempting to replace NASA. Lowell Wood, who, newspapers reported, had shouted "Nukes forever!" during Teller's talk, was accused of plotting the downfall of NASA administrator Richard Truly in order to militarize US space exploration. Chapman encouraged these ideas, arguing that weapons specialists "looking for a playground" intended only to continue their work at "taxpayer expense." Whereas previously Chapman had attempted to drum up support for a deflection program by arguing that the risk of dying in an impact exceeded that of dying in a plane, a controversial claim even among his close colleagues, he now told newspapers that the risk was "really pretty small," which meant interception schemes were "not worth the cost to society."[39]

Rather, Teller, and other advocates for an interception program belatedly attempted to defend themselves, but the damage had been done. By the summer of 1992, the impact hazard had received more attention than ever before. Yet the derisive tone of media coverage had restored what one congressional aide called the "very high giggle factor" surrounding the issue among policymakers.

Public opinion is harder to gauge. Polls suggested, though, that a growing share of Americans opposed the SDI, NASA's involvement in military projects, and spending more money on space exploration. Public support for efforts to reduce the impact risk may have further declined later in 1992 when Brian Marsden wrongly calculated that comet 109P/Swift-Tuttle could strike Earth on August 14, 2126. He was soon proved wrong and media coverage again trended toward ridicule.[40]

Truly wrote a letter to Congress confirming that NASA had completed its detection and interception workshops. He emphasized that the impact risk was dominated by large asteroids that "can be readily detected with ground-based technology, most likely decades in advance of any potential collision" using research that NASA already funded. There was no need for a new program for either detection or interception.[41]

Shortly after the publication of the completed interception report in early 1993, two Department of Defense officials declined to testify at a House space subcommittee hearing on the impact threat because, according to an anonymous government source, senior Pentagon officials did not "want to see headlines that the Air Force was chasing space rocks." The opportunities that asteroid impacts and near-impacts had opened for detection or interception programs had been thoroughly squandered.[42]

20

Avoiding Jupiter's Fate

If you ring the doorbell of a nondescript building in Toronto, Ontario, work your way up a winding staircase, and walk into a little office, you will—at the time of this writing—find a forest of telescopes. These are the treasures of the Royal Astronomical Society of Canada. Most are instruments of gleaming metal and polished glass. One, however, stands out. It towers over the others and seems hammered together out of plywood and cardboard, crudely painted a garish blue. Flimsy metal attachments, crumbling tubes, and even an ordinary jar are strapped to it with little more than duct tape and Velcro. A series of plaques adorn its base (a battered wooden box), each recording a night on which the telescope's owner discovered a comet, or at least observed one barreling toward Earth. The plaques resemble kill marks on a fighter plane.

Now over forty years old, this was the telescope long used by an American amateur astronomer and comet hunter named David Levy. It captures the essence of his craft. Its scrapyard materials reflect a lack of financial reward; its many attachments, an obsessive hunger to see more. Its sheer bulk has meaning, too. A large mirror allows a reflector to reveal dim comets as they emerge from the outer solar system. It is easy to imagine the maniacal commitment required to haul such a behemoth outdoors, night after sleepless night. All that, to glimpse the faintest smudge through the eyepiece, the telltale blur of a wayward comet.

On the night of March 23, 1993, Levy joined Gene Shoemaker, his wife, Carolyn, a comet- and asteroid hunter in her own right, and a doctoral student named Philippe Bendjoya at the Palomar Observatory. Using long-exposure photographs, the team had planned to search for a comet near Jupiter, but the sky was clouding over. They took one more picture, and to Carolyn's surprise it showed a fuzzy line. Gene suspected that the line could be a "squashed comet." When they reported their discovery, astronomers determined that Jupiter's immense gravity had ripped apart a large comet and released a cloud of dust easily visible from Earth. Astronomers referred to the surviving fragments as a single comet, and since it was the ninth discovered by Levy and the Shoemakers, they named it Shoemaker-Levy 9 (SL9). Soon scientists calculated to their astonishment that the comet would collide with Jupiter over the course of a week in July 1994.[1]

When they smashed into Jupiter, the comet fragments would set in motion environmental transformations that were readily visible from Earth. It would be the beginning of the end of the giggle factor that had long hampered efforts to mitigate the impact risk. Yet the greater seriousness with which policymakers and the US public regarded that risk would only translate into action after NASA's goals for human spaceflight finally aligned with the priorities of asteroid scientists.

A Warning from Jupiter, a Change on Earth

In December 1993 astronauts repaired the orbiting Hubble Space Telescope (HST). Now the telescope revealed that Comet Shoemaker-Levy 9 (SL9), which was expected to crash into Jupiter in about seven months, had disintegrated into twenty-one distinct fragments, the largest being about two kilometers in diameter. Here was a chance to determine whether the little worlds prioritized by the Spaceguard report could really transform a planet's atmosphere. Scientists coordinated a global observation program that would focus six spacecraft and all the world's major telescopes on Jupiter during the week of the impacts. Yet as the impacts loomed, Hal Weaver, a senior scientist on the HST team, complained, "Everything seems to be working against us." At exactly the time when scientists expected light from the collision of Fragment A to reach Earth, the Moon would obscure a star that the HST would otherwise lock onto in order to precisely target Jupiter. Then, just as the Fragment B collision became visible, the HST would enter the South Atlantic Anomaly, a region of intense radiation where the inner Van Allen radiation belt approaches Earth. Signals to the tele-

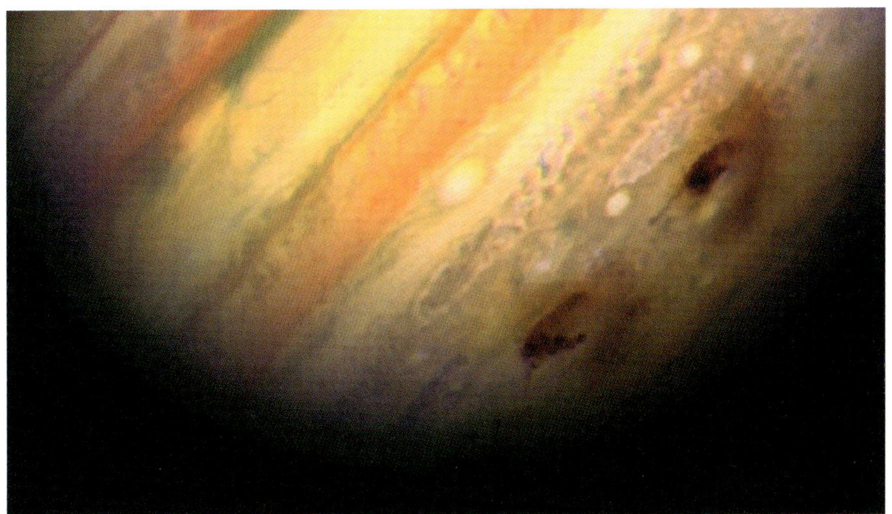

Jupiter under assault. Top: Earth-sized scars in Jupiter's atmosphere. Bottom: Astronomers at NASA's Space Telescope Science Institute in Baltimore wait to catch their first glimpse of the Fragment A collision.

scope's detectors would be scrambled in the radioactive noise. Astronomers therefore had good reason to fear that, for all their preparations, the dynamic outer space environments near Earth would foil their best attempts to observe environments further afield.[2]

In July 1994 astronomers around the world watched as Comet SL9's first fragment finally collided with Jupiter. Their pictures initially showed little hint of the flash some had expected. When the impact site finally rotated into view, however, the astonished astronomers spotted an Earth-sized scar: easily the most prominent feature on Jupiter's once-familiar disk. The impacts had launched vast quantities of dust into space, and when the dust re-entered Jupiter's atmosphere, it created gigantic bruises hot enough to temporarily blind infrared sensors on terrestrial telescopes.[3]

One by one, the comet fragments collided with Jupiter over the course of what the planetary scientist Heidi Hammel dubbed "Impact Week." By the end of the week, astronomers struggled to continue their observations during violent weather on Earth that seemed a pale imitation of the tumult on Jupiter. The great observatories atop Mauna Kea volcano in Hawaii closed their protective domes to prepare for a rare hurricane that briefly threatened the island. Dust blowing off the Sahara cloaked observatories on the Canary Islands, while Antarctic blizzards buffeted the South Pole Infrared Explorer telescope, where Jupiter was otherwise visible all day long. Thunderstorms at Cape Canaveral even forced astronauts aboard the Space Shuttle *Columbia* to stay in orbit. Yet space-based telescopes maintained their vigil, and by the end of Impact Week the HST revealed how Jupiter's zonal winds had merged the impact bruises into a ragged scar some one hundred thousand kilometers long, nearly eight times Earth's diameter. Like many astronomers, Rita Beebe was "in terror" when she realized that "a comet fragment only a mile or two across could create so huge an effect."[4]

During Impact Week, scientists planned observations, shared images, and offered interpretations in real time, using capabilities unlocked by the young internet. A single NASA website created by software engineer Ron Baalke to share impact images handled more than a million file requests (from users in fifty-nine countries) and transferred a then-extraordinary 80 gigabytes of data. NASA websites repeatedly crashed as more than 2.5 million internet users accessed them during Impact Week—a substantial share of global internet traffic.[5]

It was the world's first viral web event, and NASA took notice. Just two days after a particularly violent collision, a NASA press release announced that instead of mailing news to media outlets, the agency would from then on send announcements electronically, directly to the public. NASA soon awarded nearly $15 million (in 2024 USD) to ten organizations charged with building tools to distribute NASA images and information online. Impact Week had revealed the potential of an emerging medium that many had dismissed as a fad, and now NASA's decision gave it legitimacy as a source for news and knowledge.[6]

Journalists in other media covered the surging popularity of the internet during Impact Week but devoted far more attention to the impacts themselves. The latest images and interpretations were broadcast live on CNN and on public television in many of the top media markets in the United States. In newspaper after newspaper, articles and cartoons directly linked the impacts to the extinction of the dinosaurs and the prospects for humanity. Many of the participants in the detection and interception workshops encouraged these connections. "The fireworks happening on distant Jupiter," Carl Sagan wrote in the magazine *Parade*, raised the question "whether our civilization would survive even a much less energetic collision."[7]

Millions of people observed the Comet SL9 impacts directly, using small telescopes or large binoculars. Jupiter is bright enough to pierce light-polluted urban skies, and to the surprise of professional astronomers, its bruises were easily glimpsed using newly inexpensive or newly capable telescopes that were now available to amateurs. On July 21 alone, a thousand enthusiasts swarmed a tickets-only observing event at the Naval Observatory in Washington, DC.[8]

Amateurs, like professionals, were astonished and frightened by the changes they discerned on Jupiter's familiar face. Many astronomy enthusiasts had assumed that planetary disasters were too rare for humans to experience. It had seemed that such calamities happened billions of years ago, or in distant reaches of the galaxy. Yet now they had seen one with their own eyes. Kathy Sawyer of the *Washington Post* concluded that "citizens and scientists" alike gained "a new, more visceral awareness of Earth's place in an unsettled region where objects whiz and sometimes smash into each other."[9]

The impact risk, the planetary scientist Bevan French reflected in 1990, was a bit like the "stoplight phenomenon. You can yell and yell at the local authorities, but they usually won't put a light at a dangerous intersection until there's a terrible, deadly accident." The SL9 collisions were that accident—a "gut punch," Heidi Hammel reflected, to dispel the availability heuristic (the tendency to estimate the likelihood of events based on how easily such events could be recalled). Journalists and politicians expressed relief that the impacts had happened on Jupiter rather than Earth. "The Chicken Little Crowd," William Broad wrote in the *New York Times,* "which once drew smiles by suggesting that Earth could be devastated by killer rocks from outer space," suddenly found "its warnings and agenda taken seriously." In the House of Representatives, the Committee on Science, Space, and Technology met during Impact Week and directed NASA to work with the Department of Defense to find all asteroids larger than one kilometer in diameter whose orbits cross Earth's. Two years after it was widely ridiculed, the Spaceguard goal had been legitimized by Congress.[10]

The Impact Risk, Reduced at Last

NASA responded to the congressional directive by establishing the Near-Earth Objects Survey Group, with Gene Shoemaker again serving as chair. In June 1995 the group called for a scaled-down version of the Spaceguard program, but NASA, facing budget cuts in the new Republican Congress, recommended against funding it. Part of the problem was that officials in NASA's Office of Space Science did not believe that asteroid detection amounted to science; it seemed a public service better suited to the Department of Defense. Indeed, Lieutenant Colonel Lindley Johnson, chief of the Space Surveillance Division within the Air Force, saw an opportunity. Space surveillance had languished in the Air Force since the end of the Cold War, but the impact threat seemed to offer an opening to request new funding. Beginning in 1996, asteroid scientists experimented with newly capable CCD cameras using Air Force telescopes that had been designed to survey the skies for satellites and nuclear warheads. It was an arrangement that for a time benefited both parties: scientists gained access to a network of telescopes that allowed for an unprecedented surge in NEO discoveries, and the Air

Force justified and improved its surveillance capabilities. However, Air Force leadership preferred to focus on human enemies and continued to believe that NASA should lead what Johnson had termed planetary defense.[11]

Public interest in the impact threat scarcely ebbed in the wake of Impact Week. In 1996 and 1997 the spectacular appearance of two brilliant comets, Hyakutake and Hale-Bopp, spurred a wave of media reports on the possibility of a comet impact on Earth. In 1998, Disney and Paramount released the movies *Armageddon* and *Deep Impact*. These blockbusters, informed by asteroid and comet scientists, depicted detection and interception programs as heroic efforts to defend humanity. The impact threat was now so mainstream that the Chinese Communist Party (CCP) used it as an excuse to briefly derail negotiations aimed at drafting a treaty to ban nuclear weapons tests. In April and then July 1996, China defied pressure from the United States and Russia by detonating two nuclear devices. To the consternation of Western diplomats, China's Foreign Ministry argued that it needed to develop "peaceful" nuclear weapons that could be used to intercept a potentially threatening asteroid or comet. It was more likely that CCP officials wanted to jeopardize a major foreign policy objective of the Clinton administration in order to ease US pressure over alleged Chinese human rights violations, intellectual property theft, and nuclear proliferation. Nevertheless, in September, China joined 182 nations in signing the Comprehensive Nuclear-Test-Ban Treaty. Some worried that, by replicating the effects of a nuclear bomb, impacts could undermine efforts to monitor and enforce the treaty.[12]

Then, on December 6, 1997, an American astronomer named James Scotti used Gehrels' Spacewatch telescope to detect a kilometer-wide asteroid. Subsequent observations suggested that the asteroid would pass exceptionally close to Earth in October 2028. Brian Marsden again prematurely announced that the asteroid, called 1997 XF11, could collide with Earth. Other scientists quickly established that it could not, but Marsden's announcement inspired sensational headlines and prompted NASA to convene a meeting to establish guidelines for how information about potentially hazardous NEOs should be released to the public. Carl Pilcher, science director of NASA's Solar System Exploration program, chaired the meeting. During congressional testimony in May 1998 he formally committed NASA to meeting the goals of the 1992 Spaceguard Survey report.[13]

Meanwhile, asteroid scientists launched new survey programs, developed clearer ways of communicating impact risks to the public, and collaborated with enthusiasts to sustain popular and political support for their work. As surveillance identified more and more large asteroids and established that they could not collide with Earth, the Science Definition Team (SDT) for the proposed Large Synoptic Survey Telescope led interdisciplinary discussions that exposed the potential destructiveness of even small impacts. The threat associated with large asteroids declined as more were discovered to be in benign orbits, but the "residual hazard" posed by smaller asteroids seemed ever more menacing. The SDT had a solution: it proposed an expanded detection effort that would further reduce the impact risk by identifying all Earth-crossing asteroids down to 140 meters in diameter. Then in 2004 astronomers Roy Tucker, David Tholen, and Fabricio Bernardi identified another asteroid that seemed capable of striking Earth. This one was under 400 meters in diameter. The asteroid was named Apophis after the Egyptian god of the underworld, and headlines focused on the forecast that its close encounter with Earth would occur in 2029. Swayed in part by the discovery of this asteroid, Congress made "detecting, tracking, cataloguing, and characterizing near-Earth asteroids and comets" one of NASA's statutory responsibilities and directed NASA to undertake a survey of asteroids 140 meters across or larger. Still, NASA did not propose a program to meet that goal.[14]

Indeed, planetary defense did not attract significant funding from NASA until the agency's ambitions for human spaceflight aligned with those of asteroid scientists. After Barack Obama was elected president, the Office of Science and Technology Policy in the White House asked NASA to convene a committee to review its faltering plans for a crewed mission to Mars. The Augustine Committee, as it came to be known, recommended that NASA pursue a "flexible path" for human spaceflight: a series of stepping stones to deep space that might include a mission to an NEO. Such a mission would take advantage of surging interest in planetary defense and Obama's desire to catalyze the commercialization of space travel, which, it seemed, could include asteroid mining. In order to avoid designing an expensive new habitat to sustain astronauts in deep space, NASA ultimately proposed a mission to deflect a small asteroid—or perhaps part of a large asteroid—into lunar orbit, where astronauts could more easily pay it a visit. According to a report prepared by the Keck Institute for Space Studies at Caltech, the Asteroid Redirect Mission would be "mankind's first attempt at modifying the heavens to enable the permanent settlement of humans in space." Preparations for the mission began to funnel real money into NASA's programs for detecting and characterizing asteroids.[15]

In 2013 a small, previously undetected asteroid exploded above Chelyabinsk Oblast, Russia, with the force of perhaps thirty Hiroshima bombs, damaging over 7,000 buildings. In the United States, two days of congressional meetings in the wake of the blast helped motivate the creation of a Planetary Defense Coordination Office at NASA and accelerated the increase in funding devoted to asteroid detection and deflection programs. At the time of this writing, survey programs have detected well over thirty-one thousand NEOs; more than ten thousand of these are larger than 140 meters in diameter, and more than 90 percent are larger than a kilometer across. None will strike Earth in my lifetime, or yours. Nevertheless, there is a small but enduring risk posed by comets from the outer solar system, the lingering population of undiscovered asteroids closer to Earth, and rocks smaller than 140 meters in diameter. Polls show that planetary defense is now NASA's most popular priority.[16]

CONCLUSION

Saving Earth, Fueling the Future?

The solar system's smallest worlds have long seemed especially menacing. Fears that they herald, or perhaps cause, the destruction of life-giving environments stretch back into prehistory. In the last three decades, though, scientists have effectively measured and mitigated the impact threat, folding these tasks into the responsibilities of national governments. Today, according to the philosopher Toby Ord, "no existential risk is as well handled as that of asteroids and comets."[1]

It is tempting to conclude that a rare confluence of environmental changes and threats, all of which originated in space, spurred this remarkable reduction in risk. Close encounters with comets inspired the idea that a cosmic impact could occur, asteroid near-misses motivated the first plans to prevent such an impact, and explosions in Earth's atmosphere revealed the magnitude of the risk and the urgency of dealing with it. In the pivotal years beginning in 1989, a near-hit drew unprecedented attention to the impact risk. Cataclysmic collisions on Jupiter legitimized the risk, making efforts to mitigate it a federal responsibility in the United States. Finally, spectacular comet apparitions, false alarms, and a violent explosion over Russia enshrined the risk as one of NASA's well-funded priorities.

Yet closer study of the years from 1989 to 1994, when the American federal government began to address the impact risk, reveals that environmental changes did not by themselves lead to shifts in the course of human history. By 1994 the impact risk had become a matter of public concern partly because a small group of scientists had identified and correctly interpreted a layer of iridium in ancient rocks, then because another group found the crater made by the asteroid responsible for that iridium. The impact risk had also gained legitimacy because Martian dust storms combined with terrestrial volcanoes and forest fires to inspire new applications for novel climate models, which were used by scientists to reveal the perils of both asteroid impacts and nuclear war during an escalating arms race and a shift in US nuclear targeting. The rising seriousness with which policymakers and the public regarded the impact risk moreover registered a widespread increase in fears of, and political concern over, previously undetected vulnerabilities in the Earth system, particularly those created by global warming, the ozone hole, and nuclear winter. The impact risk could not have captured public attention without the Cold War development of technological systems with dual civilian and military applications that could monitor and manage the space environment or, in the case of the internet, channel previously unimagined flows of information. The risk also might not have sparked as much public concern without humbler advances in the telescopes available to amateur astronomers, which by 1994 were more capable and less expensive than ever before. Efforts to understand the risk and communicate it to the public depended on the coming of age of a new generation of asteroid scientists and their fruitful interaction with a still-productive old guard who had pioneered the first surveillance programs. Attempts to grapple with the impact risk also emerged out of the end of the Cold War, and the search for new threats—and new funds—by weapons specialists who were suddenly out of favor. Finally, the ability to spur public interest in the changing environment surrounding Earth built upon a foundation of popular fear and fascination that science fiction authors, filmmakers, and artists had established over many decades. All of these pieces had to be in place before cosmic environmental changes could inspire real change on Earth.

Even then, other forces undermined the development of a genuine response to the impact threat. Perhaps first among these was a widespread tendency of policymakers, journalists, and ordinary people to dismiss low-likelihood high-consequence risks. This giggle factor played a key role in discouraging both NASA and the Department of Defense from funding detection programs, let alone developing interception systems. Ironically it followed in part from the very science fiction stories and movies that raised awareness of the impact risk. The popularity of far-fetched science fiction on television and in movie theaters meant that when the genre turned to the impact risk, that risk actually seemed less realistic for some audiences.

Beyond the giggle factor, budget pressures created by federal cuts during the Reagan administration, the concept of the peace dividend at the end of the Cold War, and the Republican takeover of Congress all dissuaded NASA officials from supporting detection programs. So did NASA's decision to prioritize human spaceflight as a series of failures and cost overruns derailed robotic missions. Popular distrust of the defense establishment and allied scientists, who had repeatedly warned of dangers that in hindsight seemed overblown, also undermined public support for detection or interception programs. Waning public support for space exploration did not help, nor did a broad sense that fears of a sudden apocalypse should be consigned to the Cold War. Divisions between weapons specialists and asteroid scientists, moreover, fatally undermined any chance of confronting the impact threat in 1992. Finally, neither detection nor interception programs could fit easily within the purview of either NASA or the Department of Defense, and this encouraged both organizations to preserve their budgets by ignoring the risk.

In the United States, fears inspired by cosmic near-hits and real hits rippled through a complex social, cultural, economic, and political environment that pulled in different directions but overall, by 1994, encouraged efforts to confront the threat. Still, those efforts might have proceeded at a far slower pace were it not for the Comet SL9 impacts on Jupiter. And they might not have happened at all had they threatened powerful corporate or government interests. Because the impact hazard was not created by human beings, mitigating it did not require a radical change in the structure of society—and that was one reason it was easier to address than other potentially existential risks, such as climate change or nuclear war.

Asteroid and comet detection programs first emerged in the United States and only belatedly spread abroad. The European, Japanese, and Soviet space programs, however, led the way in dispatching robotic missions that began to characterize asteroids and comets. By cataloguing the locations of (as of this writing) more than one million asteroids and thousands of comets, survey programs provided roadmaps that guided these journeys. Today robots have collected samples from both asteroids and comets and returned their primordial materials back to Earth. Over the past two decades, NASA has undertaken a series of ambitious expeditions to these little worlds, none more spectacular than those that fulfilled Teller's vision of conducting deflection experiments in space. Observations of the Comet SL9 fragments helped scientists develop and promote *Deep Impact,* a robotic mothership and impactor that reached Comet Tempel 1 on July 4, 2005. The impactor survived eruptions of dust from the comet's surface, then smashed into the nucleus while the mothership observed the carnage. The collision shed new light on comet compositions, and confirmed that robots could autonomously impact a planetoid at a speed of some forty thousand kilometers per hour. NASA's press kit repeatedly reassured journalists that the impactor had hit Tempel 1 with a force akin to that of "a mosquito running into a 767 airliner," which meant that the comet would not shift into an orbit that might threaten Earth.[2]

The discovery of the near-Earth asteroid Apophis (Chapter 20) and Congress's subsequent mandate that a study be conducted by the National Research Council, led NASA to develop a spacecraft that would be capable of nudging a small asteroid to a different orbit. On September 26, 2022, the Double Asteroid Redirection Test (DART) mission collided with Dimorphos, the diminutive moon of a near-Earth asteroid, and gave it a noticeable jolt. In altering the environments of two little worlds, NASA confirmed that, if detection programs spotted a small, hazardous asteroid with years or decades to spare, the agency would be able to nudge it off an Earthbound track.[3]

NEO surveys have not only clarified the threat posed by asteroids and comets, but also catalogued the locations of potentially lucrative stockpiles of resources that are either hard to bring to space or rare on Earth. As in the 1970s, concerns over the sustainability of resource extraction on Earth have led entrepreneurs and engineers to explore how mining might be outsourced to space. Some argue that basic commodities could be extracted from asteroids to build and fuel large structures in space, including solar power plants that would beam clean energy to Earth. The feasibility of asteroid mining is controversial. Its effects on the global economy may be destabilizing, and its legality could be contested. Corporations explicitly devoted to asteroid mining have attracted billions in venture capital, but none have survived for long. Still, space agencies and NewSpace corporations alike are today shifting toward a model of space exploitation that will, in all likelihood, see the wholesale transformation of NEO environments through mining efforts that were first popularized in the 1970s. If so, the way would be paved by maps, created by detection programs, that reflect the influence of cosmic disasters on Earth and far beyond.[4]

Conclusion

A Shore on the Cosmic Ocean

An artist's drawing of Earth.

"Can the stability and order of the world be but a temporary
dynamic equilibrium achieved in a corner of the universe,
a short-lived eddy in a chaotic current?"
—Cixin Liu, *The Three-Body Problem* (2008)

"Within the past decade four problems have been recognized,
all of which relate to the limited size of Earth: they are
energy, food, living space, and population."
—Gerard K. O'Neill, *The High Frontier: Human
Colonies in Space* (1976)

It was the first blush of dawn on July 18, 1997. Carolyn and Gene Shoemaker scuffled in the dust of the Australian Outback, gawking at the brightening sky. Carolyn thought she had glimpsed Comet Hale-Bopp as it hurtled away from Earth. Gene had his doubts. "Do you have your glasses on?" he teased, chuckling. After a while, husband and wife clambered into a pickup truck and rattled along a dirt road toward Goat Paddock, an ancient impact crater. Then, at about 1:00 p.m., a Land Rover burst through the dust and slammed headfirst into their truck. Gene died instantly. Carolyn would need months to recover.[1]

Ironically, the geologist who more than anyone had revealed the threat posed by cosmic impacts had perished in the most ordinary of collisions. Gene's lifelong interest in those impacts had stemmed from a childhood desire to visit the Moon. By establishing geology as the leading science of NASA's Apollo program, he had hoped to earn a place among the test-pilots-turned-astronauts who landed on the Moon. But "just at the critical time when I should have been standing at the head of the line to go [to the Moon]," he remembered years later, "my adrenal cortex stopped functioning." It was a condition that would have kept his body from handling the stress of being an astronaut. Now he could do little more than teach the Apollo astronauts how to identify rocks that would be interesting to geologists.[2]

The day after Gene died, one of his former students, planetary scientist Carolyn Porco, wondered whether there might still be a way to realize her mentor's childhood dream. Porco learned that NASA's *Lunar Prospector* probe would soon be sent to collide with the Moon to search for water. Could it carry some of Gene's ashes? When Carolyn Shoemaker gave her consent, Porco made the arrangements. Within months, an ounce of Gene Shoemaker's ashes rocketed off to the Moon. Exactly two years after his death, Gene's remains smashed into Shoemaker crater at the Moon's south pole. Scientists had hoped to detect water in the impact plume, because future astronauts would need water for sustenance, protection from radiation, and rocket fuel. They did not detect water in the 1999 impact. Nevertheless, the man who proved that the Moon's craters were formed by impacts finally got his wish: he made it to the Moon, in an impact that itself created a crater.[3]

Like many others in this book, Shoemaker had devoted his life to uncovering previously hidden links between changing environments on Earth and in space. It had been the first epiphany of his career to realize that the wreckage of human violence resembled scars left by cosmic forces on Earth and across all rocky worlds. Planetary landscapes revealed a common source of environmental change: countless asteroids and comets swerving in and around the gravity wells of larger bodies, binding together the environments of the solar system. Today, scientists know that other forces once unseen and unsuspected—electromagnetism in all its wavelengths, charged particles flowing out from the Sun, microorganisms embedded within impact debris—also serve as connective tissue, uniting the environments of our corner of the universe. Through the relentless movement of matter and energy, real or perceived changes in cosmic environments shaped affairs on Earth, often by revealing the precarity of our continued existence. Earth is an island in the cosmic ocean, and sometimes the waves wash away our sandcastles.

The metaphor of space as a cosmic ocean was a favorite of Carl Sagan's. "Recently, we have waded a little out to sea," Sagan wrote, "enough to dampen our toes or, at most, wet our ankles. The water seems inviting. The ocean calls." The atoms in human bodies, Sagan explained, were all forged in the cores of long-dead stars. We come from the ocean, and, according to Sagan, "we long to return."[4]

Shoemaker's afterlife revealed how some of us are making that return. Scientists, humanists, artists, entrepreneurs, authors, and politicians have all imagined how our ocean surroundings could be reshaped to accommodate their dreams of expansion beyond Earth. Space agencies, space companies, and the space forces of powerful militaries are gaining the power to carry out those dreams, and finding that they have already, and in most cases unwittingly, made some alterations—some to our collective benefit, others to our detriment. Today our corner of the ocean is in flux. It is changing of its own accord, as it always has, but it is also changing because some of us want it to change. Where will these environmental transformations take us? What can the past reveal about the future?

Visions of the Future

Today our island seems imperiled, and for some the ocean feels irrelevant. We are using up our island, choking it with pollution. It is growing less habitable for us and for countless species on which we depend. We have gained increasingly deadly powers that threaten to spiral out of our control. We prepare to extinguish ourselves in conflicts over scraps of land. What use is the ocean if the island can no longer sustain us? Perhaps, if we can repair its problems—if we can learn to live well together—we will be worthy of the ocean. But only then.[5]

Others believe the ocean matters, yet only so far as using it benefits the island. Setting forth a little into the ocean helps us see the island more clearly. By comparing our island with other islands, we gain a sense of what ails our home—and how to fix it. We know that many of the resources we are depleting on our island exist elsewhere, and we know that acquiring them on other islands could restore our own. So we should venture as far as is useful, and not contemplate going farther until our home requires it.[6]

Still others believe that we should explore for exploration's sake, and settle new islands too, if we can. Yet when we set forth, we should do it together. Sailing the ocean is hard; surviving on its islands is harder still. It challenges us; it exposes us to experiences unavailable back home. The ocean is sublime. It can teach us what we are and where we come from; it invites a grand purpose. It calls on us to live more equitably and more sustainably than we did on our island—and, in that way, it might teach us how to live on all islands. But an ocean reserved for the privileged is not worth much to the majority.[7]

Then there are the true believers. The ocean must be traveled, wherever it may lead, for the rewards of exploration can rarely be anticipated in advance. Those with power must gain a foothold on the ocean's many shores, no matter how they do it. Our island's fate can be left to others. The privileged will create the boats; they will map the winds and settle the islands. The rest can follow, in time, according to rules already set by the few. Yet speed is of the essence, for the home-island may collapse. If we have scattered ourselves on many islands, we will survive; otherwise, we could disappear. Then there will be nothing left to contemplate the beauty of the ocean, or the island.[8]

There is overlap between these views. Like me, you might hold more than one of them. But what does the history of cosmic change teach us? Which approach to the cosmic ocean, to the environments of outer space, should we take? Owing to the growth of space companies, such as SpaceX, we seem to be embarking on a new Space Age or, to use a term now in vogue, a NewSpace age. It is time to choose a path and stick to it, to the extent that space agencies and companies can agree on a common approach in a divided world.

One lesson from the past is that we cannot turn away from space. We cannot pretend the ocean does not exist. It is not only because its waves will come whether we look for them or not; it is also because we can only understand our island by looking out toward the ocean. Indeed, how could it be otherwise? Nothing is explicable in isolation, and our island—our Earth—is no different. We have, in fact, come to depend on wading into the water. Growing satellite constellations, for example, are today essential for everything from weather forecasting to navigation, communications to weapons targeting. They themselves transform near-Earth space—the shallows near the shore—into yet another environment under our collective thumb.[9]

Nor could enthusiasts and entrepreneurs, scientists and artists stop dreaming about the deep ocean, about the environments of the Moon and beyond. There is a false dichotomy embedded within the notion that we could, an assumption that human collectives either conserve or explore, turn inward or look outward. Not only have they typically done both, but conservation can be affirmed and guided by exploration. The waves of the ocean have both revealed and helped create the precarity of the shore.

So, should we explore, exploit, and settle the deep ocean at all costs? I used to think so. My purpose in writing this book was, at first, to uncover histories that could provide new perspectives on a pending expansion of human settlement beyond Earth—to the craters of the Moon's south pole, for example, or the lava tubes beneath the Martian regolith, or even the cloud-tops of Venus. I feared that human affairs might deteriorate on Earth as the century unfolded, but I believed that they would only improve across diverse environments beyond Earth. After years of writing and teaching about climate change, I wanted to consider a hopeful part of humanity's likely future.

The past I uncovered while writing this book gave me ample reason for cynicism. As we have seen, alongside a utopian understanding of space exploration and settlement, one in which the challenges of space bring out the best in humanity, another current of thought has imagined environments beyond Earth as playgrounds in which terrestrial rules do not apply. From plans to launch a nuclear assault on the Moon to schemes for terraforming Venus, in this book I have introduced several examples of such thinking. Other proposals for space exploration and exploitation have dismissed substantial risks to environments on Earth. Returning samples from other worlds, for example, could someday contaminate Earth's biosphere with extraterrestrial microorganisms, while space mining may give corporations the capacity to extinguish human life by redirecting an asteroid toward Earth. For good reason, some space scientists argue that it is time to decolonize space exploration, according to a broad definition of colonialism that encompasses environmental plunder for the benefit of the few rather than the majority. What good is it to brave the ocean and occupy new islands if we export the problems that imperil our home?[10]

Human Ripples on the Cosmic Ocean

It is now clear to me that plans for securing a human future in the cosmic ocean should prioritize the challenges of the shore. In our solar system there is, after all, no substitute for Earth. As concern over the fate of our island swelled in the closing years of the 1960s, the American physicist Gerard O'Neill came to the same conclusion. O'Neill's students had soured on NASA's vision for the crewed exploration of the Moon and other worlds, which to them offered few benefits for ordinary people. Inspired by their ideas, O'Neill imagined what it would be like to develop a space program for Earth, and for everyone on Earth. The gravest challenges facing humanity, he reasoned, cannot simply be solved by conservation, retrenchment, or technological innovation. The resource and energy demands of billions of humans inevitably produce waste—vast quantities of it—while the world's stock of key resources is finite, and dwindling. Yet infinite energy flows through space—and limitless materials, buried in

The imagined interior of a Bernal sphere, one of the space settlements proposed by Gerard O'Neill. In the 1970s the artist, Rick Guidice, was commissioned by NASA to create a series of paintings that would help policymakers and the public imagine the possibilities for space settlement.

millions of desolate asteroids, seem just within reach. What if we could harness even a slightly bigger slice of that energy, that matter?[11]

O'Neill believed that other worlds are unsuitable for human settlement. He was more right than he knew. Robotic missions have since established that the most habitable real estate on any planet besides Earth—the upper atmosphere of Venus—would have you floating through clouds of sulfuric acid, bathed in radiation, above a scorching hellscape that literally glows in all directions. Space itself, however, may be a different story. Drawing on the ideas of other thinkers, such as the Irish communist John Desmond Bernal, O'Neill's students designed "inside-out planets": manufactured cylinders or rings, complete with artificial rivers, fields, and forests, spinning in space to create their own gravity. Building these space cities, O'Neill believed, could ensure that humanity's expansion into the solar system directly and permanently eased the human burden on Earth. Mining the Moon and asteroids would provide the resources to construct the cities. It would also enable the inhabitants of the cities to build and operate giant solar power plants in space that would beam clean energy back home. Everyone on Earth would benefit, and because the new cities would create abundant land, in time everyone could choose whether to settle in space.[12]

O'Neill's proposal for an inaugural space city, "Island One," attracted widespread support from American and European scientists, journalists, and students who sought a roadmap for an egalitarian, progressive, and sustainable future beyond Earth. O'Neill insisted that the city could be built with the technology of his day, but many engineers argued that it would depend on achieving a series of breakthroughs that would lower the cost of hauling materials into orbit and permit automated manufacturing in space. By the waning years of the 1970s, slumping energy prices and the end of the Vietnam War had eased some of the existential angst that had motivated support for O'Neill's dream. The surging importance of high-tech industries and the personal computing revolution, meanwhile, redirected techno-utopian dreams back toward Earth. O'Neill's vision for humanity's future in space faded into irrelevance.[13]

Many of the environmental problems that came into focus in the 1960s seem more pressing today, however, and the technologies required to realize O'Neill's vision may be maturing. The development of reusable rockets by SpaceX, Blue Origin, and other NewSpace companies has at last slashed the cost of reaching orbit, while the refinement of 3D printing and the rapid advance of artificial intelligence may soon unlock an era of automated construction in space.[14]

All of these developments should lead us back to the future: to a vision of the ocean in which we build in the water to support our home, for everyone's collective benefit. Today the falling cost of solar and wind power is flattening the global carbon dioxide emissions curve. Yet fossil fuels have a big advantage over renewables: they provide constant power, whereas the Sun does not always shine, and the wind does not always blow. Things are different in space. Not only does sunlight reach solar panels without traveling through Earth's obscuring atmosphere, but it can do so continuously. Building and maintaining solar power stations in space would be hard, and beaming power back to Earth could be tricky. But imagine what could be done with the roughly $12 billion USD annually set aside for human spaceflight in the United States alone.[15]

If that money were channeled into a vast campaign to develop the technology and build the infrastructure for space-based solar power, a substantial share of humanity could enjoy three broad benefits. First, the effort would accelerate the emergence of a genuine cislunar economy: a zone of commercial development that would extend from Earth to Moon. It not only would fuel the growth and proliferation of NewSpace companies, but also would inspire breakthroughs in the technology and practice of mining and building in space. It would create capabilities that could in time permit even more elaborate construction projects in space, for even more ambitious causes. If climate change progresses more rapidly than currently expected, for example, or if its impacts on populations and ecosystems are excessively damaging, then construction could shift from power stations to a solar shield that filters a tiny portion of the Sun's incoming radiation, and thereby cools Earth.[16]

Second, the construction effort could slow the pace of climate change. If ongoing improvements in launch costs and the efficiency of solar panels, for example, are sustained, then space-based solar power would soon be competitive with terrestrial renewable power sources. Eventually the automated construction of solar power stations in space, using resources available on the Moon or asteroids, could herald an era of truly pollution-free energy that leaves environments on Earth entirely unscathed. Space-based solar power may never provide all or even most of our electricity, and it might not be wise to off-world all of our power plants. Yet continuous power from space could accelerate the adoption of renewable power on Earth, making a real difference in securing a livable planet for all.[17]

Third, the effort would serve as a genuine source of inspiration for young people who care about space exploration but prioritize the fate of the only habitable planet we currently know to exist. At a recent event, well attended by a diverse crowd of Georgetown students, I proposed that building solar power plants would be a better use of NASA's resources than returning astronauts to the Moon. NASA representatives who shared the stage disagreed. Human explorers, they argued, inspire people like no robotic construction project ever could. Yet after we were done, a succession of students came to speak with me.

All agreed: forget about planting flags on the Moon or blasting billionaires to Mars. What could be more uplifting than a constellation of glittering stations, all unfurling their panel-petals, one after another to beam salvation to our imperiled planet? What could more beautifully declare that our expansion into space truly serves every member of our species?[18]

After we have stabilized Earth's climate, where might the distant future take us? It is hard to know. Advances in artificial intelligence, genetic engineering, and nanotechnology may fundamentally change what we are. A nuclear apocalypse, a catastrophic decline in biodiversity, or a continuing decline in birth rates could threaten us with extinction or so reduce our numbers that Earth offers all the space that's needed by those who have survived.[19] Yet despite the chaos and peril of our time, we can now glimpse—ever so dimly, through a haze of risk and hardship—a truly boundless horizon. Imagine a future in which we use our expertise at space construction, gained by assembling solar power plants, to build entire cities beyond Earth, spinning homes spun in turn from the infinite bounty of the asteroid belt. We might realize O'Neill's vision, a century or two late, and thereby find a way to expand sustainably while preserving our home island. In hollowed-out asteroids we could, perhaps, travel between the stars, all the while seeking to safeguard, sustain, and, ever so carefully, search for life. Because while the cosmos may be replete with environments, all available evidence suggests that life—at least complex life—is rare.

Before you dismiss this dream as fanciful, remember that what happens in space has long shaped us. Humanity's past was influenced, in part, by ripples on the cosmic ocean. More will come, no matter what we do. Now we are gaining the capacity to make our own waves. Our future may depend on how we make them.

Notes

Introduction

1 Carl Sagan, *Pale Blue Dot: A Vision of the Human Future in Space* (Ballantine Books, 1997), 4.
2 Sagan, *Pale Blue Dot,* 6.
3 Sagan, *Pale Blue Dot,* 7.
4 Sagan, *Pale Blue Dot,* 6.
5 Dagomar Degroot, "'A Catastrophe Happening in Front of Our Very Eyes': The Environmental History of a Comet Crash on Jupiter," *Environmental History* 22, no. 1 (2016): 23–49; Vaclav Smil, *Energies: An Illustrated Guide to the Biosphere and Civilization* (MIT Press, 1999), 1.
6 Toby Ord, *The Precipice: Existential Risk and the Future of Humanity* (Hachette Books, 2020), 6.
7 Patrick Moore, *The Data Book of Astronomy* (Institute of Physics, 2000), 4, 98.
8 Kenneth Lang, *The Cambridge Guide to the Solar System,* 2nd ed. (Cambridge University Press, 2011), 8, 192.
9 Benson Bobrick, *The Fated Sky: Astrology in History* (Simon and Schuster, 2005), 6.
10 Henry King, *The History of the Telescope* (Dover, 2003), 32.
11 J. L. Heilbron, *Galileo* (Oxford University Press, 2010), 148; Huib Zuidervaart, "The 'Invisible Technician' Made Visible: Telescope Making in the Seventeenth and Early Eighteenth-Century Dutch Republic," in *From Earth-Bound to Satellite: Telescopes, Skills, and Networks,* ed. A. D. Morrison-Low et al. (Brill, 2012), 41; Albert van Helden, "The Invention of the Telescope," *Transactions of the American Philosophical Society* 67, no. 4 (1977): 9.
12 Samuel K. Cohn Jr., *The Black Death Transformed: Disease and Culture in Early Renaissance Europe* (Oxford University Press, 2002); Alfred W. Crosby, *Ecological Imperialism: The Biological Expansion of Europe, 900–1900* (Cambridge University Press, 2004).
13 Elizabeth Eisenstein, *The Printing Press as an Agent of Change* (Cambridge University Press, 1980), 5; Meredith K. Ray, *Daughters of Alchemy: Women and Scientific Culture in Early Modern Italy* (Harvard University Press, 2015); John F. Richards, *The Unending Frontier: An Environmental History of the Early Modern World* (University of California Press, 2003), 19.
14 Euan Cameron, *The European Reformation,* 2nd ed. (Oxford University Press, 2012); Brad S. Gregory, *The Unintended Reformation: How a Religious Revolution Secularized Society* (Harvard University Press, 2015); Carol Pal, *Republic of Women: Rethinking the Republic of Letters in the Seventeenth Century* (Cambridge University Press, 2012), 6.

15 Robert Westman, *The Copernican Question: Prognostication, Skepticism, and Celestial Order* (University of California Press, 2011); Toby Huff, *The Rise of Early Modern Science: Islam, China, and the West,* 2nd ed. (Cambridge University Press, 2003), 350; W. G. L. Randles, *The Unmaking of the Medieval Christian Cosmos, 1500–1760: From Solid Heavens to Boundless Aether* (Ashgate, 1999), 1.

16 Owen Gingerich, *Copernicus: A Very Short Introduction* (Oxford University Press, 2016), 1; Huff, *The Rise,* 67.

17 Mark A. Waddell, *Magic, Science, and Religion in Early Modern Europe* (Cambridge University Press, 2021), 5; Allison Coudert, *Religion, Magic, and Science in Early Modern Europe and America* (Praeger, 2011), 134.

18 Classic examples include Tom Wolfe, *The Right Stuff* (Picador, 1979); John Logsdon, *The Decision to Go to the Moon* (University of Chicago Press, 1976); Walter A. McDougall, *. . . the Heavens and the Earth: A Political History of the Space Age* (Basic Books, 1985).

19 Favorite examples include David Mindell, *Digital Apollo: Human and Machine in Spaceflight* (MIT Press, 2011); Teasel Muir-Harmony, *Operation Moonglow: A Political History of Project Apollo* (Hachette UK, 2020); Asif Siddiqi, *The Red Rockets' Glare: Spaceflight and the Russian Imagination, 1857–1957* (Cambridge University Press, 2014); Monique Laney, *German Rocketeers in the Heart of Dixie* (Yale University Press, 2015); Alexander MacDonald, *The Long Space Age: The Economic Origins of Space Exploration from Colonial America to the Cold War* (Yale University Press, 2017).

20 James Fleming, *Fixing the Sky* (Columbia University Press, 2010), 137–139; Irving Langmuir, "Pathological Science," ed. and annotated by Robert N. Hall, *Physics Today* 42, no. 10 (October 1989): 36–48.

21 Sabine Höhler, *Spaceship Earth in the Environmental Age, 1960–1990* (Routledge, 2015); Peder Anker, "The Ecological Colonization of Space," *Environmental History* 10, no. 32 (2005): 239–268; Amy Nelson, "What the Dogs Did: Animal Agency in the Soviet Manned Space Flight Programme," *BJHS Themes* 2 (2017): 79–99; Kärin Nickelsen and David P. D. Munns, *Far beyond the Moon: A History of Life Support Systems in the Space Age* (University of Pittsburgh Press, 2021).

22 Jordan Bimm, "Rethinking the Overview Effect," *Quest* 21, no. 1 (2014): 39–45; Kim McQuaid, "Selling the Space Age: NASA and Earth's Environment, 1958–1990," *Environment and History* 12 (May 2006): 127; Neil M. Maher, *Apollo in the Age of Aquarius* (Harvard University Press, 2017).

23 Michael Rawson, "Discovering the Final Frontier: The Seventeenth-Century Encounter with the Lunar Environment," *Environmental History* 20 (2015): 216; Valerie A. Olsen, "NEO-ecology: The Solar System's Emerging Environmental History and Politics," in *New Natures: Joining Environmental History with Science and Technology Studies,* ed. Dolly Jørgensen, Finn Arne Jørgensen, and Sara Pritchard (University of Pittsburgh Press, 2013): 210.

24 Roger D. Launius, "Writing the History of Space's Extreme Environment," *Environmental History* 15, no. 3 (2010): 526–532; W. Henry Lambright and Anna Ya Ni, "The Environmental Frontier of Space," in *Handbook of Globalization and the Environment,* ed. Khi Thai, Dianne Rahm, and Jerrell Coggburn (CRC Press, 2007), 95; Lisa Ruth Rand, "Orbital Decay: Space Junk and the Environmental History of Earth's Planetary Borderlands" (PhD diss., University of Pennsylvania, 2016); Valerie Olson and Lisa Messeri, "Beyond the Anthropocene: Un-earthing an Epoch," *Environment and Society* 6 (2015): 28–47.

25 For a history book that uses a similar approach, see Dava Sobel, *The Planets* (Penguin, 2006).

Part I. Sun

1 Robert W. Noyes, *The Sun, Our Star* (Harvard University Press, 1982), 42.
2 Kenneth J. H. Phillips, *Guide to the Sun* (Cambridge University Press, 1992), 226.
3 Philip Judge, *The Sun: A Very Short Introduction* (Oxford University Press, 2020), 14.

1. Ice Ages Great and Small

1 Johann Jakob Scheuchzer, *Itinera Alpina Tria* (London, 1708), 9; Keith Thomas, *Man and the Natural World: Changing Attitudes in England, 1500–1800* (Allen Lane, 1983); Aaron John Henry Larsen, "Darkest Forests and Highest Mountains: The Witches' Sabbath and Landscapes of Fear in Early Modern Demonologies," *European Review of History* (2023): 1–18; Urs B. Leu, "Swiss Mountains and English Scholars: Johann Jakob Scheuchzer's Relations to the Royal Society," *Huntington Library Quarterly* 78, no. 32 (2015): 329.
2 Johann Jakob Scheuchzer, *Uresiphoites helveticus sive itinera per Helvetiae alpinas regiones facta annis 1702–11* (Leiden, 1723), 210, 234.
3 Spencer R. Weart, *The Discovery of Global Warming*, rev. expanded ed. (Harvard University Press, 2008).
4 Dagomar Degroot et al., "The History of Climate and Society: A Review of the Influence of Climate Change on the Human Past," *Environmental Research Letters* 17, no. 10 (2022): 2.
5 CenCO_2PIP Consortium, "Toward a Cenozoic History of Atmospheric CO_2," *Science* 382, no. 6675 (2023); Emily Judd et al., "A 485-Million-Year History of Earth's Surface Temperature," *Science* 385, no. 1316 (2024): 1.
6 Doug Macdougall, *Frozen Earth: The Once and Future Story of Ice Ages* (University of California, 2013); Jürgen Ehlers, Philip Hughes, and Philip L. Gibbard, *The Ice Age* (Wiley-Blackwell, 2016), 12.
7 James Woodward, *The Ice Age. A Very Short Introduction* (Oxford University Press, 2014), 13.
8 Degroot et al., "The History," 5.
9 Arlene M. Rosen et al., "Holocene Desertification, Traditional Ecological Knowledge, and Human Resilience in the Eastern Gobi Desert, Mongolia," *The Holocene* 32, no. 12 (2022): 1462.
10 Rodney Castleden, *Atlantis Destroyed* (Routledge, 2002); P. D. Nunn, *The Edge of Memory: Ancient Stories, Oral Tradition and the Post-Glacial World* (Bloomsbury, 2018); William Ryan and Walter Pitman, *Noah's Flood: The New Scientific Discoveries about the Event That Changed History* (Simon and Schuster, 1998).
11 Harvey Weiss, ed., *Megadrought and Collapse: From Early Agriculture to Angkor* (Oxford University Press, 2017); Hans Renssen, "Climate Model Experiments on the 4.2 ka Event: The Impact of Tropical Sea-Surface Temperature Anomalies and Desertification," *The Holocene* 32, no. 5 (2022): 378–389; Sugata Ray, *Climate Change and the Art of Devotion* (University of Washington Press, 2019), 16.
12 Lydia Barnett, *After the Flood: Imagining the Global Environment in Early Modern Europe* (Johns Hopkins University Press, 2019), 2.
13 Barnett, *After the Flood*, 5; Claudine Cohen, *The Fate of the Mammoth: Fossils, Myth, and History* (University of Chicago Press, 2002), 72; Hans Sloane, "An Account of Elephants Teeth and Bones Found under Ground. By Sir Hans Sloane, Bart," *Philosophical Transactions* 35 (1727): 458.
14 Barnett, *After the Flood*, 16, 132, 166.

15 Barnett, *After the Flood,* 194.

16 Dagomar Degroot, "Climate Change and Society from the Fifteenth through the Eighteenth Centuries," *WIREs Climate Change* (2018): 2.

17 Rémi Thiéblemont et al., "Solar Forcing Synchronizes Decadal North Atlantic Climate Variability," *Nature Communications* 6 (2015): 1; Chitradeep Saha and Dibyendu Nandy, "Understanding Grand Minima in Solar Activity," *arXiv*:2409.09775 (2024); Sami Solanki, Natalie Krivova, and Joanna Haigh, "Solar Irradiance Variability and Climate," *Annual Review of Astronomy and Astrophysics* 51, no. 1 (2013): 312; Francois Lapointe and Raymond S. Bradley, "Little Ice Age Abruptly Triggered by Intrusion of Atlantic Waters into the Nordic Seas," *Science Advances* 7, no. 51 (2021): eabi8230.

18 Lauren R. Marshall et al., "Volcanic Effects on Climate: Recent Advances and Future Avenues," *Bulletin of Volcanology* 84, no. 5 (2022): 54; Michael Sigl et al., "Volcanic Stratospheric Sulfur Injections and Aerosol Optical Depth during the Holocene," *Earth System Science Data* 14, no. 7 (2022): 3167; Andrea Burke et al., "High Sensitivity of Summer Temperatures to Stratospheric Sulfur Loading from Volcanoes in the Northern Hemisphere," *PNAS* 120, no. 47 (2023): e2221810120.

19 C. Pfister and H. Wanner, *Climate and Society in Europe: The Last Thousand Years* (Haupt, 2021).

20 Bradley Skopyk, *Colonial Cataclysms: Climate, Landscape, and Memory in Mexico's Little Ice Age* (University of Arizona Press, 2020); David W. Stahle et al., "Major Mesoamerican Droughts of the Past Millennium," *Geophysical Research Letters* 38, no. 5 (2011): 3; Gayatri Kathayat et al., "Protracted Indian Monsoon Droughts of the Past Millennium and Their Societal Impacts," *PNAS* 119, no. 39 (2022): e2207487119; C. Sun and Y. Liu, "Tree-Ring-Based Drought Variability in the Eastern Region of the Silk Road and Its Linkages to the Pacific Ocean," *Ecological Indicators* 96 (2019): 421–429; Heinz Wanner, Christian Pfister, and Raphael Neukom, "The Variable European Little Ice Age," *Quaternary Science Reviews* 287 (2022): 107531; Christina Karamperidou, "Extracting Paleoweather from Paleoclimate through a Deep Learning Reconstruction of Last Millennium Atmospheric Blocking," *Communications Earth & Environment* 5, no. 1 (2024): 2.

21 Christopher Friedrichs, *The Early Modern City, 1450–1750* (Routledge, 2014), 125.

22 Emmanuel Kreike, *Scorched Earth: Environmental Warfare as a Crime against Humanity and Nature* (Princeton University Press, 2022); J. R. McNeill, *Mosquito Empires: Ecology and War in the Greater Caribbean, 1620–1914* (Cambridge University Press, 2010), 4; Erica Charters, Marie Houllemare, and Peter H. Wilson, *A Global History of Early Modern Violence* (Manchester University Press, 2020).

23 Sam White, *The Climate of Rebellion in the Early Modern Ottoman Empire* (Cambridge University Press, 2011), 3; Bas van Bavel et al., *Disasters and History: The Vulnerability and Resilience of Past Societies* (Cambridge University Press, 2020), 143.

24 Timothy Brook, *The Price of Collapse: The Little Ice Age and the Fall of Ming China* (Princeton University Press, 2023), 21; Qian Liu et al., "Climate, Disasters, Wars and the Collapse of the Ming Dynasty," *Environmental Earth Sciences* 77 (2018): 2; Jianxin Cui et al., "Climatic Change and the Rise of the Manchu from Northeast China during AD 1600–1650," *Climatic Change* 156 (2019): 406.

25 Jiang Yonglin, *The Mandate of Heaven and the Great Ming Code* (University of Washington Press, 2011), 5; Chaochao Gao et al., "Volcanic Climate Impacts Can Act as Ultimate and Proximate Causes of Chinese Dynastic Collapse," *Communications Earth & Environment* 2, no. 234 (2021).

26 Wolfgang Behringer, "Climatic Change and Witch-Hunting," *Climatic Change* 43, no. 1 (1999): 336; Jacek Wijaczka, "The Impact of Climate Change on Witch Trials: Myth or Reality?," *Kwartalnik Historyczny* 129, no. 6 (2022): 5; D. P. Bell, "The Little Ice Age and the Jews," *AJS Review* 32, no. 1 (2008), 26; Abaigéal Warfield, "The Witch and the Weather," *Sixteenth Century Journal* 50, no. 4 (2019).

27 Federik Albritton Johnsson, *Enlightenment's Frontier: The Scottish Highlands and the Origins of Environmentalism* (Yale University Press, 2013), 80.

28 Sugata Ray, "Hydroaesthetics in the Little Ice Age: Theology, Artistic Cultures and Environmental Transformation in Early Modern Braj, c. 1560–70," *South Asia: Journal of South Asian Studies* 40, no. 1 (2017): 6.

29 Thomas Wickman, "Narrating Indigenous Histories of Climate Change in the Americas and Pacific," in *The Palgrave Handbook of Climate History*, ed. Sam White, Christian Pfister, and Franz Mauelshagen (Palgrave Macmillan, 2018), 387–411.

30 Dagomar Degroot, *The Frigid Golden Age: Climate Change, the Little Ice Age, and the Dutch Republic, 1560–1720* (Cambridge University Press, 2018), 258.

31 She made this remark after attending a ceremony at a shrine that, she hoped, would halt a stretch of cold, rainy weather. Madame de Sévigné to Madame de Grignan, Paris, July 24, 1675, in *The Letters of Madame de Sévigné to Her Daughter and Friends* (Roberts Brothers, 1874), https://archive.org/stream/selectionsfromraooraberich/selectionsfromraooraberich_djvu.txt.

32 Francis Bacon, *Historia Ventorum*, vol. 12 (Oxford University Press, 2007), 129; Francis Bacon, "Of Vicissitudes of Things," in *Essays or Counsels, Civil and Moral*, available at https://www.gutenberg.org/files/575/575-h/575-h.htm; Philip Jenkins, *Climate, Catastrophe, and Faith: How Changes in Climate Drive Religious Upheaval* (Oxford University Press, 2021), 111.

33 Sam White, "Unpuzzling American Climate: New World Experience and the Foundations of a New Science," *Isis* 106, no. 3 (2015): 545.

34 Johnsson, *Enlightenment's Frontier*, 75; Paul Warde, *The Invention of Sustainability: Nature and Destiny, c. 1500–1870* (Cambridge University Press, 2018), 127.

35 Anya Zilberstein, *A Temperate Empire: Making Climate Change in Early America* (Oxford University Press, 2016); Colin Coates and Dagomar Degroot, "'Les bois engendrent les frimas et les gelées': Comprendre le climat en Nouvelle-France," *Revue d'histoire de l'Amérique française* 68, no. 3 (2015): 197.

36 Alexander Koch et al., "Earth System Impacts of the European Arrival and Great Dying in the Americas after 1492," *Quaternary Science Reviews* 207 (2019): 13–36.

37 Warde, *The Invention of Sustainability*, 329.

2. Changing Stars, Changing Climates

1 Richard Holmes, *The Age of Wonder* (Vintage Books, 2008), 67; Rachel Cowgill and Peter Holman, eds., *Music in the British Provinces, 1690–1914* (Routledge, 2007), 100–111; Marilyn B. Ogilvie, *Searching the Stars: The Story of Caroline Herschel* (History Press, 2011).

2 Michael Hoskin, *William and Caroline Herschel: Pioneers in Late 18th-Century Astronomy* (Springer, 2013); Joseph Ashbrook, *The Astronomical Scrapbook* (Cambridge University Press, 1984), 245; Hoskin, *Discoverers of the Universe: William and Caroline Herschel* (Princeton University Press, 2011), 86; E. Winterburn, "Philomaths, Herschel, and the Myth of the Self-Taught Man," *Notes and Records* 68 (2014): 207–225.

3 Agnes Clerke, *A Popular History of Astronomy during the Nineteenth Century,* 4th ed. (A&C Black, 1902), 4.
4 Thomas Levenson, *The Hunt for Vulcan* (Penguin Random House, 2015), 15.
5 Hoskin, *Discoverers of the Universe,* 166; Christopher M. Linton, *From Eudoxus to Einstein: A History of Mathematical Astronomy* (Cambridge University Press, 2004), 360; Stuart Clark, *The Sun Kings* (Princeton University Press, 2007), 27.
6 William Herschel, logbook entries, June 27, 1783, to May 19, 1801, Herschel W. 3 / 1.1 No. 1, Royal Astronomy Society Archives; Clark, *The Sun Kings,* 31; V. Hoyt and Kenneth H. Schatten. "Sir William Herschel's Notebooks—Abstracts of Solar Observations," *Astrophysical Journal Supplement Series* 78 (1992): 339.
7 Jessica Riskin, *Science in the Age of Sensibility: The Sentimental Empiricists of the French Enlightenment* (University of Chicago Press, 2002), 97.
8 John Locke, *An Essay concerning Human Understanding,* bk. 4 (1690), 665; Paul Bartha, "Analogy and Analogical Reasoning," in *The Stanford Encyclopedia of Philosophy,* ed. Edward N. Zalta (Spring 2019); William Herschel, logbook entry, July 30, 1776, 9:07 p.m., Observations by William Herschel, Royal Astronomy Society Archives.
9 William Herschel, logbook entries, September 8, 1792, to May 19, 1801, Herschel W. 3 / 1.1 No. 1, Royal Astronomy Society Archives; Michael J. Crowe, *The Extraterrestrial Life Debate, 1750–1900* (Cambridge University Press, 1986), 38; Pierre Borel, *A New Treatise Proving a Multiplicity of Worlds* (Printed for John Deane, 1658); Bernard le Bovier de Fontenelle, *Conversations on the Plurality of Worlds,* trans. William Gardiner (A. Bettesworth, 1715).
10 Clark, *The Sun Kings,* 35; Jack R. White, "Herschel and the Puzzle of Infrared," *American Scientist* 100, no. 3 (2012): 218.
11 William Herschel, logbook entries, September 8, 1792, to May 19, 1801, Herschel W. 3 / 1.1 No. 1, Royal Astronomy Society Archives; Clark, *The Sun Kings,* 36.
12 William Herschel, "IX. On the method of observing the changes that happen to the fixed stars . . . ," *Philosophical Transactions of the Royal Society of London* 86 (1796): 186; Paul Warde, *The Invention of Sustainability: Nature and Destiny, c. 1500–1870* (Cambridge University Press, 2018), 291; Thomas Robert Malthus, *An Essay on the Principle of Population* (J. Johnson, 1798), 24; Alison Bashford and Joyce E. Chaplin, *The New Worlds of Thomas Robert Malthus* (Princeton University Press, 2016).
13 William Herschel, "IX. On the method," 186; Clark, *The Sun Kings,* 37.
14 William Herschel, logbook entries, January 15, 1801, to February 14, 1801, Herschel W. 3 / 1.1 No. 1, Royal Astronomy Society Archives.
15 William Herschel, logbook entries, January 15, 1801, to December 13, 1801, Herschel W. 3 / 1.1 No. 1, Royal Astronomy Society Archives.
16 William Herschel, logbook entries, April 19, 1779, to July 20, 1816, Herschel W. 3 / 1.1 No. 1 and No. 2, Royal Astronomy Society Archives; D. E. Parker, T. P. Legg, and C. K. Folland, "A New Daily Central England Temperature Series, 1772–1991," *International Journal of Climatology* 12 (1992): 317–342.
17 William Herschel, logbook entries, September 18, 1796, to January 31, 1800, Herschel W. 3 / 1.1 No. 1, Royal Astronomy Society Archives.
18 William Herschel, "XVI. Additional observations tending to investigate the symptoms of the variable emission of the light and heat of the sun . . . ," *Philosophical Transactions of the Royal Society of London* 91 (1801): 354–362; Clark, *The Sun Kings,* 37; William Herschel, logbook entries, January 15, 1801, to March 15, 1806, Herschel W. 3 / 1.1 No. 1, Royal Astronomy Society Archives.

19 See David Nash et al., "Climate Indices in Historical Climate Reconstructions," *Climate of the Past* 17 (2021).

20 Clark, *The Sun Kings,* 38.

21 Deborah R. Coen, *Climate in Motion: Science, Empire, and the Problem of Scale* (University of Chicago Press, 2018), 14; James Rodger Fleming, *Historical Perspectives on Climate Change* (Oxford University Press, 2005), 11; Rudolf Wolf, "Neue Untersuchungen uber die Periode der Sonnenflecken und ihre Bedeutung," *Mittheilungen der Naturforschenden Gesellschaft in Bern* 255 (1852): 250, 251.

22 Vimal Mishra et al., "Drought and Famine in India, 1870–2016," *Geophysical Research Letters* 46, no. 4 (2019): 2076; Mike Davis, *Late Victorian Holocausts: El Niño Famines and the Making of the Third World* (Verso, 2002), 339; Sunil Amrith, *Unruly Waters: How Rains, Rivers, Coasts and Seas Have Shaped Asia's History* (Basic Books, 2018), 64; Simon Naylor, *The Observatory Experiment: Meteorology in Britain and Its Empire* (Cambridge University Press, 2024).

23 Clark, *The Sun Kings,* 138; Pierre Sokolsky, *The Clock in the Sun: How We Came to Understand Our Nearest Star* (Columbia University Press, 2024), 124.

24 William Stanley Jevons, letters, February 14, 1878, to March 3, 1878, in *Letters & Journal of W. Stanley Jevons,* ed. Herriet A. Jevons (Macmillan, 1886); Harro Maas, *William Stanley Jevons and the Making of Modern Economics* (Cambridge University Press, 2005); Theodore M. Porter, *The Rise of Statistical Thinking, 1820–1900* (Princeton University Press, 2020); Sokolsky, *The Clock,* 162; "The Predicted Drought," *South Australian Register,* October 3, 1889.

25 Matthias Heymann and Dania Achermann, "From Climatology to Climate Science in the Twentieth Century," in *The Palgrave Handbook of Climate History,* ed. Sam White, Christian Pfister, and Franz Mauelshagen (Palgrave Macmillan, 2018), 233–249; Sokolsky, *The Clock,* 176, 194.

3. Hidden Connections and Solar Science

1 Stuart Clark, *The Sun Kings* (Princeton: Princeton University Press, 2007), 79.

2 Richard Carrington, "Sunday July 18," Carrington Letters, Carrington 58:3, Royal Astronomical Society.

3 Minutes of Council, vol. IV, November 1856 to June 1866 , and Friday November 11, 1859, and Richard Carrington, "Sunspot Observations of Richard Christopher Carrington (1826–75)," Royal Astronomical Society Archives; Clark, *The Sun Kings,* 15.

4 Kenneth Lang, *The Cambridge Guide to the Solar System,* 2nd ed. (Cambridge University Press, 2011), 98; Philip Judge, *The Sun: A Very Short Introduction* (Oxford University Press, 2020), 5; Markus Aschwanden, "The Sun," in *Encyclopedia of the Solar System,* 3rd ed., ed. Tilman Spohn, Doris Breuer, and Torrence Johnson (Elsevier, 2014), 235; Merav Opher, Abraham Loeb, and J. E. G. Peek, "A Possible Direct Exposure of the Earth to the Cold Dense Interstellar Medium 2–3 Myr Ago," *Nature Astronomy* (2024): 1.

5 Lang, *Cambridge Guide to the Solar System,* 262.

6 Lang, *Cambridge Guide to the Solar System,* 103.

7 Z. Svestka, "Solar Activity," in *Dynamic Sun,* ed. B. N. Dwivedi and E. N. Parker (Cambridge University Press, 2013), 252; J. T. Gosling, "The Solar Wind," in Spohn, Breuer, and Johnson, *Encyclopedia of the Solar System,* 268.

8 Judge, *The Sun,* 38; Delores J. Knipp et al., "The May 1967 Great Storm and Radio Disruption Event," *Space Weather* 14, no. 9 (2016): 618.

9 Aschwanden, "The Sun," 249; Space Studies Board and National Research Council, *Severe Space Weather Events: Understanding Societal and Economic Impacts: A Workshop Report* (National Academies Press, 2009), 101; *The 1989 System Disturbances Disturbance Analysis Working Group Report,* North American Electric Reliability Corporation (1990), 37; D. H. Boteler, "A 21st Century View of the March 1989 Magnetic Storm," *Space Weather* 17, no. 10 (2019): 1433.

10 Alan Gurney, *Compass: A Story of Exploration and Innovation* (W. W. Norton, 2004), 8.

11 Clark, *The Sun Kings,* 50; Monika Korte and Mioara Mandea, "Geomagnetism: From Alexander von Humboldt to Current Challenges," *Geochemistry, Geophysics, Geosystems* 20 (2019): 3801–3820; Robert W. Noyes, *The Sun, Our Star* (Harvard University Press, 1982), 193.

12 Pablo Ortega et al., "A Model-Tested North Atlantic Oscillation Reconstruction for the Past Millennium," *Nature* 523, no. 7558 (2015): 74; Nicholas P. McKay and Darrell S. Kaufman, "An Extended Arctic Proxy Temperature Database for the Past 2,000 Years," *Scientific Data* 1, no. 1 (2014): 3; William Scoresby, "Journal for 1816," in *The Arctic Whaling Journals of William Scoresby the Younger,* ed. C. Ian Jackson (Hakluyt Society, 2008), 236.

13 Vidar Enebakk, "Hansteen's Magnetometer and the Origin of the Magnetic Crusade," *British Journal for the History of Science* 47, no. 4 (2014): 590; Michael Bravo, "Geographies of Exploration and Improvement: William Scoresby and Arctic Whaling, 1782–1822," *Journal of Historical Geography* 32, no. 3 (2006): 512–538.

14 Clark, *The Sun Kings,* 68; Enebakk, "Hansteen's Magnetometer," 591, 600.

15 Matthew Goodman, "Follow the Data: Administering Science at Edward Sabine's Magnetic Department, Woolwich, 1841–57," *Notes and Records: The Royal Society Journal of the History of Science* (2018); Edward Sabine, "On the Periodical Laws Discoverable in the Mean Effects of the Larger Magnetic Disturbances," *Philosophical Transactions of the Royal Society of London* 142 (1852): 103–124.

16 Clark, *The Sun Kings,* 69; Leon Golub and Jay Pasachoff, *The Sun* (Reaktion Books, 2017), 149; Minutes of Council, vol. IV, November 1856 to June 1866, and Friday November 11, 1859, Royal Astronomical Society Archives.

17 See Jürgen Osterhammel, *The Transformation of the World: A Global History of the Nineteenth Century* (Princeton University Press, 2015); Simone M. Müller, *Wiring the World: The Social and Cultural Creation of Global Telegraph Networks* (Columbia University Press, 2016).

18 Osterhammel, *The Transformation,* 649.

19 Clark, *The Sun Kings,* 131.

20 Clark, *The Sun Kings,* 136; Iwan Rhys Morus, "The Sciences," in *A Companion to Nineteenth-Century Britain,* ed. Chris Williams (Wiley, 2004), 460.

21 Annie S. D. Maunder and E. Walter Maunder, *The Heavens and Their Story* (Robert Culley, 1910), 103.

22 Maunder and Maunder, *The Heavens,* 105.

23 Clark, *The Sun Kings,* 153.

24 Mary Bruck, *Women in Early British and Irish Astronomy* (Springer, 2009), 157.

25 D. E. Packer, "New Photographic Discovery—The Solar Corona Photographed in Daylight," *Popular Astronomy* 3, no. 7 (March 1896); E. Miller, "The Corona of the Sun as Seen by E. Miller, May 3d, 1899," *Popular Astronomy* 8, no. 2 (1899); W. B. Featherstone, "Photographing the Corona," *Popular Astronomy* 8, no. 5 (1900); "Origin of a Disturbed Region Observed in the Corona of 1901, May 17–18," *Popular Astronomy* 10, no. 8 (1902); Walter M. Mitchell, "The Auroral Phenomenon of August 21, 1903," *Popular Astronomy* 11, no. 10 (1903); William J. S.

Lockyer, "Magnetic Storms, Aurorae, and Solar Phenomena." *Popular Astronomy* 11, no. 10 (1903); E. Walter Maunder, "The Solar Origin of Terrestrial Magnetic Disturbances," *Popular Astronomy* 13, no. 2 (1905); Hisashi Hayakawa et al., "The Extreme Space Weather Event in 1903 October / November: An Outburst from the Quiet Sun," *Astrophysical Journal Letters* 897, no. 1 (2020): L10.

26 Clark, *The Sun Kings*, 162; Maunder and Maunder, *The Heavens*, 190.

4. Solar Storm and Existential Risk

1 Michael B. Oren, *Six Days of War: June 1967 and the Making of the Modern Middle East* (Presidio Press, 2003), 62; Dale Andrade, *America's Last Vietnam Battle: Halting Hanoi's 1972 Easter Offensive* (University Press of Kansas, 1995); Eric Schlosser, *Command and Control* (Penguin, 2013), 356; Wilmot Hess, *The Radiation Belt and Magnetosphere* (Blaisdell, 1968), 186.

2 Delores J. Knipp et al., "The May 1967 Great Storm and Radio Disruption Event," *Space Weather* 14, no. 9 (2016): 616; P. J. Citrone, "Paper Session I-B—USAF Space Weather Support," The Space Congress Proceedings, http://commons.erau.edu/space-congress-proceedings/proceedings-1995-32nd/april-25-1995/20.

3 Knipp et al., "May 1967 Great Storm," 627; M. Messerotti, "Solar Radio Observations in Trieste Contributed to Avert a Nuclear War in 1967," *Journal of the Italian Astronomical Society* (2023): 40.

4 J. R. McNeill and Peter Engelke, *The Great Acceleration: An Environmental History of the Anthropocene since 1945* (Harvard University Press, 2016).

5 Matteo Leone and Nadia Robotti, "Guglielmo Marconi, Augusto Righi and the Invention of Wireless Telegraphy," *European Physical Journal* 46, no. 1 (2021): 19; Timothy C. Campbell, *Wireless Writing in the Age of Marconi* (University of Minnesota Press, 2006), xi.

6 C. S. Gillmore, "Threshold to Space: Early Studies of the Ionosphere," in *Space Science Comes of Age*, ed. P. Hanle and V. del Chamberlain (Smithsonian Institution Press, 1981), 101; J. E. Taylor, "Characteristics of Electric Earth-Current Disturbances, and Their Origin," *Proceedings of the Royal Society of London* 71, no. 467–476 (1903): 225–227; J. E. Taylor, "Wireless Telegraphy in Relation to Interferences and Perturbations," *Journal of the Institution of Electrical Engineers* 47, no. 208 (1911): 120; Priya Satia, "War, Wireless, and Empire: Marconi and the British Warfare State, 1896–1903," *Technology and Culture* 51, no. 4 (2010): 829–853; Aitor Anduaga, *Wireless and Empire: Geopolitics, Radio Industry, and Ionosphere in the British Empire, 1918–1939* (Oxford University Press, 2009).

7 Hisashi Hayakawa et al., "The Extreme Solar and Geomagnetic Storms on 20–25 March 1940," *Monthly Notices of the Royal Astronomical Society* (2021); Becaja Caldwell, Eoin McCarron, and Seth Jonas, "An Abridged History of Federal Involvement in Space Weather Forecasting," *Space Weather* 15, no. 10 (2017): 1224; J. J. Love and P. Coïsson, "The Geomagnetic Blitz of September 1941," *Eos* 97 (2016), https://doi.org/10.1029/2016EO059319.

8 Robert McMahon, "US National Security from Eisenhower to Kennedy," in *The Cambridge History of the Cold War*, vol. 1, ed. Melvyn Leffler and Odd Arne Westad (Cambridge University Press, 2010), 310.

9 Knipp et al., "May 1967 Great Storm," 616.

10 Elisheva R. Coleman, Samuel A. Cohen, and Michael S. Mahoney, "Greek Fire: Nicholas Christofilos and the Astron Project in America's Early Fusion Program," *Journal of Fusion Energy* 30 (2011): 241.

11 Mark Wolverton, *Burning the Sky: Operation Argus and the Untold Story of the Cold War Nuclear Tests in Outer Space* (Abrams, 2018), 23; Toshihiro Higuchi, *Political Fallout: Nuclear Weapons Testing and the Making of a Global Environmental Crisis* (Stanford University Press, 2020), 175.

12 Caldwell, McCarron, and Jonas, "An Abridged History," 1228; Gian Luca Delzanno, Joseph E. Borovsky, and Evgeny Mishin, "Active Experiments in Space: Past, Present, and Future," *Frontiers in Astronomy and Space Sciences* 7 (2020): 5; T. I. Gombosi et al., "Anthropogenic Space Weather," *Space Science Reviews* 212, no. 3 (2017): 1006.

13 J. W. Findlay, "West Ford and the Scientists," *Proceedings of the IEEE* 52, no. 5 (1964): 455–460.

14 "Protests Continue Abroad," *New York Times,* October 23, 1961, 12; "Space Needle Plan Is Scored by Russia," *New York Times,* September 22, 1961, 2; John Finney, "Needle Antennas Stir Space Furor," *New York Times,* July 30, 1961, 48. See Lisa Ruth Rand, "Falling Cosmos: Nuclear Reentry and the Environmental History of Earth Orbit," *Environmental History* 24, no. 1 (2019): 78–103.

15 R. M. Winglee and E. M. Harnett, "Radiation Mitigation at the Moon by the Terrestrial Magnetosphere," *Geophysical Research Letters* 34, no. 21 (2007); Caldwell, McCarron, and Jonas, "An Abridged History," 1228.

16 Ying D. Liu et al., "Observations of an Extreme Storm in Interplanetary Space Caused by Successive Coronal Mass Ejections," *Nature Communications* 5, no. 1 (2014): 1.

17 Delores J. Knipp et al., "On the Little-Known Consequences of the 4 August 1972 Ultra-Fast Coronal Mass Ejecta," *Space Weather* 16, no. 11 (2018): 1641; Office of the Chief of Naval Operations, *The Mining of North Vietnam, 8 May 1972 to 14 January 1973* (US Department of the Navy, 1975).

18 V. D. Albertson and J. M. Thorson, "Power System Disturbances during a K-8 Geomagnetic Storm: August 4, 1972," *IEEE Transactions on Power Apparatus and Systems* 4 (1974): 1025.

19 Joe Allen et al., "Effects of the March 1989 Solar Activity," *Eos* 70, no. 46 (1989): 1479; D. H. Boteler, "A 21st Century View of the March 1989 Magnetic Storm," *Space Weather* 17, no. 10 (2019); *The 1989 System Disturbances Disturbance Analysis Working Group Report,* North American Electric Reliability Corporation (1990), 37; Sebastien Guillon et al., "A Colorful Blackout," *IEEE Power and Energy Magazine* 14, no. 6 (2016): 59–71; Léonard Bolduc, "GIC Observations and Studies in the Hydro-Québec Power System," *Journal of Atmospheric and Solar-Terrestrial Physics* 64, no. 16 (2002): 1793.

20 Boteler, "A 21st Century View," 1438; *The 1989 System Disturbances,* 42; Space Studies Board and National Research Council, *Severe Space Weather Events* (National Academies Press, 2009), 18; L. V. Medford et al., "Transatlantic Earth Potential Variations during the March 1989 Magnetic Storms," *Geophysical Research Letters* 16, no. 10 (1989): 1145.

21 *The 1989 System Disturbances,* 37.

22 P. R. Barnes et al., *Electric Utility Industry Experience with Geomagnetic Disturbances,* Report no. ORNL-6665, Oak Ridge National Laboratory, 1991, https://info.ornl.gov/sites/publications/Files/Pub57561.pdf.

23 Boteler, "A 21st Century View," 1438; Space Studies Board et al., *Severe Space Weather Events,* 1; Caldwell, McCarron, and Jonas, "An Abridged History," 1232; L. Van der Zel, *Monitoring and Mitigation of Geomagnetically Induced Currents,* Electric Power Research Institute (EPRI) report no. 1015938 (December 2008), 1–2; John Kappenman, *Geomagnetic Storms and Their Impacts on the US Power Grid* (Metatech, 2010), 1–2.

24 Paul Cannon et al., *Extreme Space Weather* (Royal Academy of Engineering, 2013), 30, 33; Space Studies Board et al., *Severe Space Weather Events,* 2.

25 Space Studies Board et al., *Severe Space Weather Events,* 82; Cannon et al., *Extreme Space Weather,* 32; *Report of the Commission to Assess United States National Security Space Management and Organization,* March 28, 2001, https://www.govinfo.gov/content/pkg/CHRG-107shrg81578/html/CHRG-107shrg81578.htm.

26 J. P. Eastwood et al., "The Economic Impact of Space Weather: Where Do We Stand?," *Risk Analysis* 37, no. 2 (2017): 211; David Webb and Joe Allen, "Spacecraft and Ground Anomalies Related to the October-November 2003 Solar Activity," *Space Weather* 2, no. 3 (2004).

27 Ying D. Liu et al., "Observations of an Extreme Storm," 2.

28 Troy Cline, "Interview of Dr. John Kappenman, October 1, 2013," Capital Reporting Company, https://smd-cms.nasa.gov/wp-content/uploads/2023/09/SWLH_John_Kappenman_transcript.pdf.

29 Hiroyuki Maehara et al., "Superflares on Solar-Type Stars," *Nature* 485, no. 7399 (May 2012): 478–481F; Fusa Miyake et al., "A Signature of Cosmic-Ray Increase in AD 774–775 from Tree Rings in Japan," *Nature* 486, no. 7402 (2012): 242; Nicolas Brehm et al., "Tree-Rings Reveal Two Strong Solar Proton Events in 7176 and 525 BCE," *Nature Communications* 13, no. 1 (2022): 1196; Ulf Büntgen et al., "Tree Rings Reveal Globally Coherent Signature of Cosmogenic Radiocarbon Events in 774 and 993 CE," *Nature Communications* 9, no. 1 (2018): 3605; Valeriy Vasilyev et al., "Sun-Like Stars Produce Superflares Roughly Once per Century," *Science* 386 (2024): 1301–1305.

30 Space Studies Board et al., *Severe Space Weather Events,* 78; Greg M. Lucas et al., "A 100-Year Geoelectric Hazard Analysis for the US High-Voltage Power Grid," *Space Weather* 18, no. 2 (2020): e2019SW002329; Edward W. Cliver et al., "Extreme Solar Events," *Living Reviews in Solar Physics* 19, no. 1 (2022): 30.

31 North America Electric Reliability Corporation (NERC), *Effects of Geomagnetic Disturbances on the Bulk Power System* (2012), 46, Patrick Picher et al., "Study of the Acceptable DC Current Limit in Core-Form Power Transformers," *IEEE Transactions on Power Delivery* 12, no. 1 (1997): 257–265.

32 Mike Hapgood et al., "Development of Space Weather Reasonable Worst-Case Scenarios for the UK National Risk Assessment," *Space Weather* 19, no. 4 (2021).

33 James Jinks and Peter Hennessy, *The Silent Deep: The Royal Navy Submarine Service since 1945* (Penguin, 2015).

34 Graham Spinardi, *From Polaris to Trident: The Development of US Fleet Ballistic Missile Technology* (Cambridge University Press, 1994), 82.

35 Gombosi et al., "Anthropogenic Space Weather," 1025.

Part I Conclusion

1 Marlin Schuetz et al., "Optical SETI Observations of the Anomalous Star KIC 8462852," *Astrophysical Journal Letters* 825, no. 1 (2016): L5; Freeman Dyson, "Search for Artificial Stellar Sources of Infrared Radiation," *Science* 131, no. 3414 (1960): 1667.

2 Tabetha Boyajian et al., "Planet Hunters IX. KIC 8462852–Where's the Flux?," *Monthly Notices of the Royal Astronomical Society* 457, no. 4 (2016): 3988; Miguel Martinez et al., "Orphaned Exomoons," *Monthly Notices of the Royal Astronomical Society* 489, no. 4 (2019): 5119.

Part II. Venus

1 Intergovernmental Panel on Climate Change (IPCC), "Summary for Policymakers," in Working Group 1, IPCC, *Climate Change 2021: The Physical Science Basis* (Cambridge University Press, 2021), 4.

2 Duncan H. Forgan, *Solving Fermi's Paradox* (Cambridge University Press, 2019), 5.

3 Forgan, *Solving Fermi's Paradox,* 7.

4 Forgan, *Solving Fermi's Paradox,* 11.

5 Forgan, *Solving Fermi's Paradox,* 47.

6 Suzanne E. Smrekar, Ellen R. Stofan, and Nils Mueller, "Venus: Surface and Interior," in *Encyclopedia of the Solar System,* 3rd ed., ed. Tilman Spohn, Doris Breuer, and Torrence Johnson (Elsevier, 2014), 324; Tereza Constantinou, Oliver Shorttle, and Paul B. Rimmer, "A Dry Venusian Interior Constrained by Atmospheric Chemistry," *Nature Astronomy* (2024).

7 David A. Rothery, Neil McBride, and Iain Gilmour, *An Introduction to the Solar System,* 3rd ed. (Cambridge University Press, 2018), 188; Frederic W. Taylor and Donald M. Hunten, "Venus: Atmosphere," in Spohn, Breuer, and Johnson, *Encyclopedia of the Solar System,* 321.

8 Leonard G. Wilson, "Uniformitarianism and Actualism," in *The History of Science and Religion in the Western Tradition: An Encyclopedia,* ed. Gary B. Ferngren, Edward J. Larson, and Darrel W. Amundsen (Routledge, 2003), 493. See Peter Braunstein and Michael William Doyle, eds., *Imagine Nation: The American Counterculture of the 1960's and 70's* (Routledge, 2013).

5. Measuring the Universe

1 James Cook, *The Journals of Captain Cook* (Penguin, 2000), 40.

2 William Sheehan and John E. Westfall, *The Transits of Venus* (Prometheus Books, 2004), 169.

3 Cook, *The Journals,* 164; Anne Salmond, *Two Worlds: First Meetings between Māori and Europeans, 1642–1772* (University of Hawai'i Press, 1992), 250.

4 J. L. Heilbron, *Galileo* (Oxford University Press, 2010), 167.

5 Paolo Palmieri, "Galileo and the Discovery of the Phases of Venus," *Journal for the History of Astronomy* 32, no. 2 (2001): 110; Heilbron, *Galileo,* 171.

6 Sheehan and Westfall, *The Transits of Venus,* 64.

7 Allan Chapman, "Horrocks, Crabtree and the 1639 Transit of Venus," *Astronomy & Geophysics* 45, no. 5 (2004): 7.

8 Wilbur Applebaum, *Venus Seen on the Sun: The First Observation of a Transit of Venus by Jeremiah Horrocks* (Brill, 2012), xi.

9 Sheehan and Westfall, *The Transits of Venus,* 14, 224; S. Vince, *A Treatise on Practical Astronomy* (J. Archdeacon Printer, 1790), 191.

10 David K. Love, *Edmond Halley: The Many Discoveries of the Most Curious Astronomer Royal* (Prometheus, 2023), 68; Edmond Halley, "A New Method of Determining the Parallax of the Sun," *Philosophical Transactions of the Royal Society of London* 29 (1716): 454–464.

11 Alfred Crosby, *The Measure of Reality: Quantification and Western Society, 1250–1600* (Cambridge University Press, 1997), 19. See also Jessica Marie Otis, *By the Numbers: Numeracy, Religion, and the Quantitative Transformation of Early Modern England* (Oxford University Press, 2024).

12 Sheehan and Westfall, *The Transits of Venus,* 142.

13 Sheehan and Westfall, *The Transits of Venus,* 152; Patrick Moore, *The Planet Venus* (Faber and Faber, 1959), 115.
14 Samuel Dunn, "A Determination of the Exact Moments of Time," *Philosophical Transactions of the Royal Society of London* (London, 1770); Sheehan and Westfall, *The Transits of Venus,* 157.
15 Sheehan and Westfall, *The Transits of Venus,* 224.
16 Jessica Ratcliff, *The Transit of Venus Enterprise in Victorian Britain* (University of Pittsburgh Press, 2008), 149; Alex Soojung-Kim Pang, *Empire and the Sun* (Stanford University Press, 2002).
17 See, for example, Aileen Fyfe and Bernard Lightman, *Science in the Marketplace: Nineteenth-Century Sites and Experiences* (University of Chicago Press, 2019).

6. Venus as a Changing Earth

1 Stephen Gillett, "Venus," in *The Greenwood Encyclopedia of Science Fiction and Fantasy,* ed. Gary Westfahl (Greenwood Press, 2005).
2 Inge Keil, "Johann Wiesel's Telescopes and His Clientele," in *From Earth-Bound to Satellite,* ed. A. D. Morrison-Low et al. (Brill, 2011), 21; Helge Kragh, *The Moon That Wasn't* (Birkhauser, 2008), 5; Johannes Kepler, *Kepler's Conversation with Galileo's Sidereal Messenger,* trans. Edward Rosen (Johnson Reprint, 1965), 47; David Dunér, "Venusians: The Planet Venus in the 18th-Century Extraterrestrial Life Debate," *Journal of Astronomical Data* 19 (2013): 147.
3 Kragh, *The Moon,* 150; T. W. Webb, *Celestial Objects for Common Telescopes* (Longmans, Green, and Co., 1893), 52.
4 Kragh, *The Moon,* 142; Michael J. Crowe, *The Extraterrestrial Life Debate, 1750–1900* (Cambridge University Press, 1986), 251.
5 Francesco Bianchini, *Observations concerning the Planet Venus,* trans. Sally Beaumont and Peter Fay (Springer, 1996), 56, 93; John Jerome Schroeter [Johann Hieronymus Schröter], "New Observations and Further Proof . . . ," *Philosophical Transactions of the Royal Society* 85 (1795): 169; Luís Tirapicos and Thomas Horst, "Francesco Bianchini (1662–1729) and the Origins of Planetary Globes," *Nuncius* 35, no. 2 (2020): 251.
6 Jean le Rond d'Alembert, "Planete," in *Encyclopédie,* vol. 12 (Neufchastel, 1765), 705; Bianchini, *Observations,* 157; Francesco Bianchini, *Hesperi et Phosphori* (1728) (Linda Hall Library).
7 Agnes Clerke, *The Planet Venus* (Witherby and Co., 1893), 40, 64; William Sheehan and John E. Westfall, *The Transits of Venus* (Prometheus Books, 2004), 139; Fredric Taylor, *The Scientific Exploration of Venus* (Cambridge University Press, 2014), 9.
8 John Jerome Schroeter [Johann Hieronymus Schröter], "Observations on the Atmospheres of Venus and the Moon. . . . ," *Philosophical Transactions of the Royal Society* 82 (1792): 337; William Herschel, "Observations on the Planet Venus," in *The Scientific Papers of Sir William Herschel,* vol. 1 (Royal Society and Royal Astronomical Society, 1912), 441–452; Schroeter [Schröter], "New Observations and Further Proof," 169; Johann Hieronymus Schröter, *Aphroditographische Fragmente, zur genauern Kenntniss des Planeten Venus* (Helmstedt, 1796), 194.
9 Clerke, *The Planet Venus,* 38; Mary Brück, *Agnes Mary Clerke and the Rise of Astrophysics* (Cambridge University Press, 2002).
10 Katrin Kleemann, *A Mist Connection: An Environmental History of the Laki Eruption of 1783 and Its Legacy* (De Gruyter, 2023), 187; Robert R. Herrick and Scott Hensley, "Surface

Changes Observed on a Venusian Volcano during the Magellan Mission," *Science* (2023): eabm7735.

11 Patrick Moore, *The Planet Venus* (Faber and Faber, 1959), 48, 53, 75; David Harry Grinspoon, *Venus Revealed: A New Look Below the Clouds of Our Mysterious Twin Planet* (Perseus, 1997), 44, 95.

12 Crowe, *The Extraterrestrial Life Debate,* 73; Immanuel Kant, *Universal Natural History and Theory of the Heavens,* trans. Stanley L. Jaki (Scottish Academic Press, 1981), 144.

13 Camille Flammarion, *La pluralité des mondes habités* (Mallet-Bachelier, 1862), 418; Richard Proctor, *Our Place among Infinities,* 2nd ed. (Henry S. King and Co., 1876), 52.

14 Clarence J. Glacken, *Traces on the Rhodian Shore* (University of California Press, 1967), 575; Bernard le Bovier de Fontenelle, *Conversations on the Plurality of Worlds,* trans. William Gardiner (A. Bettesworth, 1715), 101; Kant, *Universal Natural History.*

15 Dunér, "Venusians," 150; Vincent Roy-Di Piazza, "Ghosts from Other Planets," *Annals of Science* 77, no. 4 (2020): 469.

16 Crowe, *The Extraterrestrial Life Debate,* 321, 377, 394, 471; Thomas Dick, *Celestial Scenery, or the Wonders of the Planetary System Displayed* (Sumner and Goodman, 1848), 270; Auguste Comte, "Cours de philosophie positive," from *The Extraterrestrial Life Debate: Antiquity to 1915, a Source Book,* ed. Michael J. Crowe (University of Notre Dame Press, 2008), 316; Enoch Fitch Burr, *Ecce Coelum; or, Parish Astronomy* (1867); Richard Proctor, *Other Worlds than Ours* (1870).

17 Clerke, *The Planet Venus,* 25.

18 Clerke, *The Planet Venus,* 30; Crowe, *The Extraterrestrial Life Debate,* 352.

19 "Folder 3. Venus Drawings, 1909," box 1 large, Working Papers, 1883–1916, MS.1, Percival Lowell Papers, Lowell Observatory Archive; Percival Lowell, "Detection of Venus' Rotation Period and of the Fundamental Physical Features of the Planet's Surface," *Popular Astronomy* 4, no. 6 (1896): 281; William Sheehan, *Planets and Perception: Telescopic Views and Interpretations, 1609–1909* (University of Arizona Press, 1988), 230; E. M. Antoniadi, "Notes of the Rotation Period of Venus," *Royal Astronomical Society Monthly Notices* 58 (1898): 314; Percival Lowell, "The Markings on Venus," *Astronomische Nachrichten* 160 (1902), 130.

20 Charles Edward Housden, *Is Venus Inhabited?* (Longmans, Green, 1915), 15.

21 Housden, *Is Venus Inhabited?,* 36; Charles Edward Housden, *The Riddle of Mars* (Longmans, Green, 1914).

22 Clerke, *The Planet Venus,* 33; Moore, *The Planet Venus,* 129.

23 Brian Stableford, *Science Fact and Science Fiction: An Encyclopedia* (Routledge, 2006); John Ellard Gore, *Worlds of Space* (London, 1894), 12; W. G. Colgrove, *A Ready Reference Handbook of the Solar System* (Colgrove, 1933); Svante Arrhenius, *Destinies of the Stars* (G. P. Putnam's Sons, 1918), 252, 254.

24 Moore, *The Planet Venus,* 50, 67, 126; Carl Sagan, "Venus," *International Science and Technology* (March 1963): 7.

25 Carl Sagan, "The Planet Venus," *Science* 133 (1961): 849; Daniel Yergin, *The Prize: The Epic Quest for Oil, Money and Power* (Free Press, 2008).

7. Worlds in Collision

1 Patrick Moore, *The Planet Venus* (Faber and Faber, 1959), 99; Michael J. Crowe, *The Extraterrestrial Life Debate, 1750–1900* (Cambridge University Press, 1986), 204; Gruithuisen, "Bewohnern," 1, 4 (1833), 53.

2 Moore, *The Planet Venus,* 102; Agnes Clerke, *The Planet Venus* (Witherby and Co., 1893), 18; John A. Paterson, "The Astronomy of 1897," *Transactions of the Royal Astronomical Society of Canada* (1897): 108.

3 David Harry Grinspoon, *Venus Revealed: A New Look Below the Clouds of Our Mysterious Twin Planet* (Perseus, 1997), 44, 127; Fredric Taylor, *The Scientific Exploration of Venus* (Cambridge University Press, 2014),12, 76, 90.

4 Immanuel Velikovsky, *Worlds in Collision* (Paradigma, 2009), 19.

5 T. J. J. See, "Sirius in Ancient Times," *Popular Astronomy,* 2, no. 5 (1895): 193.

6 Velikovsky, *Worlds in Collision,* 41, 117.

7 Velikovsky, *Worlds in Collision,* 125, 135.

8 Velikovsky, *Worlds in Collision,* 66, 75, 83, 110, 371.

9 Velikovsky, *Worlds in Collision,* 149, 211, 261, 299.

10 Velikovsky, *Worlds in Collision,* 14, 360; V. Bargmann and Lloyd Motz, "On the Recent Discoveries concerning Jupiter and Venus," *Science* 138, no. 3547 (1962): 1350; Walter Sullivan, "Science: 'The Velikovsky Affair,'" *New York Times,* October 2, 1966; Stephen M. Hudspeth, "Creativity and Acceptance in the Sciences." *Yale Scientific Magazine* 41, no. 7 (April 1967): 30; Immanuel Velikovsky, "My Challenge to Conventional Views in Science," *Pensée* 4, no. 2 (1974): 12.

11 Philip Wylie and Edwin Balmer, *When Worlds Collide* (University of Nebraska Press, 1999); Horace Kallen, "Shapley, Velikovsky and the Scientific Spirit." *Pensée* 2, no. 2 (1972): 36.

12 Eric Larrabee, "The Day the Sun Stood Still," *Harper's,* January 1950; Kallen, "Shapley, Velikovsky and the Scientific Spirit," 37.

13 Harlow Shapley to Editorial Department, The Macmillan Company, January 18, 1950; James Putnam to Harlow Shapley, January 24, 1950; Harlow Shapley to James Putnam, January 25, 1950; Immanuel Velikovsky to Harlow Shapley, March 31, 1947; George Brett to Harlow Shapley, February 1, 1950; and Harlow Shapley to George Brett, February 9, 1950; all in Papers of Harlow Shapley, 1906–1966, Papers, c. 1921–1965, box 1A, HUG 4773.10, Harvard University Archives.

14 Kallen, "Shapley, Velikovsky and the Scientific Spirit," 37; Lynn Rose, "The Censorship of Velikovsky's Interdisciplinary Synthesis," *Pensée* 2, no. 2 (1972): 29.

15 Harlow Shapley to T. O. Thackrey, June 6, 1950; Harlow Shapley to H. F. Latham, July 7, 1950; and Harlow Shapley to Ellsworth Huntington, September 16, 1935; all in Papers of Harlow Shapley, 1906–1966, Papers, c. 1921–1965, box 1A, HUG 4773.10, Harvard University Archives.

16 H. S. Latham to Harlow Shapley, June 8, 1950, and Harlow Shapley to H. F. Latham, June 14 1950, both in Papers of Harlow Shapley, 1906–1966, Papers, c. 1921–1965, box 1A, HUG 4773.10, Harvard University Archives; George Kolodiy, "Velikovsky: Paradigms in Collision," *Bulletin of the Atomic Scientists* 31, no. 2 (1975): 37; Alfred De Grazia, "The Scientific Reception System and Dr. Velikovsky," *American Behavioral Scientist* 7, no. 1 (1963): 46; David Stove, "The Scientific Mafia," *Pensée* 2, no. 2 (1972): 7; "A Scientific Approach to Velikovsky," *Yale Scientific Magazine* 41, no. 7 (1967), 5, box 802, folder 9, and Donald Goldsmith, "Preface for Velikovsky-Symposium Volume," box 803, folder 4, both in Seth MacFarlane Collection of the Carl Sagan and Ann Druyan Archive, Library of Congress; Harold Latham, *My Life in Publishing* (E. P. Dutton, 1965).

17 Cecilia Payne-Gaposchkin, "Worlds in Collision," *Popular Astronomy* 58 (1950); Derek York, "Velikovsky: Made-to-Order Wandering Worlds," *The Globe and Mail,* February 16, 1981; Kallen, "Shapley, Velikovsky and the Scientific Spirit," 37; Robert Cowen, "Velikovsky: Tardy Justice," *Christian Science Monitor,* March 12, 1974.

18 Isaac Asimov, "Worlds in Confusion," *Magazine of Fantasy and Science Fiction,* October 1969, 5; Michael Gordin, "The Unseasonable Grooviness of Immanuel Velikovsky," in *Groovy Science,* ed. David Kaiser and W. Patrick McCray (University of Chicago Press, 2016), 210.

19 Asimov, "Worlds in Confusion," 6; De Grazia, "The Scientific Reception System"; "1967: A New Report on the Velikovsky Controversy," *Yale Scientific Magazine* 41, no. 7 (1967): 2.

20 Michael Gordin, *The Pseudoscience Wars: Immanuel Velikovsky and the Birth of the Modern Fringe* (University of Chicago Press, 2012), 9; Thomas Kuhn, *The Structure of Scientific Revolutions* (University of Chicago Press, 2012); Karl Hufbauer, "From Student of Physics to Historian of Science: T. S. Kuhn's Education and Early Career, 1940–1958," *Physics in Perspective* 14 (2012): 459.

21 Gordin, "The Unseasonable Grooviness," 213; Lewis Mumford, *The Myth of the Machine: The Pentagon of Power* (Harcourt Brace Jovanovich, 1970); Audra Wolfe, *Competing with the Soviets: Science, Technology, and the State in Cold War America* (Johns Hopkins University Press, 2013), 3.

22 Norman W. Storer, "The Sociological Context of the Velikovsky Controversy," AAAS Symposium, February 25, 1974; Hudspeth, "Creativity and Acceptance," 30; Rose, "Censorship of Velikovsky's Interdisciplinary Synthesis," 45; Cowen, "Velikovsky: Tardy Justice."

23 Fred Warshofsky, "When the Sky Rained Fire," *Reader's Digest,* December 1975, 235; Kallen, "Shapley, Velikovsky and the Scientific Spirit," 39; James Hazelwood, "Theory of Venus and the Earth," *Oakland Tribune,* February 26, 1974, 2.

24 Grinspoon, *Venus Revealed,* 64; Frederic W. Taylor and Donald M. Hunten, "Venus: Atmosphere," in *Encyclopedia of the Solar System,* 3rd ed., ed. Tilman Spohn, Doris Breuer, and Torrence Johnson (Elsevier, 2014), 307; D. E. Jones, "The Microwave Temperature of Venus," *Planetary and Space Science* 5, no. 2 (1961): 167.

25 Bargmann and Motz, "On the Recent Discoveries," 1351; "A Record of Success," *Pensée,* 2, no. 2 (1972), 11; Hudspeth, "Creativity and Acceptance," 28; Anthony Leviero, "Radio Finds Venus Hotter than 212 F," *New York Times,* June 5, 1956; "Temperature of Venus," *New York Times,* October 11, 1959; "300 F. Reading on Venus Indicates No Life There," *New York Times,* February 21, 1963; "Image of Venus: 'Glowing Earth,'" *New York Times,* February 21, 1963; Immanuel Velikovsky, "H. H. Hess and My Memoranda," *Pensée* 2, no. 3 (1972): 22; William Kellogg and Carl Sagan, "The Atmospheres of Mars and Venus," National Academy of Sciences, 1961; Jet Propulsion Laboratory, *Mariner Mission to Venus* (McGraw-Hill, 1963), 112.

26 Gordin, "The Unseasonable Grooviness," 217; Karl Abraham, "'Never-Proved Wrong' Scientist Makes a New Prediction Here," *Evening Bulletin,* March 24, 1966.

27 *Yale Scientific Magazine* 41, no. 7 (April 1967): 14; "A Record of Success," 42; William Mullen, "The Center Holds," *Pensée* 2, no. 2 (1972): 32; BBC, "Worlds in Collision," January 11, 1973; "The Future of a Publishing Idea," *Pensée* 2, no. 3 (1972), 4; Ian Maciver, "Velikovsky's Critics and Catastrophism," *Pensée* 2, no. 3 (1972), 43; P. P. M. Meincke, "Notes on Dr. Immanuel Velikovsky," *Pensée* 2, no. 3 (1972), 42; "Courses on Velikovsky," *Pensée* 3, no. 1 (1973), 37; "Velikovsky at Princeton," *Pensée* 3, no. 1 (1973), 39; "A Report on the Velikovsky Affair," box 802, folder 12, Seth MacFarlane Collection of the Carl Sagan and Ann Druyan Archive, Library of Congress; Edward Edelson, "The Maverick Doctor Who Leads a Comet Cult," *Sunday News,* March 10, 1974; "Sciences in Collision." *Newsweek,* February 25, 1974; Gordin, "The Unseasonable Grooviness," 212.

28 Stephen L. Talbott to Carl Sagan, November 15, 1972, and Carl Sagan to Stephen L. Talbott, December 18, 1972, both in box 802, folder 9, Seth MacFarlane Collection of the Carl Sagan

and Ann Druyan Archive, Library of Congress; "Velikovsky's Challenge to Science," *Pensée* 4, no. 2 (1974): 23; "The Genesis of a Symposium," *Pensée* 4, no. 2 (1974): 28; "Velikovsky: AAAS Forum for a Mild Collision," *Science* 183, no. 4129 (April 19, 1975): 1059–1062; Edelson, "The Maverick"; Goldsmith, "Preface"; York, "Velikovsky."

29 Carl Sagan, "Not the Knowing but the Finding Out," *Christian Science Monitor,* November 16, 1965; William Poundstone, *Carl Sagan: A Life in the Cosmos* (Holt Paperbacks, 2000); Keay Davidson, *Carl Sagan: A Life* (Wiley, 2000), 12; Carl Sagan, "Direct Contact among Galactic Civilizations by Relativistic Interstellar Spaceflight," *Planetary and Space Science* 11 (1963): 485; Sagan, "On the Origin and Planetary Distribution of Life," *Radiation Research* 15, no. 2 (1961): 174–192; "Yet Another Chapter . . . ," *Pensée* 4, no. 1 (Winter 1973–1974): 57.

30 Velikovsky, "My Challenge to Conventional Views," 14; J. Derral Mulholland, "Movements of Celestial Bodies," AAAS Symposium, February 25, 1974; Gordin, "The Unseasonable Grooviness," 220.

31 Carl Sagan, "An Analysis of 'Worlds in Collision,'" AAAS Symposium, February 25, 1974, box 803, folder 1, Seth MacFarlane Collection of the Carl Sagan and Ann Druyan Archive, Library of Congress; Rupert Wildt, "Note on the Surface Temperature of Venus," *Astrophysical Journal* 91 (1940): 266–268.

32 Joel N. Shurkin, "Blaze of Glory for an Old Man," *Philadelphia Inquirer,* February 27, 1974; Robert Cooke, "Theory on Collision of Planets Sets Up a Few Earthy Ripples, by Jupiter," *Boston Globe,* February 26, 1974; Walter Sullivan, "Writer Collides with Scientists," *New York Times,* February 26, 1974; Miranda Robertson, "Velikovsky in the Open," *Nature* 248 (1974); Adrianne Marcus, "Worlds in Collision: The Struggle at the Science Convention," *Pacific Sun,* March 7–13, 1974.

33 Ruth Velikovsky Sharon, *The Glory and the Torment: The Life of Dr. Immanuel Velikovsky* (Paradigma, 2010), Carl Sagan, "Episode 4: Heaven and Hell," *Cosmos,* PBS, 1980; Robert Jastrow, "Velikovsky, a Star-Crossed Theoretician of the Cosmos," *New York Times,* December 2, 1979.

34 Dorothy Vitaliano, "Geomythology: Geological Origins of Myths and Legends," in *Myth and Geology,* ed. L. Piccardi and W. B. Masse (Geological Society, 2007), 1.

8. Remaking Venus and Earth

1 Carl Sagan, "Extraterrestrial Objects of Possible Biological Interest: Notebook," 1957, and Sagan, *The Radiation Balance of Venus,* Jet Propulsion Laboratory Technical Report no. 32-24, September 15, 1960, both in box 804, folder 4, Seth MacFarlane Collection of the Carl Sagan and Ann Druyan Archive, Library of Congress; Sagan, "The Planet Venus: Recent Observations Shed Light on the Atmosphere, Surface, and Possible Biology of the Nearest Planet," *Science* 133, no. 3456 (1961): 849–858; William L. Laurence, "Venus Mystery: Wild Variance in the Temperature of the Planet Is Reported," *New York Times,* January 7, 1962.

2 Carl Sagan, "The Trouble with Venus," *IAU Symposium* 40 (1969): 118; "Secret Planet Lowers Its Mask: Press Conference of Soviet and Foreign Journalists anent 'Venera-4,'" box 804, folder 4, Seth MacFarlane Collection of the Carl Sagan and Ann Druyan Archive, Library of Congress; Walter Sullivan, "New Study Points to Life on Venus," *New York Times,* December 8, 1964; Ralph Lorenz, *Exploring Planetary Climate* (Cambridge University Press, 2019), 69; Brian Harvey, *Russian Planetary Exploration* (Springer, 2006), 97.

3 Joshua Lederberg, "Exobiology: Approaches to Life beyond the Earth," *Science* 132, no. 3424 (1960): 400; Keay Davidson, *Carl Sagan: A Life* (Wiley, 2000), 89.

4 Dagomar Degroot, "One Small Step for Man, One Giant Leap for Moon Microbes?," *Isis* 114, no. 2 (2023): 272.

5 Audra J. Wolfe, "Germs in Space: Joshua Lederberg, Exobiology, and the Public Imagination, 1958–1964," *Isis* 93, no. 2 (2002): 185; Steven J. Dick, *The Biological Universe* (Cambridge University Press, 1996), 5, 18, 348, 351; Davidson, *Carl Sagan*, 90; Michael Meltzer, *When Biospheres Collide: A History of NASA's Planetary Protection Programs* (NASA, 2011), 24, 38; Joshua Lederberg, "Sputnik + 30," *Journal of Genetics* 66 (1987): 217.

6 Geoffrey Landis, "The Sultan of the Clouds," *Asimov's Science Fiction,* September 2010; Sagan, "The Planet Venus," 857; Harold M. Shmeck Jr., "Habitable Venus Scientist's Goal," *New York Times,* March 27, 1961; Fredric Taylor, *The Scientific Exploration of Venus* (Cambridge University Press, 2014),153, 271, 274; David Harry Grinspoon, *Venus Revealed: A New Look Below the Clouds of Our Mysterious Twin Planet* (Perseus, 1997), 123; Magnus Larsson and Alex Kaiser, "Cloud Ten," in *Inner Solar System: Prospective Energy and Material Resources,* ed. V. Badescu and K. Zacny (Springer, 2015), 452.

7 Sagan, "The Trouble with Venus," 126; Harold Morowitz and Carl Sagan, "Life in the Clouds of Venus?," *Nature* 215, no. 5107 (1967): 1260; Sagan, "The Planet Venus," 857.

8 Martin Beech, *Terraforming: The Creating of Habitable Worlds* (Springer, 2009), 10.

9 J. Williamson [Will Stewart], "Collision Orbit," *Astounding Science Fiction,* July 1942; Beech, *Terraforming,* 7; Taylor, *Scientific Exploration of Venus,* 277.

10 Anne Ehrlich and Paul Ehrlich, *The Population Bomb* (Ballantine Books, 1968), 1; Sabine Höhler, *Spaceship Earth in the Environmental Age, 1960–1990* (Routledge, 2015), 2; Patrick McCray, *The Visioneers: How a Group of Elite Scientists Pursued Space Colonies, Nanotechnologies, and a Limitless Future* (Princeton University Press, 2013), 25; Pierre Desrochers and Christine Hoffbauer, "The Post War Intellectual Roots of the Population Bomb," *Electronic Journal of Sustainable Development* 1, no. 3 (2009): 73.

11 Ehrlich and Ehrlich, *The Population Bomb;* Adrian Berry, "Venus: The Next Frontier?," *Daily Telegraph Magazine,* April 30, 1971; Gerard K. O'Neill, *The High Frontier: Human Colonies in Space* (1977), 15.

12 Joseph Seckbach to Carl Sagan, April 11, 1969, box 804, folder 7, Seth MacFarlane Collection of the Carl Sagan and Ann Druyan Archive, Library of Congress; J. Pournelle, *A Step Farther Out* (Star Books, 1980); A. Berry, *The Next Ten Thousand Years* (Coronet Books, 1976); James Hansen and J. W. Hovenier, "Interpretation of the Polarization of Venus," *Journal of Atmospheric Sciences* 31, no. 4 (1974): 1137–1160; Morowitz and Sagan, "Life in the Clouds of Venus?," 1060; Beech, *Terraforming,* 194; Robert E. Evenson and Douglas Gollin, "Assessing the Impact of the Green Revolution, 1960 to 2000," *Science* 300, no. 5620 (2003): 758.

13 F. R. Gross, "Buoyant Probes into the Venus Atmosphere," *Journal of Spacecraft and Rockets* 3, no. 4 (1966): 582; J. F. Baxter, *Final Report: Buoyant Venus Station Feasibility Study,* vol. 1, NASA report no. CR-66404, 1967; Stanley Sadin and Ronald Frank, "Buoyant Venus Station Requirements," *Journal of Spacecraft and Rockets* 7, no. 8 (1970): 905.

14 Michael Carroll, *Drifting on Alien Winds* (Springer, 2011), 223; Geoffrey A. Landis, *Settling Venus: A City in the Clouds?,* AIAA report no. 2020-4152, ASCEND 2020, November 16–18, 2020.

15 K. M. Kiran Babu and Rajkumar S. Pant, "A Review of Lighter-than-Air Systems for Exploring the Atmosphere of Venus," *Progress in Aerospace Sciences* 112 (2020): 100587;

Dale C. Arney and Christopher A. Jones, *High Altitude Venus Operational Concept (HAVOC): An Exploration Strategy for Venus,* AIAA report no. 2015-4612, AIAA SPACE 2015 Conference and Exposition, August 21–September 2, 2015; Larsson and Kaiser, "Cloud Ten," 461.

16 J. E. Oberg, *New Earths* (New American Library, 1981); S. J. Adelman, "Can Venus Be Transformed into an Earth-Like Planet?," *Journal of the British Interplanetary Society* 35 (1982): 3; P. Sargent, *Venus of Dreams* (Bantam Spectra Books, 1986); Charles Cockell, "Life on Venus," *Planetary and Space Science* 47 (1999): 1488; Geoffrey A. Landis, "Colonization of Venus," paper presented at the Conference on Human Space Exploration, Space Technology & Applications International Forum, Albuquerque, February 2–6, 2003; M. J. Fogg, "The Terraforming of Venus," *Journal of the British Interplanetary Society* 40 (1987): 556; Geoffrey Landis, *Terraforming Venus,* AIAA report no. 2011-7215, AIAA SPACE 2011 Conference & Exposition, September 27–29, 2011, 2.

17 Naomi Oreskes, "The Scientific Consensus on Climate Change," *Science* 306, no. 5702 (2004): 1686. See Spencer R. Weart, *The Discovery of Global Warming,* rev. expanded ed. (Harvard University Press, 2008).

18 William Kellogg, "Predicting the Climate," in *Man's Impact on the Climate,* ed. William Henry Matthews, William Kellogg, and G. D. Robinson (MIT Press, 1971).

19 Weart, *Discovery of Global Warming,* 73; Matthew Shindell, *The Life and Science of Harold C. Urey* (University of Chicago Press, 2019), 139.

20 Jean Jouzel, "A Brief History of Ice Core Science over the Last 50 Yr," *Climate of the Past* 9, no. 6 (2013): 2525–2547; Henry Nielsen and Kristian Hvidtfeldt Nielsen, *Camp Century: The Untold Story of America's Secret Arctic Military Base under the Greenland Ice* (Columbia University Press, 2021), 10.

21 Willi Dansgaard, *Frozen Annals: Greenland Ice Sheet Research* (Niels Bohr Institute, 2005).

22 Willi Dansgaard et al., "One Thousand Centuries of Climatic Record from Camp Century on the Greenland Ice Sheet," *Science* 166 (1969): 377; James Woodward, *The Ice Age: A Very Short Introduction* (Oxford University Press, 2014), 125

23 Yuk Yung to Carl Sagan, October 14, 1981, box 805, folder 4, Seth MacFarlane Collection of the Carl Sagan and Ann Druyan Archive, Library of Congress; Yuk L. Yung and W. B. DeMore, "Photochemistry of the Stratosphere of Venus: Implications for Atmospheric Evolution," *Icarus* 51, no. 2 (1982): 199–247; Lawrence Colin et al., "Future Exploration of Venus," in *Comparative Climatology of Terrestrial Planets,* ed. Stephen J. Mackwell et al. (University of Arizona Press, 2014).

24 Leah Aronowsky, "Gas Guzzling Gaia, or: A Prehistory of Climate Change Denialism," *Critical Inquiry* 47, no. 2 (2021): 308; James Lovelock, *Gaia: A New Look at Life on Earth* (Oxford University Press, 1979), 10.

25 James Hansen, "Climatic Change: Understanding Global Warming," box 660, folder 4, Seth MacFarlane Collection of the Carl Sagan and Ann Druyan Archive, Library of Congress; Aronowsky, "Gas Guzzling Gaia," 313.

26 James Hansen, *Storms of My Grandchildren* (Bloomsbury, 2008); Nathaniel Rich, *Losing Earth: A Recent History* (Farrar, Straus and Giroux, 2019), 30.

27 Grinspoon, *Venus Revealed,* 84.

28 James Hansen et al., "Climate Impact of Increasing Atmospheric Carbon Dioxide," *Science* 213, no. 4511 (1981): 958; Rich, *Losing Earth,* 31.

29 National Research Council (NRC) et al., "Carbon Dioxide and Climate: A Scientific Assessment" (National Academies Press, 1979), 2, 6.

30 NRC et al., "Carbon Dioxide and Climate," 6; S. C. Sherwood et al., "An Assessment of Earth's Climate Sensitivity Using Multiple Lines of Evidence," *Reviews of Geophysics* 58, no. 4 (2020): e2019RG000678.

31 John Noble Wilford, "Craft to Study Heat Effect of Venus," *New York Times,* May 22, 1978; "Venus Shot," *U.S. News,* May 29, 1978; John Noble Wilford, "'Pioneers' to Begin Probing," *New York Times,* November 28, 1978; Bill Densmore, Associated Press, December 14, 1978; John Noble Wilford, "Space Probe Supports Venus 'Greenhouse' Theory," *New York Times,* December 15, 1978; Thomas O'Toole, "Fiery Heat, Fierce Winds Keep Venus a 'Hell in the Heavens,'" *Washington Post,* December 15, 1978; Walter Sullivan, "Probes Show Venus in Terrifying Light," *New York Times,* January 25, 1979; Robert C. Cowen, "Space Probes Unravel Venus's 'Greenhouse' Mystery," *Christian Science Monitor,* April 10, 1981; Tom Murphy, "Scientists Describe Venus Environment as Hellish," Associated Press, November 6, 1981.

32 Robert Strand, United Press International, December 4, 1980; Walter Sullivan, "Study Finds Warming Trend That Could Raise Sea Levels," *New York Times,* August 22, 1981; "Heating Up the Atmosphere," *New York Times,* August 29, 1981; James Morton Turner and Andrew Isenberg, *The Republican Reversal* (Harvard University Press, 2018), 6.

33 "Statement of Dr. Carl Sagan, Laboratory for Planetary Studies, Cornell University," Joint Hearing on Carbon Dioxide and the Greenhouse Effect; US House of Representatives, February 28, 1984; John Perry, "Much Ado about CO_2," *Parade* (1984).

34 Wm A Mac Grefor to Carl Sagan, October 3, 1988, box 661, folder 2 Seth MacFarlane Collection of the Carl Sagan and Ann Druyan Archive, Library of Congress; Crispin Tickell, "The Experiment That Could Become Too Hot to Handle," *The Times* (London), August 17, 1982.

35 Jerry Grey, "We're Not Condemned to Carbon-Based Fuels," *New York Times,* May 26, 1988; Bill Green, "Earth to NASA," *New York Times,* August 27, 1989; "Venus May Tell Us How to Escape Its Fate," *New York Times,* September 21, 1989.

36 Rich, *Losing Earth,* 165.

37 Rolf Müller, "A Brief History of Stratospheric Ozone Research," *Meteorologische Zeitschrift* 18, no. 1 (2009): 5.

38 K. Madhava Sarma and Stephen O. Andersen, *Protecting the Ozone Layer: The United Nations History* (Earthscan , 2002), 8; Franck Montmessin et al., "A Layer of Ozone Detected in the Nightside Upper Atmosphere of Venus," *Icarus* 216, no. 1 (2011): 82; Steven C. Wofsy and Michael B. McElroy, "HOx, NOx, and ClOx: Their Role in Atmospheric Photochemistry," *Canadian Journal of Chemistry* 52, no. 8 (1974): 1582.

39 Mario J. Molina and F. Sherwood Rowland, "Stratospheric Sink for Chlorofluoromethanes: Chlorine Atom-Catalysed Destruction of Ozone," *Nature* 249, no. 5460 (1974): 810; Paul W. Barnes et al., "Interactive Effects of Changes in UV Radiation and Climate on Terrestrial Ecosystems, Biogeochemical Cycles, and Feedbacks to the Climate System," *Photochemical & Photobiological Sciences* (2023): 42.

40 S. A. Abbasi and Tasneem Abbasi, *Ozone Hole: Past, Present, Future* (Springer, 2017).

Part II Conclusion

1 Janusz J. Petkowski et al., "Astrobiological Potential of Venus Atmosphere Chemical Anomalies and Other Unexplained Cloud Properties," *Astrobiology* 24, no. 4 (2024): 343.

2 Jane S. Greaves et al., "Phosphine Gas in the Cloud Decks of Venus," *Nature Astronomy* 5, no. 7 (2021): 655; G. L. Villanueva et al., "No Evidence of Phosphine in the Atmosphere of Venus from Independent Analyses," *Nature Astronomy* 5, no. 7 (2021): 631–635.

3 Fredric Taylor, *The Scientific Exploration of Venus* (Cambridge University Press, 2014),118, 219.

Part III. Moon

1 "Loon." Loon LLC. Available at: https://loon.com.

2 Oliver Morton, *The Moon: A History for the Future* (Economist Books, 2019), 21.

3 Morton, *The Moon,* 21; Matija Ćuk et al., "Tidal Evolution of the Moon from a High-Obliquity, High-Angular-Momentum Earth," *Nature* 539 (2016): 402.

9. Muses of the Moon

1 Gabriella Bernardi, *The Unforgotten Sisters: Female Astronomers and Scientists before Caroline Herschel* (Springer, 2016), 98; Enola Pellegrini, "Maria Clara Eimmart (1676–1707), une femme astronome à Nuremberg," *Arts, civilisation et histoire de l'Europe* 20 (2023): 12.

2 Diane Woodin, "Visions of Urania: Women, Art, and Astronomy in Eighteenth-Century Europe" (PhD diss., University of North Carolina, 2018), 30; Gabriella Bernardi, "Domestic Astronomy in the Seventeenth and Eighteenth Centuries," in *The Palgrave Handbook of Women and Science since 1660,* ed. Claire G. Jones, Alison E. Martin, and Alexis Wolf (Palgrave, 2022), 270; Sabina Lessmann, "Susanna Maria von Sandrart: Women Artists in 17th-Century Nuremberg," *Woman's Art Journal* 14 (1993): 10.

3 David Rothery, *Moons: A Very Short Introduction* (Oxford University Press, 2015), 23.

4 Maggie Aderin-Pocock, *The Book of the Moon: A Guide to Our Closest Neighbor* (Abrams Image, 2019), 24.

5 Alexandra Loske and Robert Massey, *Moon: Art, Science, Culture* (Octopus, 2018); Edgar Williams, *Moon: Nature and Culture* (Reaktion Books, 2014); Scott L. Montgomery, *The Moon and the Western Imagination* (University of Arizona Press, 1999).

6 Ewen Whitaker, *Mapping and Naming the Moon* (Cambridge University Press, 2003), 13.

7 Oliver Morton, *The Moon: A History for the Future* (Economist Books, 2019), 43.

8 Leonardo da Vinci, *The Codex Leicester: Notebook of a Genius* (Powerhouse, 2001), 34; Johannes Kepler, *Kepler's Somnium: The Dream, or Posthumous Work on Lunar Astronomy* (Courier, 2003); Svante Arrhenius, *Destinies of the Stars* (G. P. Putnam's Sons, 1918), 244.

9 Galileo Galilei, *The Starry Messenger* (Venice, 1610); Bill Leatherbarrow, *The Moon* (Reaktion Books, 2018), 28; William Sheehan and Thomas Dobbins, *Epic Moon: A History of Lunar Exploration in the Age of the Telescope* (Willmann-Bell, 2001), 7.

10 J. L. Heilbron, *Galileo* (Oxford University Press, 2010), 153; Michael J. Crowe, *The Extraterrestrial Life Debate, 1750–1900* (Cambridge University Press, 1986), 84.

11 John Wilkins, *The Discovery of a World in the Moone* (London, 1638); Michael Rawson, "Discovering the Final Frontier: The Seventeenth-Century Encounter with the Lunar Environment," *Environmental History* 20 (2015): 199; Francis Godwin, *The Man in the Moone, or a discourse of a voyage thither* (London, 1657).

12 Heilbron, *Galileo,* 200.

13 Rebekah Higgitt, "Longitude," in *Encyclopedia of the History of Science,* https://doi.org/10.34758/esnh-7544; Richard de Grijs, "A (Not So) Brief History of Lunar Distances: Lunar

Longitude Determination at Sea before the Chronometer," *arXiv preprint,* arXiv:2007.14504 (2020).

14 Sheehan and Dobbins, *Epic Moon,* 15, 20; Dava Sobel, *Longitude* (Bloomsbury, 2010), 26.

15 Morton, *The Moon,* 48; Sheehan and Dobbins, *Epic Moon,* 36.

16 Leatherbarrow, *The Moon,* 34; William Sheehan, *Planets and Perception: Telescopic Views and Interpretations, 1609–1909* (University of Arizona Press, 1988), 44; Sheehan and Dobbins, *Epic Moon,* 19.

17 Christiaan Huygens, *The Celestial Worlds Discovered: Or, Conjectures Concerning the Inhabitants, Plants and Productions of the Worlds in the Planets* (Timothy Childe, 1698), 3, 110, 132; Joseph Ashbrook, *The Astronomical Scrapbook* (Cambridge University Press, 1984), 239; Sheehan and Dobbins, *Epic Moon,* 27; Whitaker, *Mapping and Naming the Moon,* 75.

18 Morton, *The Moon,* 55; Ashbrook, *The Astronomical Scrapbook,* 242.

19 William Herschel, "May 28, 1776, to July 30, 1776," Observations by William Herschel, Royal Astronomy Society Archives.

20 William Herschel, "September 11, 1780, to October 11, 1783," Observations by William Herschel, Royal Astronomy Society Archives; Arrhenius, *Destinies of the Stars,* 246; Katrin Kleemann, *A Mist Connection: An Environmental History of the Laki Eruption of 1783 and Its Legacy* (De Gruyter, 2023).

21 William Herschel, "May 21, 1785, to October 22, 1790," Observations by William Herschel, Royal Astronomy Society Archives; Leatherbarrow, *The Moon,* 42.

22 James R. Voelkel, *Johannes Kepler and the New Astronomy* (Oxford University Press, 2001), 71; Sheehan and Dobbins, *Epic Moon,* 7.

23 William Herschel, "July 30, 1776," Observations by William Herschel, Royal Astronomy Society Archives.

24 William Herschel, "January 25, 1779, to July 10, 1780," Observations by William Herschel, Royal Astronomy Society Archives; Michael Hoskin, *Discoverers of the Universe: William and Caroline Herschel* (Princeton University Press, 2011), 42.

25 Leatherbarrow, *The Moon,* 44.

26 Whitaker, *Mapping and Naming the Moon,* 105.

27 W. M. W. Payne, "The Moon," *Popular Astronomy* 2, no. 2 (1894), 72; Agnes Clerke, *A Popular History of Astronomy during the Nineteenth Century,* 4th ed. (A&C Black, 1902), 289; Richard Proctor, *The Moon: Her Motions, Aspect, Scenery, and Physical Condition,* 2nd ed. (Longmans, Green, and Co., 1878), 145.

28 Alexander Mikaberidze, *The Napoleonic Wars: A Global History* (Oxford University Press, 2020), 230; Leatherbarrow, *The Moon,* 44.

29 Myles W. Jackson, *Spectrum of Belief: Joseph von Fraunhofer and the Craft of Precision Optics* (MIT Press, 2000), 7.

30 Gruithuisen, *Entdeckung vieler deutlicher Spuren der Mondbewohner, besonders eines colossalen Kunstgebäudes derselben* (Joseph Lindauer, 1824), 54.

31 Gruithuisen, *Entdeckung,* 40; Vincent S. Foster, *Modern Mysteries of the Moon* (Springer, 2015), 4; R. Baum, "The Man Who Found a City in the Moon," *Journal of the British Astronomical Association* 102, no. 3 (1992): 159.

32 Herbert Sadler, "Schröter," Observations of Herbert Sadler (1856–1898), Royal Astronomical Society Archive; Crowe, *The Extraterrestrial Life Debate,* 208.

33 Richard Adams Locke, *Some Account of the Great Astronomical Discoveries Lately Made by Sir John Herschel at the Cape of Good Hope* (Effingham Wilson, 1836), 18; Crowe, *The Extraterres-*

trial Life Debate, 211; Aaron Parrett, *The Translunar Narrative in the Western Tradition* (Ashgate, 2004), 84.

34 Locke, *Some Account,* 85; Crowe, *The Extraterrestrial Life Debate,* 212.

35 Richard Proctor, "The Moon; What We Know About It," *New York Times,* October 26, 1874; Richard Albert Edward Brooks, ed., *The Diary of Michael Floy Jr., Bowery Village, 1833–1837* (Yale University Press, 1941); Allan Nevins, ed., *The Diary of Philip Hone, 1828–1851* (Dodd, Mead, 1927), 173; David A. Copeland, "A Series of Fortunate Events: Why People Believed Richard Adams Locke's 'Moon Hoax,'" *Journalism History* 33, no. 3 (2007): 140; Steven W. Ruskin, "A Newly-Discovered Letter of JFW Herschel concerning the 'Great Moon Hoax,'" *Journal for the History of Astronomy* 33, no. 1 (2002): 73.

36 Edgar Allan Poe, "Richard Adams Locke," *Literary America,* 1848, available at www.eapoe.org/works/misc/litamlra.htm; Brian Thornton, "The Moon Hoax: Debates about Ethics in 1835 New York Newspapers," *Journal of Mass Media Ethics* 15, no. 2 (2000): 90; István Kornél Vida, "The 'Great Moon Hoax' of 1835," *Hungarian Journal of English and American Studies* (2012): 431.

37 Matthew Goodman, *The Sun and the Moon* (Basic Books, 2008), 12; Mario Castagnaro, "Lunar Fancies and Earthly Truths: The Moon Hoax of 1835 and the Penny Press," *Nineteenth-Century Contexts* 34, no. 3 (2012): 253–268; Ulf Jonas Bjork, "'Sweet Is the Tale': A Context for the New York Sun's Moon Hoax," *American Journalism* 18, no. 4 (2001): 14.

10. The Promise and Peril of Lunar Life

1 Mihkel Joeveer, "Mädler, Johann Heinrich von," in *The Biographical Encyclopedia of Astronomers* (Springer, 2007): 723; Edmund Neison, *The Moon: And the Condition and Configuration of Its Surface* (London, 1876), 105; Ewen Whitaker, *Mapping and Naming the Moon* (Cambridge University Press, 2003), 121; Agnes Clerke, *The System of the Stars* (Longmans, Green, and Co., 1890), 204.

2 Patrick Moore, "The Linné Controversy: A Look into the Past," *Journal of the British Astronomical Association* 87 (1977): 363; James Nasmyth and James Carpenter, *The Moon Considered as a Planet, a World, and a Satellite,* 2nd ed. (John Murray, Albemarle Street, 1874), 150.

3 Michael J. Crowe, *The Extraterrestrial Life Debate, 1750–1900* (Cambridge University Press, 1986); Agnes Clerke, *A Popular History of Astronomy during the Nineteenth Century,* 4th ed. (A&C Black, 1902), 314.

4 Crowe, *The Extraterrestrial Life Debate,* 394; Clerke, *Popular History of Astronomy,* 313; Richard Proctor, *The Moon: Her Motions, Aspect, Scenery, and Physical Condition* (Longmans, Green, 1873), 193; William Pickering, *The Moon* (Doubleday, Page and Co., 1903), 39; William Rutter Dawes, "Journal of Telescopic Astronomical Observations Made at Haddenham near Thame," February 14, 1867, Royal Astronomical Society Archive.

5 W. R. Birt, "Telescopic Work for Moonlight Evenings," *English Mechanic,* November 26, 1870; Iwan Rhys Morus, "The Sciences," in *A Companion to Nineteenth-Century Britain,* ed. Chris Williams (Wiley, 2004), 461.

6 W. R. Birt, *English Mechanic,* December 30, 1870, to June 28, 1878; C. Gaudibert, *English Mechanic,* April 25, 1873, to January 8, 1873; G. Knutt, *English Mechanic,* January 5, 1877; J. W. Durrad, *English Mechanic,* May 10, 1872, to July 5, 1872; T. W. Webb, *English Mechanic,* May 21, 1875; F. B. Allison, *English Mechanic,* November 28, 1881, to March 17, 1883; author unknown, "χ of Schroter N. of Plato," *English Mechanic,* 1877–1878; D. J. Smith, "A History of

English Mechanics," *English Mechanic,* January 18, 1924; and "Cleomedes," September 26, 1881, and R. S. Lesuteluiys, November 1, 1881, *Selenographical Club: Lunar Drawings with Notes,* 2 vols., 1880–1883, both in Royal Astronomical Society Archives.

7 Richard Proctor, "The Moon Crater, Hyginus," *New York Times,* January 18, 1879; Richard Proctor, "Changes in the Moon," *New York Times,* February 9, 1879; Joshua Nall, "Constructing Canals on Mars: Event Astronomy and the Transmission of International Telegraphic News," *Isis* 108, no. 2 (2017): 284, 290; Pickering, *The Moon,* v, 59; Dava Sobel, *The Glass Universe: How the Ladies of the Harvard Observatory Took the Measure of the Stars* (Penguin Books, 2016); Catherine Nisbett Becker, "Professionals on the Peak," *Science in Context* 22, no. 3 (2009): 490; William Pickering, Letters from Arequipa to his mother, 1891–1893, HUG 1691.4, Harvard University Archives.

8 Percival Lowell, "July 10, 1894," and "August 10, 1894," observation logs, Lowell Observatory Archives; Pickering, *The Moon,* 61.

9 Pickering, Letters from Arequipa to his mother, 1891–1893, William Pickering, General Folder, HUG 1691, Harvard University Archives; Pickering, *The Moon,* 19, 41, 56; "Our Astronomical Column," *Nature* (1902): 41.

10 Philipp Fauth, *The Moon in Modern Astronomy* (D. van Nostrand Co., 1909), 158; William Sheehan and Thomas Dobbins, *Epic Moon: A History of Lunar Exploration in the Age of the Telescope* (Willmann-Bell, 2001), 280; Roger Launius, *Reaching for the Moon* (Yale University Press, 2019), 43; John A. Paterson, "The Astronomy of 1897," *Transactions of the Royal Astronomical Society of Canada* (1897): 110.

11 Launius, *Reaching for the Moon,* 89–111; William E. Burrows, *This New Ocean: The Story of the First Space Age* (Modern Library, 1999), 287, 293; Asif A. Siddiqi, *Challenge to Apollo: The Soviet Union and the Space Race* (NASA, 2000), 398.

12 Dagomar Degroot, "One Small Step for Man, One Giant Leap for Moon Microbes?," *Isis* 114, no. 2 (2023): 272; W. Vishniac, "The Wolf Trap," in *Concepts for Detection of Extraterrestrial Life,* ed. Freeman Henry Quimby (NASA, 1964), 39.

13 National Research Council, *A Review of Space Research* (National Academies Press, 1962), 9–13; Michael Meltzer, *When Biospheres Collide: A History of NASA's Planetary Protection Programs* (NASA, 2011); Carl Sagan, "Biological Contamination of the Moon," *PNAS* 46, no. 4 (1960): 400; Sagan, "Indigenous Organic Matter on the Moon," *PNAS* 46, no. 4 (1960): 395; Sagan, "Organic Matter on the Moon" (National Academy of Sciences, 1961), box 1002, folder 4, Seth MacFarlane Collection of the Carl Sagan and Ann Druyan Archive, Library of Congress; Keay Davidson, *Carl Sagan: A Life* (Wiley, 2000), 51.

14 Susan Mangus and William Larsen, *Lunar Receiving Laboratory Project History,* NASA report CR-2004-208938, 2, 33; 1964 NASA Authorization: Hearings before the Committee on Science and Astronautics, US House of Representatives. 88th Congress First Session on H.R. 5466, March 4 and 5, 1963, 1088.

15 Meltzer, *When Biospheres Collide,* 186; "LRL (Lunar Receiving Laboratory) Equipment and System Problems" HIS 35638, Apollo 076-22; "Request for Approval of Project #72-9056 LRL (Lunar Receiving Laboratory) Biomedical Support Facility," HIS 35747, Apollo 076-22, Johnson Space Center (JSC) Archives; Howard Eckles, "Dept (Department) of Interior Observations on the LRL (Lunar Receiving Laboratory) Procedures Relative to Invertebrate and Fish Species Feb 12–14," HIS 37092, Back Contamination and Quarantine, Apollo 076-24, JSC History Collection, Neumann Library.

16 NASA MSC. May 12, 1969; "Integrated Quarantine Operations Plan," NASA MSC, May 15 1969; "Spacecraft Quarantine and Release Plan," NASA MSC, May 21, 1969, NASA HQ Archives.

17 Michael Crichton, *The Andromeda Strain* (Vintage Books, 2017); John Pickering, "LRL (Lunar Receiving Laboratory) Bioprotocol Readiness Review and Certification Procedures," HIS 36891, Back Contamination and Quarantine, Apollo 076-23, JSC History Collection, Neumann Library; "Apollo 11 Recovery and Quarantine News Briefing, June 16, 1969," and "Apollo 11 Mission Director's Briefing for News Media, June 16, 1969," NASA HQ Archives.

18 Victor Cohn, "Lunar Contamination: Growing Worry," *Washington Post,* May 28, 1969; "Is the Earth Safe from Lunar Contamination?," *Time,* June 13, 1969; Thomas O'Toole, "Photo Worker Gets Moon Dust on Hands," *Washington Post,* July 27, 1969; "Defenseless?," *Time,* August 4, 1969; "Inside the Box," *Newsweek,* August 4, 1969; George Getze, "Certain Earth Elements Killed by Moon Dust," *Los Angeles Times,* March 12, 1970; Ronald Kotulak, "Soil of Moon May Be Boon to Man," *Chicago Tribune,* May 3, 1970; "Menace in Moon Soil?," *Time,* March 30, 1970; LRL Summary Report No. 3 and 18, NASA HQ Archives.

19 "Earth Germ Survives Three Years in Camera on Moon," *Los Angeles Times,* May 23, 1970; F. J. Mitchell and W. L. Ellis, "Surveyor III: Bacterium Isolated from Lunar-Retrieved TV Camera," in *Lunar and Planetary Science Conference Proceedings,* vol. 2 (MIT Press, 1971), 2721.

20 LRL Summary Report Nos. 4-18, NASA HQ Archives.

21 David Yates Graham, interview by Dagomar Degroot, June 28, 2023.

22 Toby Ord, *The Precipice: Existential Risk and the Future of Humanity* (Hachette Books, 2020), 93.

11. Lunar Changes, Human Ambitions

1 Robert A. McCutcheon, "The 1936–1937 Purge of Soviet Astronomers," *Slavic Review* 50, no. 1 (1991): 100–117; William Sheehan and Thomas Dobbins, "Kozyrev, Nikolai Alexandrovich," in *Biographical Encyclopedia of Astronomers,* ed. Thomas Hockey et al. (Springer, 2014): 1242–1244; Aleksandr I. Solzhenitsyn, *The Gulag Archipelago,* trans. Thomas P. Whitney (Harper and Row, 1973), 480, 482.

2 N. A. Kozyrev, "Volcanic Activity on the Moon," *International Geology Review* 1, no. 10 (1959); Patrick Moore, "Transient Lunar Phenomena: A Review, 1967," *Journal of the British Astronomical Association* 78 (1967): 139; Don Wilhelms, *To a Rocky Moon: A Geologist's History of Lunar Exploration* (University of Arizona Press, 1993), 30; Walter Sullivan, "News of Volcano on Moon Upheld," *New York Times,* January 20, 1959, 20; "Russians Reaffirm Moon Volcano View," *New York Times,* November 20, 1958, 7; N. A. Kozyrev, "Volcanic Phenomena on the Moon," *Nature* 198, no. 4884 (1963): 979–980; A. A. Mills, "Transient Lunar Phenomena and Electrostatic Glow Discharges," *Nature* 225, no. 5236 (1970): 929.

3 Alan H. Anderson Jr., "Apparent Activity on Moon's Surface Draws Rising Scientific Interest," *Goddard News,* July 11, 1966.

4 Robert O'Connell and Anthony Cook, "Revisiting the 1963 'Aristarchus Events,'" *Journal of the British Astronomical Association* 123, no. 4 (2013): 202; K. E. Chilton, "Transient Lunar Phenomena," *Journal of the Royal Astronomical Society of Canada* 63 (1969): 203; Zdeněk Kopal and Robert W. Carder, *Mapping of the Moon: Past and Present* (Springer Science & Business Media, 2013), 115, 151, 222; Gerald Schaber, *The U.S. Geological Survey, Branch of Astrogeology* (US Geological Survey, 2005), 19.

5 John J. Gilheany, "Text of Paper Delivered to National Amateur Astronomers Convention, Denver, 31 August 1964," and "Transfer of FY 64 Funds from Headquarters to Goddard Space Flight Center," NASA HQ Archives.

6 Jaylee Burley and Barbara M. Middlehurst, "Apparent Lunar Activity: Historical Review," *PNAS* 55, no. 5 (1966): 1007; B. M. Middlehurst and J. M. Burley, "Chronological Listing of Lunar Events," Goddard Space Flight Center, 1966, X-641-66-178; "Moon Blink Net Organized for Lunar Study," NASA Press Release No. 65-370, December 20, 1964.

7 *Monthly Progress Report on Alphonsus (Operation Moon Blink),* July 20, 1964, Trident Engineering Associates Inc., Washington, DC; "Moon Blink Sees Red in Lunar Crater," NASA Press Release, November 1, 1964; "NASA Moon-Blink Project Searching for Lunar Volcanoes," *Missile / Space Daily,* June 10, 1964; Charles L. Ricker, "Lunar Transient Phenomena: The ALPO Programme," *Journal of the British Astronomical Association* 78 (1968): 217–219; K. E. Chilton, "Transient Lunar Phenomena," *Journal of the Royal Astronomical Society of Canada* 63 (1969): 203; W. Patrick McCray, *Keep Watching the Skies! The Story of Operation Moonwatch and the Dawn of the Space Age* (Princeton University Press, 2008), 20.

8 Anderson, "Apparent Activity"; Moore, "Transient Lunar Phenomena," 142; Bill J. Austin, "RCAA Joins Lunar Search"; "Trident Briefing on Project Moon-Blink, November 22, 1965," Trident Engineering Associates Inc.; Mitchell A. Kapland to Dr. James B. Edson, June 10, 1965; G. A. Guter to J. J. Gilheany, August 6, 1965; ARGUS / ASTRONET Progress Report no. 6-15-65; all in NASA HQ Archives. See also "Letters," *Sky and Telescope* 351 (June 1964).

9 Trident Engineering Associates, *Project Moon-Blink* (NASA, 1966) 13; "Astronomers Photograph 'Blinks' of Color on Moon's Surface," NASA Press Release no. 65-370, November 30, 1965, NASA HQ Archives; Winifred Sawtell Cameron, "Comparative Analyses of Observations of Lunar Transient Phenomena," *Icarus* 16, no. 2 (1972): 339–387.

10 Philip Stooke, *The International Atlas of Lunar Exploration* (Cambridge University Press, 2008), 22; Roger Launius, *Reaching for the Moon* (Yale University Press, 2019), 58; Asif Siddiqi, *Deep Space Chronicle: A Chronology of Deep Space and Planetary Probes* (NASA, 2002), 51; *Report to the Congress from the President of the United States: United States Aeronautics and Space Activities,* 1964, White House Central Files, Lyndon B. Johnson Presidential Library.

11 Wilhelms, *To a Rocky Moon,* 134; Stooke, *The International Atlas,* 26, 31, 52.

12 Wilhelms, *To a Rocky Moon,* 144, 165; Stooke, *The International Atlas,* 52, 56.

13 Stooke, *The International Atlas,* 129, 160; Wilhelms, *To a Rocky Moon,* 176.

14 James C. Cornell Jr., *Strange, Sudden, and Unexpected* (Smithsonian Institution Press, 1972); Louis Schneider, "Lunar International Observers Network Operation during the Apollo 10 Mission," Lockheed Electronics Company, June 27, 1969, 4, available at https://ntrs.nasa.gov/archive/nasa/casi.ntrs.nasa.gov/19700020725.pdf; Barbara M. Middlehurst, *Operation LION: Report for Period of the Flight of Apollo 11,* Lockheed Electronics Co., September 8, 1969, available at https://ntrs.nasa.gov/archive/nasa/casi.ntrs.nasa.gov/19700020566.pdf.

15 Schneider, "Lunar International Observers Network"; Chilton, "Transient Lunar Phenomena," 203.

16 Schneider, "Lunar International Observers Network."

17 Middlehurst, *Operation LION,* 10.

18 *Apollo 11 Technical Air-to-Ground Voice Transcription* (Manned Spacecraft Center, 1969), 146–147, available at https://www.hq.nasa.gov/alsj/a11/a11transcript_tec.pdf; Arlin Crotts,

The New Moon: Water, Exploration, and Future Habitation (Cambridge University Press, 2014), 266, 487; Middlehurst, *Operation LION,* 10.

19 Middlehurst, *Operation LION,* 10; Donald A. Beattie, *Taking Science to the Moon* (Johns Hopkins University Press, 2003), 260.

20 J. D. Burke, Study Leader, Advanced Lunar Studies, to Martin W. Molloy, February 20, 1970, NASA HQ Archives; Stooke, *The International Atlas,* 246; Neil M. Maher, *Apollo in the Age of Aquarius* (Harvard University Press, 2017).

21 Apollo Program Director to Special Assistant to the Director Space Station Task Force, Deputy Associate Administrator for Space Science and Applications (Science) to Associate Administrator for Space Science and Applications, April 13, 1970, and NASA Administrator to Chairman of Committee on Aeronautical and Space Sciences, US Senate, August 11, 1970, box 007436, NASA HQ Archives; "Scientists Decry Moon Flight Cut," *New York Times,* September 4, 1970; Earl Swift, *Across the Airless Wilds: The Lunar Rover and the Triumph of the Final Moon Landings* (HarperCollins, 2021).

22 Richard J. Pike, "The Lunar Crater Linne," *Sky and Telescope* 46 (1973): 366; R. R. Vondrak, "Dispersal of Gases Released at the Lunar Surface," in *Lunar and Planetary Science Conference,* vol. 7 (1976), 37; J. E. Geake and A. A. Mills, "Possible Physical Processes Causing Transient Lunar Events," *Physics of the Earth and Planetary Interiors* 14, no. 3 (1977): 300; K. A. Howard, "Lunar Avalanches," in *Lunar and Planetary Science Conference,* vol. 4 (1973).

23 Audouin Dollfus, "Langrenus: Transient Illuminations on the Moon," *Icarus* 146, no. 2 (2000): 430; Peter H. Schultz, Matthew I. Staid, and Carlé M. Pieters, "Lunar Activity from Recent Gas Release," *Nature* 444, no. 7116 (2006): 184; Arlin Crotts, "Transient Lunar Phenomena: Regularity and Reality," *Astrophysical Journal* 697, no. 1 (2009): 1.

12. Exploiting the Moon

1 Joshua Lederberg, "Sputnik + 30," *Journal of Genetics* 66 (1987): 217.

2 Lederberg, "Sputnik + 30," 217; Leonard Reiffel, "Sagan Breached Security by Revealing US Work on a Lunar Bomb Project," *Nature* 405, no. 6782 (2000): 13.

3 Justin Allen Holcomb, Rolfe David Mandel, and Karl William Wegmann, "The Case for a Lunar Anthropocene," *Nature Geoscience* (2023): 1–3.

4 Roger Launius, *Reaching for the Moon* (Yale University Press, 2019), 46.

5 Philip Stooke, *The International Atlas of Lunar Exploration* (Cambridge University Press, 2008), 10, 15, 53, 60; Max Frankel, "Soviet Rocket Hits Moon after 35 Hours," *New York Times,* September 14, 1959.

6 Stooke, *The International Atlas,* 87, 95; Launius, *Reaching for the Moon,* 50.

7 Stooke, *The International Atlas,* 116, 140, 207, 212, 224, 240, 265, 289, 300, 327, 347.

8 William E. Burrows, *This New Ocean: The Story of the First Space Age* (Modern Library, 1999), 220.

9 Boris Chertok, *Rockets and People: Creating a Rocket Industry,* vol. 2, ed. Asif Siddiqi (NASA History Division, 2006), 440; Stooke, *The International Atlas,* 10.

10 Chertok, *Rockets and People,* 440; Asif A. Siddiqi, *Challenge to Apollo: The Soviet Union and the Space Race* (NASA, 2000), 2:169.

11 J. Reiffel, *A Study of Lunar Research Flights,* vol. 1 (Air Force Special Weapons Center, 1959), 2, 3, 124, 180, 189, 253, National Security Archive, George Washington University.

12 Reiffel, "A Study," 134, 242.

13 US Army, *Project Horizon* (1959), vol. 1: *Summary and Supporting Considerations,* 2, 53, 60, vol. 2: *Technical Considerations & Plans,* 57, National Security Archive, George Washington University.

14 US Army, *Project Horizon,* 1:1, 75, 2:9, 144, 234, 293; Stooke, *The International Atlas,* 14.

15 Rod Pyle, *Amazing Stories of the Space Age* (Prometheus Books, 2017), 31.

16 US Army, *Project Horizon,* 2:30.

17 US Army, *Project Horizon,* 1:4; David Munns and Kärin Nickelsen. "To Live among the Stars: Artificial Environments in the Early Space Age," *History and Technology* 33, no. 3 (2017): 273; Leah Aronowsky, "Of Astronauts and Algae: NASA and the Dream of Multispecies Spaceflight," *Environmental Humanities* 9, no. 2 (2017): 359–377.

18 US Army, *Project Horizon,* 1:2, 2:11, 46.

19 US Air Force, "Lunar Expedition Plan: LUNEX," Headquarters Space Systems Division, Air Force Systems Command, May 1961; Carl Berger, "The Air Force in Space Fiscal Year 1961," USAF Historical Division Liaison Office, April 1966, National Security Archive, George Washington University.

20 USAF Directorate of Space Planning and Analysis, *Military Lunar Base Program or S.R. 183 Lunar Observatory Study,* vol. 1, Air Force Ballistic Missile Division, April 1960, National Security Archive, George Washington University.

21 USAF Directorate of Space Planning and Analysis, *Military Lunar Base Program,* vol. 1.

22 "Lunar Expedition Plan: LUNEX"; David Compton, *Where No Man Has Gone Before* (NASA History Office, 1989); Grzegorz Racki and Christian Koeberl, "In Search of Historical Roots of the Meteorite Impact Theory," *Meteoritics & Planetary Science* 54, no. 10 (2019): 2604–2630.4

23 Stooke, *The International Atlas,* 393; John Westfall, "Mapping Luna Incognita," *Journal of the Association of Lunar and Planetary Observers* 34 (1990): 149.

24 Leonard David, *Moon Rush: The New Space Race* (National Geographic, 2019), 30, 120.

25 Stooke, *The International Atlas,* 406.

26 David, *Moon Rush,* 173.

27 David, *Moon Rush,* 180.

28 David, *Moon Rush,* 174; Christian Davenport, *The Space Barons: Elon Musk, Jeff Bezos, and the Quest to Colonize the Cosmos* (Public Affairs, 2019), 274.

29 David, *Moon Rush,* 119; E. Heggy et al., "Bulk Composition of Regolith Fines on Lunar Crater Floors: Initial Investigation by LRO / Mini-RF," *Earth and Planetary Science Letters* 541 (2020): 116274.

30 Maggie Aderin-Pocock, *The Book of the Moon: A Guide to Our Closest Neighbor* (Abrams Image, 2019), 204.

31 Charles R. Phillips, *The Planetary Quarantine Program: Origins and Achievements, 1956–1973,* Report no. NASA-SP-4902, Scientific and Technical Information Office, NASA, 1974.

32 Michael Meltzer, *When Biospheres Collide: A History of NASA's Planetary Protection Programs* (NASA, 2011), 46.

33 Meltzer, *When Biospheres Collide,* 46.

34 Ronald Kotulak, "Man Has Contaminated Moon, NASA Chief Dr. Paine Says," *Chicago Tribune,* November 27, 1969; Meltzer, *When Biospheres Collide,* 51; J. B. Barengoltz et al., *A Proposed New Policy for Planetary Protection,* JPL Publication 81–90, NASA, 1981, available at https://ntrs.nasa.gov/api/citations/19820003222/downloads/19820003222.pdf.

35 Keren Shahar and Dov Greenbaum, "Lessons in Space Regulations from the Lunar Tardigrades of the Beresheet Hard Landing," *Nature Astronomy* 4, no. 3 (2020): 208.

36 Andrew C. Schuerger et al., "A Lunar Microbial Survival Model for Predicting the Forward Contamination of the Moon," *Astrobiology* (2019); GengXin Xie et al., "The First Biological Experiment on Lunar Surface for Humankind: Device and Results," *Acta Astronautica* 214 (2024): 216–223.

Part III Conclusion

1 Alice Gorman, *Dr. Space Junk vs the Universe: Archaeology and the Future* (MIT Press, 2019), 161; Leonard David, *Moon Rush: The New Space Race* (National Geographic, 2019), 195.

2 "Treaty on Principles Governing the Activities of States in the Exploration and Use of Outer Space," UN Office for Outer Space Affairs, available at https://www.unoosa.org/oosa/en/ourwork/spacelaw/treaties/outerspacetreaty.html; "Agreement Governing the Activities of States on the Moon and Other Celestial Bodies," UN Office for Outer Space Affairs, available at https://www.unoosa.org/oosa/en/ourwork/spacelaw/treaties/moon-agreement.html; Joseph Reynolds, "Legal Implications of Protecting Historic Sites in Space," in *Archaeology and Heritage of the Human Movement into Space*, ed. Beth Laura O'Leary and Peter Joseph Capelotti (Springer, 2014), 114.

3 "The Artemis Accords," NASA, available at https://www.nasa.gov/wp-content/uploads/2022/11/Artemis-Accords-signed-13Oct2020.pdf.

Part IV. Mars

1 Adrian Howkins, *Frozen Empires: An Environmental History of the Antarctic Peninsula* (Oxford University Press, 2016), 5.

2 Stephen Pyne, *The Ice: A Journey to Antarctica*, 2nd ed. (University of Washington Press, 2017), 316.

3 William A. Cassidy, *Meteorites, Ice, and Antarctica: A Personal Account* (Cambridge University Press, 2012), 35; David S. McKay et al., "Search for Past Life on Mars: Possible Relic Biogenic Activity in Martian Meteorite ALH84001," *Science* 273, no. 5277 (1996): 924–930; Sarah Johnson, *The Sirens of Mars: Searching for Life on Another World* (Crown, 2020), 75.

4 Robert Markley, *Dying Planet: Mars in Science and the Imagination* (Duke University Press, 2005), 324.

5 Markley, *Dying Planet*, 323–337.

13. Unveiling a New World

1 Dagomar Degroot, *The Frigid Golden Age: Climate Change, the Little Ice Age, and the Dutch Republic, 1560–1720* (Cambridge University Press, 2018), 231.

2 Robert Markley, *Dying Planet: Mars in Science and the Imagination* (Duke University Press, 2005), 34.

3 Markus Hotakainen, *Mars from Myth and Mystery to Recent Discoveries* (Springer, 2009), 22.

4 Paul G. Abel, *Visual Lunar and Planetary Astronomy* (Springer Science & Business Media, 2013), 70.

5 Michael J. Crowe, *The Extraterrestrial Life Debate, 1750–1900* (Cambridge University Press, 1986), 480; Fred Price, *The Planet Observer's Handbook*, 2nd ed. (Cambridge University Press, 2000), 136.

6 Gerald North, *Advanced Amateur Astronomy* (Cambridge University Press, 1997), 27; William Sheehan, *Planets and Perception: Telescopic Views and Interpretations, 1609–1909* (University of Arizona Press, 1988), 2.

7 F. W. Taylor, *Planetary Atmospheres* (Oxford University Press, 2010), 186; Peter Read and Stephen Lewis, *The Martian Climate Revisited* (Springer, 2004), 116.

8 R. J. Lillis et al., "Early Mars Chronology," *Lunar and Planetary Institute Contributions* 1791 (2014): 1348.

9 Taylor, *Planetary Atmospheres,* 187.

10 Christiaan Huygens à Constantyn Huygens, frère, 14 Juin 1673, available at http://www.dbnl.org/tekst/huygoo3oeuvo7_01/huygoo3oeuvo7_01_0158.php?q=mars#hl1; Markley, *Dying Planet,* 43, 47; Camille Flammarion, *The Planet Mars,* trans. Patrick Moore (Springer, 2014), 29, 38; Price, *The Planet Observer's Handbook,* 148; William Sheehan, *The Planet Mars: A History of Observation and Discovery* (University of Arizona Press, 1996), appendix 2.

11 Flammarion, *The Planet Mars,* 51; William Herschel, "September 24, 1779, to July 22 1781," MSS Herschel W. 3.1.5, Mars, Royal Astronomical Society Archive.

12 William Herschel, "June 19, 1779, to December 1 1785," RAS MSS Herschel W. 3.1.5, Mars, Royal Astronomical Society Archive; Markley, *Dying Planet,* 47; Yves Balkanski et al., "Mortality Induced by PM 2.5 Exposure Following the 1783 Laki Eruption Using Reconstructed Meteorological Fields," *Scientific Reports* 8, no. 1 (2018): 1; William Herschel, "On the Remarkable Appearances at the Polar Regions of the Planet Mars," *Philosophical Transactions of the Royal Society of London* 74 (1784): 260.

13 Flammarion, *The Planet Mars,* 55; Markley, *Dying Planet,* 49.

14 Flammarion, *The Planet Mars,* 78; David G. Horvath et al., "Evidence for Geologically Recent Explosive Volcanism in Elysium Planitia, Mars," *Icarus* 365 (2021): 114499.

15 Wilhelm Beer and Johann Heinrich Mädler, *Beiträge zur physischen Kenntniss der himmlischen Körper im Sonnensysteme* (Bernhard Friedrich Voigt, 1841), 112.

16 Beer and Mädler, *Beiträge,* 113; Flammarion, *The Planet Mars,* 99; Sheehan, *Planets and Perception,* 45; Sheehan, *The Planet Mars,* 49.

17 Flammarion, *The Planet Mars,* 134; Mr. Lockyer, "Observations on the Planet Mars," *Memoirs of the Royal Astronomical Society* 32 (1862–1863): 183; R. J. McKim, "The Opposition of Mars, 1999," *Journal of the British Astronomical Association* 117, no. 6 (1999): 315.

18 John Phillips, "On the Telescopic Appearance of the Planet Mars," *Proceedings of the Royal Society of London* 12 (1862–1863): 436.

19 Phillips, "On the Telescopic Appearance," 436; Lockyer, "Observations on the Planet Mars," 179; C. Linsser, "Correspondenznachrichten aus Russland," in *Wochenschrift für Astronomie, Meteorologie und Geographie,* ed. Eduard Heis (H. W. Schmidt, 1864), 120; Flammarion, *The Planet Mars,* 74.

20 Flammarion, *The Planet Mars,* 172; John Herschel, *Outlines of Astronomy,* 2nd ed. (Longman, Brown, Green and Longmans, 1849), 263; John Phillips, "Further Observations on the Planet Mars," *Proceedings of the Royal Society of London* 14 (1865): 45.

21 Jean Chrétien Ferdinand Hoefer, *Histoire de l'astronomie depuis ses origines jusqu'à nos jours* (Hachette, 1873), 563; Flammarion, *The Planet Mars,* 105; Herschel, *Outlines,* 510.

22 Richard Proctor, *Other Worlds than Ours* (Appleton, 1882), 99; Howard Carlton, *Cosmology and the Scientific Self in the Nineteenth Century* (Springer, 2022), 114.

14. Aliens on an Aged Earth

1 E. M. Antoniadi, *The Planet Mars,* trans. Patrick Moore (W. & J. Mackay, 1975), 59; Camille Flammarion, *The Planet Mars,* trans. Patrick Moore (Springer, 2014), 204: C. H. Davis, "Drawings of Mars and Jupiter," *Monthly Notices of the Royal Astronomical Society* 36 (1875): 15; R. W. Zurek and L. J. Martin, "Interannual Variability of Planet-Encircling Dust Storms on Mars," *Journal of Geophysical Research* 98, no. E2 (1993): 3247; G. F. Chambers, "Correspondence—The Planet Mars," *Astronomical Register* 13 (1875): 22; Richard J. McKim, William P. Sheehan, and Randall Rosenfield, "Etienne Leopold Trouvelot and the Planet-Encircling Martian Dust Storm of 1877," *Journal of the British Astronomical Association* 119, no. 6 (2009): 351.

2 William Sheehan, *Planets and Perception: Telescopic Views and Interpretations, 1609–1909* (University of Arizona Press, 1988), 265; Andrea Bernagozzi, Antonella Testa, and Pasquale Tucci, "Observing Mars with Schiaparelli's Telescope," *Third European Workshop on Exo-Astrobiology* 545 (2004): 157.

3 Giovanni Schiaparelli, *Corrispondenza su Marte,* vol. 1: *1877–1889* (Nistri Lischi, 1965); Sheehan, *Planets and Perception,* 101.

4 K. Maria D. Lane, *Geographies of Mars: Seeing and Knowing the Red Planet* (University of Chicago Press, 2011), 24.

5 Nathaniel E. Green, "Mars and the Schiaparelli Canals," *The Observatory* 3 (1879): 252; Edward Maunder, "The Canals on Mars," *The Observatory* 11 (1888): 348; Michael J. Crowe, *The Extraterrestrial Life Debate, 1750–1900* (Cambridge University Press, 1986), 490; Steven J. Dick, *The Biological Universe* (Cambridge University Press, 1996), 85; Neil English, *Chronicling the Golden Age of Astronomy* (Springer, 2018), 333; Matthew Shindell, *For the Love of Mars: A Human History of the Red Planet* (University of Chicago Press, 2023), 101.

6 Flammarion, *The Planet Mars,* 273; Sheehan, *Planets and Perception,* 109; Crowe, *The Extraterrestrial Life Debate,* 490; W. F. Denning, "The Physical Appearance of Mars in 1886," *Nature* (1886): 105; Green, "Mars and the Schiaparelli Canals," 252; Maunder, "The Canals," 348.

7 Fred Price, *The Planet Observer's Handbook,* 2nd ed. (Cambridge University Press, 2000), 154; Flammarion, *The Planet Mars,* 263; William Pickering, "Visual Observation of the Surface of Mars," *Sidereal Messenger* 9 (1890), 370; William Pickering, "The Physical Aspect of the Planet Mars," *Science* 289 (1888): 83.

8 Giovanni Schiaparelli, *Il Pianeta Marte* (Casa Editrice Dottor Francesco Vallardi, 1893), 4; Green, "Mars and the Schiaparelli Canals," 252; Maunder, "The Canals," 348; Dick, *The Biological Universe,* 68; Robert Markley, *Dying Planet: Mars in Science and the Imagination* (Duke University Press, 2005), 83; W. W. Payne, "The Planet Mars," *Popular Astronomy* 3, no. 7 (1896): 346; Camille Flammarion, *La Planète Mars et ses conditions d'habitabilité* (Librairie Scientifique, 1892), 418.

9 Joshua Nall, "Constructing Canals on Mars: Event Astronomy and the Transmission of International Telegraphic News," *Isis* 108, no. 2 (2017): 300.

10 "Observations of Mars in South America," *New York Herald,* August 10, 1892, 3; "Large Areas of Blue," *Boston Daily Globe,* August 10, 1892, 4; "Many Changes Noted in Fiery Mars," *Chicago Tribune,* August 10, 1892, 1.

11 "A Square Look at Mars," *New York Times,* 1892, 5; "The Double 'Canals' of Mars," *New York Herald,* August 20, 1892, 6; "Melting Snow in Mars," *New York Herald,* September 1, 1892, 7; Pickering, "The Physical Aspect," 83.

12 Nall, "Constructing Canals on Mars," 282, 301; Markley, *Dying Planet,* 60.

13 Sheehan, *Planets and Perception,* 173.

14 Sheehan, *Planets and Perception,* 173.

15 Ken Croswell, *Magnificent Mars* (Simon and Schuster, 2003), 14; Sheehan, *The Planet Mars,* 202; Timothy J. Yamamura, "Fictions of Science, American Orientalism, and the Alien / Asian of Percival Lowell," in *Dis-Orienting Planets: Racial Representations of Asia in Science Fiction,* ed. Isiah Lavender (University Press of Mississippi, 2017), 92.

16 W. F. Denning, "The Relative Powers of Large and Small Telescopes in Showing Planetary Detail," *Nature* 52 (1895): 233.

17 Sheehan, *Planets and Perception,* 7.

18 Sheehan, *Planets and Perception,* 174.

19 Lowell, *Mars and Its Canals* (Macmillan, 1906), 7.

20 A. E. Douglass, "The Lowell Observatory and Its Work," *Popular Astronomy* 2, no. 9 (1895): 395; Percival Lowell, "Atmosphere: In Its Effect on Astronomical Research," lecture, spring 1897, Lowell Observatory Archives.

21 "Douglass, A. E., February to May 24, 1894," folders 1–3, box 4, Correspondence, 1876–1916, and "Early Reports on Astronomical Work in Arizona," and "Solar Activity and Rainfall." folders 1 and 4, box 13, Working Papers, 1883–1916, MS.1, Percival Lowell Papers, Lowell Observatory Archives.

22 Percival Lowell, "Mars: The Canals I," *Popular Astronomy* 2, no. 6 (1895): 258.

23 A. E. Douglass, "The Lick Review of 'Mars,'" *Popular Astronomy* 4, no. 4 (1896): 200.

24 Percival Lowell, *Mars as the Abode of Life* (Macmillan, 1908), 149, 155, 157; Lowell, "Mars: Oases," *Popular Astronomy* 2, no. 8 (1895): 348; Lowell, "Some of the Double Canals of Mars," *Popular Astronomy* 5, no. 5 (1897): 235; Lowell, "Markings in the Syrtis Major," *Popular Astronomy* 4, no. 6 (1896): 291; "Douglass, A. E., July to December, 1894," folder 6, box 4, Correspondence, 1876–1916, MS.1, Percival Lowell Papers, Lowell Observatory Archive.

25 Percival Lowell, "Mars: The Polar Snows," *Popular Astronomy* 2, no. 2 (1894): 54; Sheehan, *The Planet Mars,* 109; Jennifer Putnam and William Sheehan, "A Complicated Relationship," *Journal of Astronomical History and Heritage* 24, no. 1 (2021): 170.

26 Molesworth, vol. 47, *Report on the Observations of Mars,* Royal Astronomical Society (RAS) MSS Molesworth, 1882–1905; "Mars Is Coming," *Tacoma Times,* 1909.

27 Norriss Hetherington, "Percival Lowell: Professional Scientist or Interloper?," *Journal of the History of Ideas* 42, no. 1 (1981): 159.

28 "Mars—Oases and Canals, 1912–1914," folder 25, box 2; "Canals Visibility, 1894–1901," folder 18, box 5; "Comments on Lampland's Photographs of the Canals on Mars," folder 1, box 5-A; "Mars Canals and Oases, 1907" and "Mars Canal Charts," folders 5 and 8-A, box 1 Large, Working Papers, 1883–1916; "Douglass, A. E., January, 1895," folder 1, box 4, Correspondence, 1876–1916; all in MS.1, Percival Lowell Papers, Lowell Observatory Archive.

29 "Mars, 1903," folder 5, box 2, Working Papers, 1883–1916, MS.1, Percival Lowell Papers, Lowell Observatory Archive; Steven J. Dick, *Life on Other Worlds* (Cambridge University Press, 2001), 28.

30 K. Maria D. Lane, "Geographers of Mars: Cartographic Inscription and Exploration Narrative in Late Victorian Representations of the Red Planet," *Isis* 96, no. 4 (2005): 492; Joshua Nall, *News from Mars: Mass Media and the Forging of a New Astronomy, 1860–1910* (University of Pittsburgh Press, 2019); Sheehan, *Planets and Perception.*

31 Lowell, *Mars as the Abode of Life,* 106.

32 Percival Lowell, "Mars," *Popular Astronomy* 4, no. 3 (1896): 113; Lowell, "North Polar Rifts and Arctic Canals on Mars," *Popular Astronomy* 10, no. 3 (1902): 114; Lowell, "Mars Canals and Poles Plates, Part Two, 1894," folder 12, box 5-A, Working Papers, 1883–1916, MS. 1, Percival Lowell Papers, Lowell Observatory Archives.

33 Lowell, *Mars as the Abode of Life,* 168, 210; Lowell, *Mars and Its Canals,* 39; "Du Maurier's View," *The Enterprise,* August 27, 1897.

34 James Cutts, Karl Blasius, and W. James Roberts, "Evolution of Martian Polar Landscapes," *Journal of Geophysical Research: Solid Earth* 84, no. B6 (1979): 2975; Ronald Greely, *Planetary Landscapes,* 2nd ed. (Springer Science & Business Media, 1994), 153; Philip B. James et al., "North Polar Dust Storms in Early Spring on Mars," *Icarus* 138, no. 1 (1999): 64; Percival Lowell, "Mars: Spring Phenomena," *Popular Astronomy* 2, no. 3 (1894); 100.

35 Crowe, *The Extraterrestrial Life Debate,* 498; H. Perrotin, "Observations de la planète Mars," *Comptes rendus de l'Académie des Sciences* 115 (1892): 379.

36 "Think Martians Want to Talk," *Chicago Tribune,* August 4, 1894; "The Signals from Mars," *Popular Astronomy* 3, no. 1 (1895): 47.

37 Justice Stahn, "Letter to Prof. Douglas, Dec 25th, 1900," and Daniel E. Parks, "Letter to Prof. A. E. Douglass, January 17th, 1901," both in Correspondence, 1876–1916; "Cloud on Mars' Terminator, 1903," folder 3, box 5-A, Working Papers, 1883–1916, MS.1; all in Percival Lowell Papers, Lowell Observatory Archives; Lowell, *Mars and Its Canals,* 102; W. W. Campbell, "An Explanation of the Bright Projections Observed on the Terminator of Mars," *Astronomical Society of the Pacific* 6 (1894): 102; Percival Lowell, "Mars: Atmosphere," *Popular Astronomy* 2, no. 4 (1894): 158; Lowell, "Projections from Limb of Mars," *Popular Astronomy* 8, no. 1 (1901).

38 Richard Grove and George Adamson, *El Niño in World History* (Palgrave Macmillan, 2018), 95.

39 Joëlle Gergis and Anthony Fowler, "A History of ENSO Events since A.D. 1525," *Climatic Change* 92 (2009): 343–387; Gaurav Srivastava, Arindam Chakraborty, and Ravi S. Nanjundiah, "Multidecadal See-Saw of the Impact of ENSO on Indian and West African Summer Monsoon Rainfall," *Climate Dynamics* 52, no. 11 (2019): 6633–6649.

40 Philip Gooding, "ENSO, IOD, Drought, and Floods in Equatorial Eastern Africa, 1876–1878," Stephen J. Rockel, "A Forgotten Drought and Famine in East Africa, 1883–1885," and Theresa Ventura, "'A Drought So Extraordinary,'" all in *Droughts, Floods, and Global Climatic Anomalies in the Indian Ocean World,* ed. Philip Gooding (Springer International, 2022), 259, 289, 345; Mike Davis, *Late Victorian Holocausts: El Niño Famines and the Making of the Third World* (Verso, 2002), 44.

41 Davis, *Late Victorian Holocausts,* 339; Gayatri Kathayat et al., "Protracted Indian Monsoon Droughts of the Past Millennium and Their Societal Impacts," *PNAS* 119, no. 39 (2022): 5.

42 Lowell, *Mars as the Abode of Life,* 119; J. R. Holt, "The Polar Snows and Non-Glaciation of Mars," *Popular Science* 2, no. 2 (1894): 79; Michael J. Everhart, *Oceans of Kansas: A Natural History of the Western Interior Sea* (Indiana University Press, 2017), 16; Meredith McKittrick, *Green Lands for White Men: Desert Dystopias and the Environmental Origins of Apartheid* (University of Chicago Press, 2024); James Belich, *Replenishing the Earth: The Settler Revolution and the Rise of the Angloworld* (Oxford University Press, 2011), 394.

43 David Blackbourn, *The Conquest of Nature: Water, Landscape, and the Making of Modern Germany* (W. W. Norton, 2007), 5; Jürgen Osterhammel, *The Transformation of the World: A Global History of the Nineteenth Century* (Princeton University Press, 2015), 689.

44 Robert Crossley, *Imagining Mars: A Literary History* (Wesleyan University Press, 2011), 36; David A. Weintraub, *Life on Mars* (Princeton University Press, 2020), 89; Osterhammel, *Transformation of the World,* 691.

45 "Douglass, A. E., March to May, 1895," and "Douglass, A. E., April to November, 1895," folders 3 and 4, box 4, Correspondence, 1876–1916, MS.1, Percival Lowell Papers, Lowell Observatory Archives; A. E. Douglass, "The Lowell Observatory in Mexico," *Popular Astronomy* 4, no. 9 (1897), 490.

46 William Graves Hoyt, *Lowell and Mars* (University of Arizona Press, 1976), 59; Lowell, *Mars and Its Canals,* 37.

15. Meeting the Martians

1 Howard Sutton, "Charles Cros, the Outsider," *French Review* 39, no. 4 (1966): 515.

2 Charles Cros, "Moyens de communication aves les planètes," *Bibliotheca Augustana,* available at https://www.hs-augsburg.de/~harsch/gallica/Chronologie/19siecle/Cros/cro_plan.html.

3 Florence Raulin-Cerceau, "The Pioneers of Interplanetary Communication: From Gauss to Tesla," *Acta Astronautica* 67, no. 11–12 (2010): 1391; Sutton, "Charles Cros," 515.

4 F. C. Durant, *Report of Meetings of Scientific Advisory Panel on Unidentified Flying Objects Convened by Office of Scientific Intelligence* (CIA, 1953); Jason Wright, Chelsea Haramia, and Gabriel Swiney, "Geopolitical Implications of a Successful SETI Program," *Space Policy* 63 (2023): 101517.

5 John A. Paterson, "The Astronomy of 1897," *Transactions of the Royal Astronomical Society of Canada* (1897): 114.

6 Vesto Slipher, "Chapters for Mars Book, 1928–1931," box 11, Working Papers, MS.2, V. M. Slipher Papers, Lowell Observatory Archives.

7 Percival Lowell, *Mars as the Abode of Life* (Macmillan, 1908), 110; K. Maria D. Lane, "Geographers of Mars: Cartographic Inscription and Exploration Narrative in Late Victorian Representations of the Red Planet," *Isis* 96, no. 4 (2005): 499.

8 Lowell, *Mars as the Abode of Life,* 109. See Julia Adeney Thomas, ed., *Altered Earth: Getting the Anthropocene Right* (Cambridge University Press, 2022).

9 Lowell, "Notes on Origin of Species, 1916," folder 8, box 4-A, Working Papers, 1883–1916, MS.1, Percival Lowell Papers, Lowell Observatory Archives; Lowell, *Mars as the Abode of Life,* 109; Gregory Claeys, "The 'Survival of the Fittest' and the Origins of Social Darwinism," *Journal of the History of Ideas* 61, no. 2 (2000): 224.

10 Stephen Baxter, "Martian Chronicles: Narratives of Mars in Science and SF," *Foundation* 0 (1996), 5; Garrett P. Serviss, *Edison's Conquest of Mars* (Frank Leslie's Popular Monthly, 1898).

11 Kurd Lasswitz, *Auf zwei Planeten* (Jazzybee, 2016); William Fischer, *The Empire Strikes Out: Kurd Lasswitz, Hans Dominik, and the Development of German Science Fiction* (Popular Press, 1984), 81.

12 Richard Stites, *Utopian Vision and Experimental Life in the Russian Revolution* (Oxford University Press, 1989): 4; Alexander Bogdanov, *Red Star: The First Bolshevik Utopia* (Indiana University Press, 1984).

13 Brookings Institution, *Proposed Studies on the Implications of Peaceful Space Activities for Human Affairs* (NASA, 1960), 183.

14 Michael J. Crowe, *The Extraterrestrial Life Debate, 1750–1900* (Cambridge University Press, 1986), 400; F. Drake and D. Sobel, *Is Anyone Out There? The Scientific Search for Extraterres-*

trial Intelligence (New York, 1992), 172; Christopher Keep, "Life on Mars? Hélène Smith, Clairvoyance, and Occult Media," *Journal of Victorian Culture* 25, no. 4 (2020): 542.

15 Francis Galton, "Intelligible Signals between Neighbouring Stars," *Fortnightly Review* 60 (1896): 664; Crowe, *Extraterrestrial Life Debate,* 400; Camille Flammarion, *Omega: The Last Days of the World* (Cosmopolitan, 1894), 16.

16 Marc Seifer, *Wizard: The Life and Times of Nikola Tesla* (Citadel, 1998).

17 Steven J. Dick, *The Biological Universe* (Cambridge University Press, 1996), 401.

18 Nikola Tesla, "The Problem of Increasing Human Energy," *Century Illustrated Magazine,* June 1900, 209.

19 "Tesla's New Wonder," *Audubon County Journal,* January 10, 1901; "Talking with the Planets," *Collier's Weekly,* February 9, 1901.

20 "Some Planet Affected His Machine," *New York Sun,* January 3, 1901; Nikola Tesla, "Letter to the Editor of the New York Sun," *New York Sun,* January 9, 1901; Kenneth L. Corum and James F. Corum, "Nikola Tesla and the Planetary Radio Signals," 5th International Tesla Conference, Belgrade, 1996, 6; Seifer, *Wizard.*

21 F. Bagenal et al., "Introduction," in *Jupiter: The Planet, Satellites and Magnetosphere* (Cambridge University Press, 2004), 4.

22 Corum and Corum, "Nikola Tesla," 5; Nicholas Schneider and Fran Bagenal, "Io's Neutral Clouds, Plasma Torus, and Magnetospheric Interaction," in *Io after Galileo,* ed. Rosaly Lopes and John Spencer (Praxis, 2007), 266.

23 Margaret Cheney and Robert Uth, *Tesla, Master of Lightning* (Barnes and Noble, 1999), 134; "A Giant Eye to See Around the World," *Albany Telegram,* February 25, 1923; "Theory a Failure," *Billings Gazette,* June 14, 1907.

24 Dick, *The Biological Universe,* 404; Michael Brown, "Radio Mars," *Transmitting the Past: Historical and Cultural Perspectives on Broadcasting* (University of Alabama Press, 2005).

25 Dick, *The Biological Universe,* 408; "To Communicate with Mars," *Blue Grass Blade,* 1909; "Hello, Mars—This Is the Earth!," *Popular Science,* September 1919, 74.

26 Carl Sagan, *Contact* (Simon and Schuster, 2016); Cixin Liu, *The Three-Body Problem,* trans. Ken Liu (Tor Books, 2014); Jaime Green, *The Possibility of Life: Science, Imagination, and Our Quest for Kinship in the Cosmos* (Harlequin, 2023); Nathalie A. Cabrol, *The Secret Life of the Universe: An Astrobiologist's Search for the Origins and Frontiers of Life* (Simon and Schuster UK, 2024).

27 Raulin-Cerceau, "The Pioneers," 1391; METI International, available at http://meti.org/mission; Paolo Musso, "A Language Based on Analogy to Communicate Cultural Concepts in SETI," *Acta Astronautica* 68, no. 3 (2011): 489.

28 Keith Cooper, *The Contact Paradox* (Bloomsbury, 2021), 295; Dick, *The Biological Universe,* 408.

29 Amelia Urry, "The Astronomer at the End of the World: Visions of Scientific Authority in Camille Flammarion's Apocalyptic Fiction," *Apocalyptica* 1, no. 1 (2022): 68; Iwan Rhys Morus, "The Sciences," in *A Companion to Nineteenth-Century Britain,* ed. Chris Williams (Wiley, 2004), 462.

30 Robert Markley, *Dying Planet: Mars in Science and the Imagination* (Duke University Press, 2005), 124; Roslyn D. Haynes, *H. G. Wells: Discoverer of the Future* (NYU Press, 1980), 46; H. G. Wells, "Intelligence on Mars," in *H. G. Wells: Early Writings in Science and Science Fiction,* ed. Robert Philmus and David Y. Hughes (University of California Press, 1975).

31 H. G. Wells, *The War of the Worlds* (Plain Label Books, 1898), 4; Alan Bewell, *Romanticism and Colonial Disease* (Johns Hopkins University Press, 2003), xiv.

32 "Science Finds Life Growing Harder for the People on Mars," *Washington Times,* August 27, 1922.

33 A. Brad Schwartz, *Broadcast Hysteria: Orson Welles's War of the Worlds and the Art of Fake News* (Farrar, Straus and Giroux, 2015).

34 "Radio Listeners in Panic," *New York Times,* October 31, 1938; "Geologists at Princeton Hunt 'Meteor' in Va.," *New York Times,* October 31, 1938; Schwartz, *Broadcast Hysteria,* 8; Hadley Cantril, *The Invasion from Mars: A Study in the Psychology of Panic* (Transaction, 1940), vii; Joy Elizabeth Hayes and Kathleen Battles, "Exchange and Interconnection in US Network Radio," *International Studies in Broadcast and Audio Media* 9, no. 1 (2011): 51.

35 "FCC to Scan Script of 'War' Broadcast," *New York Times,* November 1, 1938; Schwartz, *Broadcast Hysteria,* 14.

16. Reshaping Planetary Climates

1 "Science Magazine, 1895–1909," folder 52, box 13, Correspondence, 1876–1916, MS.1, Percival Lowell Papers, Lowell Observatory Archives; George Webb, *Tree Rings and Telescopes: The Scientific Career of A. E. Douglass* (University of Arizona Press, 1983), 49.

2 Z. Svestka, "Solar Activity," in *Dynamic Sun,* ed. B. N. Dwivedi and E. N. Parkers (Cambridge University Press, 2013); George Sarton, "When Was Tree-Ring Analysis Discovered?," *Isis* 45 (December 1954): 384; Stefan Norrgård and Samuli Helama, "Dendroclimatic Investigations and Cross-Dating in the 1700s," *Canadian Journal of Forest Research* 51, no. 2 (2021): 267; Webb, *Tree Rings,* 105.

3 A. E. Douglass, "A Method of Estimating Rainfall by the Growth of Trees," *Bulletin of the American Geographical Society* 46, no. 5 (1914): 321; Webb, *Tree Rings,* 103.

4 Valerie Trouet, *Tree Story: The History of the World Written in Rings* (Johns Hopkins University Press, 2020), 11.

5 Ellsworth Huntington, "Changes of Climate and History," *American Historical Review* 18, no. 2 (1913): 213; Kevin Anchukaitis, "Tree Rings Reveal Climate Change Past, Present, and Future," *Proceedings of the American Philosophical Society* 161, no. 3 (2017): 260.

6 "Associated Press, 1911–1916," folder 39, box 1; "Daily Mail, London, 1907," folder 13, box 3; "Goddard, M., Undated," folder 19, box 6; "New York Herald, 1903," "New York Times, 1905–1915," and "New York World, 1903," folders 40 to 42, box 10; all in Correspondence, 1876–1916, MS.1, Percival Lowell Papers, Lowell Observatory Archives.

7 "Douglass, A. E., 1900–1901," folder 6, box 4; "Illustrated London News, 1903," folder 29, box 7; "Maunder, E. Walter, 1903," folder 73, box 9; all in Correspondence, 1876–1916, MS.1, Percival Lowell Papers, Lowell Observatory Archives; Webb, *Tree Rings,* 47.

8 "Lampland, Carl Otto, 1903–1907," folder 31, box 8, and "Antoniadi, E. M., 1909," folder 27, box 1, Correspondence, 1876–1916, MS.1, Percival Lowell Papers, Lowell Observatory Archives; Robert Markley, *Dying Planet: Mars in Science and the Imagination* (Duke University Press, 2005), 77; Oliver Morton, *Mapping Mars: Science, Imagination, and the Birth of a World* (Picador, 2002), 38; Percival Lowell, "Planetary Photography," *Nature* 77 (1908): 403.

9 "Lowell on His Death—Memorial by E. C. Slipher, 1916," folder 19, box 1, Personal, 1806–2016, MS.1, Percival Lowell Papers, Lowell Observatory Archives; Markley, *Dying Planet,* 159, 174; William M. Sinton, "Spectroscopic Evidence for Vegetation on Mars," *Astrophysical Journal* 126 (1957): 231; Earl C. Slipher, *A Photographic History of Mars, 1905–1961* (Lowell Observatory, 1962).

10 Paolo Ulivi and David M. Harland, *Robotic Exploration of the Solar System,* pt. 1 (Springer Praxis Books, 2007), 30, 41; Markley, *Dying World,* 235.

11 Johnson, *The Sirens of Mars,* 76; Edward Clifton Ezell and Linda Newman Ezell, *On Mars: Exploration of the Red Planet, 1958–1978,* NASA SP-4212 (NASA), 1984; Ulivi and Harland, *Robotic Exploration,* 99.

12 Markley, *Dying Planet,* 114, 239; V. G. Perminov, "The Difficult Road to Mars: A Brief History of Mars Exploration in the Soviet Union," *Monographs in Aerospace History* 15 (1999): 59.

13 V. G. Perminov, "Difficult Road to Mars," 59; Ulivi and Harland, *Robotic Exploration,* 105; Markley, *Dying Planet,* 239.

14 Ulivi and Harland, *Robotic Exploration,* 120.

15 Kevin Schindler, *Lowell Observatory* (Arcadia, 2016); A. E. Douglass, "Scales of Seeing," *Popular Astronomy* 4, no. 4 (1898): 193; Douglass, "The Effect of Mountains on the Quality of the Atmosphere," *Popular Astronomy* 7, no. 7 (1899): 354; Douglass, "Atmosphere, Telescope, and Observer," *Popular Astronomy* 5, no. 2 (1897): 65.

16 Keay Davidson, *Carl Sagan: A Life* (Wiley, 2000), 69; Stephen Baxter, "Martian Chronicles: Narratives of Mars in Science and SF," *Foundation* 0 (1996), 5.

17 Lawrence Badash, *A Nuclear Winter's Tale: Science and Politics in the 1980s* (MIT Press, 2009), 54; Richard Turco, "Appendix II: History of the TTAPS Nuclear Winter Study," 1, Seth MacFarlane Collection of the Carl Sagan and Ann Druyan Archive, Library of Congress; Owen Toon, James Pollack, and Carl Sagan, "Physical Properties of the Particles Composing the Martian Dust Storm of 1971–1972," *Icarus* 30 (1977): 663.

18 James Pollack et al., "Volcanic Explosions and Climatic Change," *Journal of Geophysical Research* 81, no. 6 (1976): 1071; Wilfried Mausbach, "Nuclear Winter," in *Nuclear Threats, Nuclear Fear and the Cold War of the 1980s,* ed. Eckart Conze, Martin Klimke, and Jeremy Varon (Cambridge University Press, 2016), 29.

19 Erik Conway, Donald K. Yeomans, and Meg Rosenburg, *A History of Near-Earth Objects Research* (NASA Office of Communications, 2022), 87.

20 Badash, *A Nuclear Winter's Tale,* 56; Conway, Yeomans, and Rosenburg, *Near-Earth Objects Research,* 88.

21 Committee on the Atmospheric Effects of Nuclear Explosions, *The Effects on the Atmosphere of a Major Nuclear Exchange* (National Academy Press, 1985), 174;Wilfried Mausbach, "Nuclear Winter," in Conze, Klimke, and Varon, *Nuclear Threats, Nuclear Fear,* 29; Richard Turco et al., "The Climatic Effects of Nuclear War," *Scientific American* 251, no. 2 (1984): 33.

22 Carl Sagan, "Nuclear War and Climatic Catastrophe," *Foreign Affairs* 62, no. 2 (1983): 257–292; See also Paul Ehrlich, Carl Sagan, and Donald Kennedy, *The Cold and the Dark: The World after Nuclear War* (W. W. Norton, 1985).

23 Badash, *A Nuclear Winter's Tale,* 163.

24 Badash, *A Nuclear Winter's Tale,* 315; Paul Rubinson, "The Global Effects of Nuclear Winter: Science and Antinuclear Protest in the United States and the Soviet Union during the 1980s," *Cold War History* 14, no. 1 (2014): 66.

25 Markley, *Dying Planet,* 238–266; Dirk Schulze-Makuch, "We May Be Looking for Martian Life in the Wrong Place," *Nature Astronomy* (2024): 1.

26 Markley, *Dying Planet,* 305; Justin A. Holcomb et al., "The Emerging Archaeological Record of Mars," *Nature Astronomy* 8, no. 12 (2024): 1490.

27 Jacob Haqq-Misra, *Sovereign Mars: Transforming Our Values through Space Settlement* (University Press of Kansas, 2022), 2.

28 Markley, *Dying Planet,* 297.

29 Christopher Oze et al., "Perchlorate and Agriculture on Mars," *Soil Systems* 5, no. 3 (2021): 37.

30 Jeffrey S. Kargel, *Mars—A Warmer, Wetter Planet* (Springer, 2004); James Lovelock and Michael Allaby, *The Greening of Mars* (St. Martin's Press, 1984), 44, 101; Kim Stanley Robinson, *Red Mars* (Bantam Books, 1993); Samaneh Ansari et al., "Feasibility of Keeping Mars Warm with Nanoparticles," *Science Advances* 10, no. 32 (2024): eadn4650.

31 David Seargent, *Weird Astronomical Theories of the Solar System and Beyond* (Springer, 2015), 21; SpaceX, "Mars," available at http://www.spacex.com/mars.

Part IV Conclusion

1 Guillem Anglada-Escudé et al., "A Terrestrial Planet Candidate in a Temperate Orbit around Proxima Centauri," *Nature* 536, no. 7617 (2016): 437–440.

2 Cecilia Garraffo, Jeremy J. Drake, and Ofer Cohen, "The Space Weather of Proxima Centauri b," *arXiv preprint* arXiv:1609.09076 (2016); Shane Smith et al., "A Radio Technosignature Search towards Proxima Centauri Resulting in a Signal of Interest," *Nature Astronomy* 5, no. 11 (2021): 1148.

Part V. Comets and Asteroids

1 Donald Yeomans, *Comets: A Chronological History of Observation, Science, Myth, and Folklore* (Wiley, 1991), 52; Gary W. Kronk, *Cometography: A Catalog of Comets,* vol. 2: *1800–1899* (Cambridge University Press, 2003), 268; Antonella Gasperini, Daniele Galli, and Laura Nenzi, "The Worldwide Impact of Donati's Comet on Art and Society in the Mid-19th Century," *Proceedings of the International Astronomical Union* 5, no. S260 (2009): 340; Camille Flammarion, *Mémoires biographiques et philosophiques d'un astronome* (Ernest Flammarion, 1911), 3; William Graves Hoyt, *Lowell and Mars* (University of Arizona Press, 1976), 15.

2 N. D. James, "Comet C / 1996 B2 (Hyakutake): The Great Comet of 1996," *Journal of the British Astronomical Association* 108 (1998): 157–171; Natalie Starkey, *Catching Stardust: Comets, Asteroids and the Birth of the Solar System* (Bloomsbury Sigma, 2018), 46; Josep M. Trigo-Rodriguez, *Asteroid Impact Risk* (Springer, 2022), 4; Rodney Gomes et al., "Origin of the Cataclysmic Late Heavy Bombardment Period of the Terrestrial Planets," *Nature* 435, no. 7041 (2005): 466; Kleomeris Tsiganis et al., "Origin of the Orbital Architecture of the Giant Planets of the Solar System," *Nature* 435, no. 7041 (2005): 459.

3 Andrew Karam, *Comets: Nature and Culture* (Reaktion Books, 2017), 14.

4 David A. Rothery, Neil McBride, and Iain Gilmour, *An Introduction to the Solar System,* 3rd ed. (Cambridge University Press, 2018), 259.

17. Warnings in the Heavens

1 Camille Flammarion, *Omega: The Last Days of the World* (Cosmopolitan, 1894), 10, 23; *Lick Observatory Bulletin* 5 (1908): 59; William Huggins, "Preliminary Note on the Photographic Spectrum of Comet b 1881," *Proceedings of the Royal Society of London* 33, no. 216–219 (1882): 1; Hugh Frank Newall, *The Spectroscope and Its Work* (Society for Promoting Christian Knowledge, 1910), 83; Amelia Urry, "The Astronomer at the End of the World: Visions of Scientific Authority in Camille Flammarion's Apocalyptic Fiction," *Apocalyptica* 1, no. 1 (2022): 69.

2 John Robert Christianson, *On Tycho's Island: Tycho Brahe and His Assistants, 1570–1601* (Cambridge University Press, 2000), 8.
3 Tabitta van Nouhuys, *The Ages of Two-Faced Janus* (Brill, 1998), 3; Victor Thoren, "The Comet of 1577," *Archives internationales d'histoire des sciences* 29, no. 104 (1979): 54.
4 Roger Beck, *A Brief History of Ancient Astrology* (Wiley, 2008); Van Nouhuys, *Two-Faced Janus*, 4, 573; John Sellars, *Stoicism* (Routledge, 2014), 3; Peter Barker and Bernard R. Goldstein, "The Role of Comets in the Copernican Revolution," *Studies in History and Philosophy of Science Part A* 19, no. 3 (1988): 300.
5 Donald Yeomans, *Comets: A Chronological History of Observation, Science, Myth, and Folklore* (Wiley, 1991), 56; Gary W. Kronk, *Cometography: A Catalog of Comets*, vol. 1: *Ancient–1799* (Cambridge University Press, 1999), 331.
6 Yeomans, *Comets*, 83; Robert Hooke, "Cometa," in *Early Science in Oxford*, vol. 8, ed. R. T. Guther (Oxford, 1931), 243.
7 Richard S. Westfall, *The Life of Isaac Newton* (Cambridge University Press, 2015).
8 Kronk, *Cometography*, 1:375; Andrew Karam, *Comets: Nature and Culture* (Reaktion Books, 2017), 52; Allan Chapman, "Edmond Halley's Use of Historical Evidence," *Notes and Records of the Royal Society of London* 48, no. 2 (1994): 168.
9 Yeomans, *Comets*, 57; James Rodger Fleming, *Historical Perspectives on Climate Change* (Oxford University Press, 2005), 11.
10 Edmond Halley, "Some Considerations," *Philosophical Transactions of the Royal Society* 33, no. 383 (1724): 118.
11 Yeomans, *Comets*, 164; William Whiston, *A New Theory of the Earth* (John Whiston, 1737), 195.
12 Lorraine Daston, *Classical Probability in the Enlightenment* (Princeton University Press, 1998); Ian Hacking, *The Emergence of Probability* (Cambridge University Press, 2006); Edmond Halley, "A Synopsis of the Astronomy of Comets" (Printed for John Senex, 1705), 2; K. Zachmann, "Risk in Historical Perspective," in *Risk—A Multidisciplinary Introduction*, ed. C. Klüppelberg, D. Straub, and I. M. Welpe (Springer, 2014).
13 Voltaire, "Lettre sur la prétendue comète," in *Œuvres complètes de Voltaire* (Garnier, 1879), 52; Joseph Jérôme Le Français de Lalande, "Réflexions sur les comètes: Qui peuvent approcher de la terre" (Gilbert, 1773), 39; P. Stewart, "Science and Superstition: Comets and the French Public in the 18th Century," *American Journal of Physics* 54 (1986), 16.
14 Ilaria Ampollini, "1773: France Starts to Discuss How to Communicate Risk," *Public Understanding of Science* 27, no. 8 (2018): 1005.
15 Barry Hankins, *The Second Great Awakening and the Transcendentalists* (Greenwood, 2004), 12.
16 Gary W. Kronk, *Cometography: A Catalog of Comets*, vol. 2: *1800–1899* (Cambridge University Press, 2003), 59; Karam, *Comets*, 114; William Miller, *Wm. Miller's Apology and Defence* (Joshua V. Himes, 1945); F. Arago, *Tract on Comets* (Hilliard, Gray and Co., 1832); Edgar Allan Poe, *The Conversation of Eiros and Charmion* (Lindhardt og Ringhof, 2023).
17 Andy Bruno, *Tunguska: A Siberian Mystery and Its Environmental Legacy* (Cambridge University Press, 2022), 28.
18 Bruno, *Tunguska*, 9.
19 Ruth Freitag, "Elliptical Designs: Halley's Comet as a Medium for Advertising Messages," *U.S. Library of Congress Quarterly Journal* 40, no. 3 (1983): 266–277; Camille Flammarion, "La Recontre de la comête," *Bulletin de la Société astronomique* 24 (1910): 59; Richard J. Goodrich, *Comet Madness* (Rowman and Littlefield, 2023), 37, 57.

20 "Comet's Poisonous Tail," *New York Times,* February 8, 1910; "Chicago Is Terrified," *New York Times,* May 18, 1910; David Peck Todd, *Halley's Comet* (American Book Co., 1910); H. J. P. Arnold, "Halley's Comet in 1910," *Spaceflight* 28 (1986); Goodrich, *Comet Madness,* 37, 57.

21 Agustín Sánchez-Lavega et al., "The Origin of Jupiter's Great Red Spot," *Geophysical Research Letters* 51, no. 12 (2024): 1; W. Ernest Cooke, "A Few Notes on Jupiter," *Transactions and Proceedings and Report of the Royal Society of South Australia* 9 (1887): 9; H. C. Russell "Recent Changes in the Surface of Jupiter," *The Observatory* 4 (1881): 324; William Noble, "Note on Two Sketches of Jupiter," *Monthly Notices of the Royal Astronomical Society* 40, no. 2 (1879): 86.

22 F. J. Hargreaves, B. M. Peek, and T. E. R. Phillips, "The Revival of Activity in the Southern Hemisphere of Jupiter," *Monthly Notices of the Royal Astronomical Society* 89, no. 2 (December 1928): 210; W. F. Denning, "Dark Spots in the N. Tropical Region of Jupiter," *Popular Astronomy* 7, no. 1 (1899): 14; E. C. Slipher, "Recent Changes and Extraordinary Coloration in Jupiter's Cloud Forms," *Publications of the American Astronomical Society* 9 (1939): 167.

18. Encountering Near-Earth Objects

1 William Graves Hoyt, *Coon Mountain Controversies* (University of Arizona Press, 1987), 74.

2 Hoyt, *Coon Mountain Controversies,* 82.

3 Hoyt, *Coon Mountain Controversies,* 296.

4 C. J. Cunningham, *Discovery of the First Asteroid, Ceres* (Springer, 2016); J. Bode, *Einleitung zur Kenntnis des Gestirnten Himmels,* 2nd ed. (Dieterich Harmsen, 1772); J. Titius von Wittenberg, *Betrachtung uber die Natur* (Herrn Karl Bonnet, 1766).

5 Erik Conway, Donald K. Yeomans, and Meg Rosenburg, *A History of Near-Earth Objects Research* (NASA Office of Communications, 2022), 22.

6 Ursula Marvin, "Meteorites in History," in *The History of Meteoritics and Key Meteorite Collections,* ed. A. J. Bowden, Gerald Joseph Home McCall, and Richard John Howarth (Geological Society, 2006): 34; John Burke, *Cosmic Debris* (University of California Press, 1991), 39.

7 Conway, Yeomans, and Rosenburg, *Near-Earth Objects Research,* 23; Hoyt, *Coon Mountain Controversies,* 47.

8 Warren D. Cummings, *Evolving Theories on the Origin of the Moon* (Springer, 2019), 1; James Powell, *Mysteries of Terra Firma: The Age and Evolution of the Earth* (Simon and Schuster, 2007), 175; Kathleen Mark, *Meteorite Craters* (University of Arizona Press, 1995), 29; G. K. Gilbert, "The Moon's Face: A Study of the Origin of Its Features," in *Bulletin XII* (Philosophical Society of Washington, 1893).

9 Conway, Yeomans, and Rosenburg, *Near-Earth Objects Research,* 46; Ralph Belknap Baldwin, *The Face of the Moon* (University of Chicago Press, 1949).

10 Milo D. Nordyke, "Nuclear Craters and Preliminary Theory of the Mechanics of Explosive Crater Formation," *Journal of Geophysical Research* 66, no. 10 (1961): 3439; Conway, Yeomans, and Rosenburg, *Near-Earth Objects Research,* 46.

11 Eugene Shoemaker, *Impact Mechanics at Meteor Crater, Arizona,* Report no. 59–108, US Geological Survey, 1959; Eugene Shoemaker and Robert Hackman, "Stratigraphic Basis for a Lunar Time Scale," *The Moon* 14 (1962): 289; Conway, Yeomans, and Rosenburg, *Near-Earth Objects Research,* 47.

12 L. R. Magnolia, *Interplanetary Matter* (STL Technical Library, 1963), 43; "Scientist Says Man May Divert Asteroid to Shatter Continent," *Washington Post,* January 19, 1962.

13 "Nuclear End to Asteroid Threat Hinted," *Chicago Tribune,* July 28, 1966; Walter Sullivan, "Asteroid's Collision with Earth in '68 Is Doubted," *New York Times,* July 31, 1966; Robert S. Richardson, "The Discovery of Icarus," *Scientific American* 212, no. 4 (1965): 106; Brian G. Marsden, "Comets and Asteroids: Searches and Scares," *Advances in Space Research* 33, no. 9 (2004): 1514.

14 Sullivan, "Asteroid's Collision with Earth"; "Observatory Denies Asteroid Perils Earth," *Washington Post,* July 28, 1966; Walter Sullivan, "The Threat of the Wandering Asteroids," *New York Times,* June 9, 1968.

15 L. A. Kleiman, *Project Icarus* (MIT Press, 1968), 158.

16 Kleiman, *Project Icarus,* 152.

17 David Morrison, "The Contemporary Hazard of Comet Impacts," in *Comets and the Origin and Evolution of Life,* ed. P. J. Thomas, R. D. Hicks, C. F. Chyba, and C. P. McKay (Springer, 2006), 287; Kleiman, *Project Icarus,* 161; "Systems Engineering: Avoiding an Asteroid," *Time,* June 16, 1967.

18 "JPL Radar Icarus Tiny, Very Rough," Office of Public Information, Jet Propulsion Laboratory, November 22, 1968.

19 Hannes Alfven to Homer E. Newell, November 7, 1969; Homer E. Newell to Hannes Alfven, November 17, 1969; Arnold W. Frutkin to Associate Administrator, Office of Space Science and Applications, November 19, 1969; Donald P. Hearth to Hannes Alfven, November 26, 1969; and Wernher von Braun to Ernest Stuhlinger, May 25, 1971; all NASA HQ 10104; "Study of Multiple Asteroid Flyby Missions," NASA Contract NAS 2-6866, November 17, 1972; "Comets and Asteroids: A Strategy for Exploration," NASA TM X-64677, May 1972; *The 1973 Report and Recommendations of the NASA Science Advisory Committee on Comets and Asteroids,* NASA, October 1973; Hannes Alfvén to George M. Low, April 10, 1975, NASA HQ 10003; Hannes Alfvén and Gustaf Arrhenius, "Mission to an Asteroid," *Science* 167, no. 3915 (1970): 139.

20 Su-Shu Huang, *Velocity Modification* (NASA, 1961); Conway, Yeomans, and Rosenburg, *Near Earth Objects Research,* 246.

21 Donella Meadows et al., *The Limits to Growth* (Universe Books, 1972), 24; Meg Jacobs, *Panic at the Pump* (Macmillan, 2016); Michael Gaffey and Thomas McCord, "A Source of Natural Resources Outside Earth" (MIT, 1976), 2, NASA HQ 10003.

22 Gaffey and McCord, "Source of Natural Resources," 39.

23 "Star Dreck," *West,* March 22, 1992; D. Morrison, *The Spaceguard Survey: Report of the NASA International Near-Earth- Object Detection Workshop* (JPL / NASA, 1992), 22, NASA HQ 10003; Conway, Yeomans, and Rosenburg, *Near-Earth Objects Research,* 58.

24 Arthur C. Clarke, *Rendezvous with Rama* (Harcourt Brace, 1973); Larry Niven and Jerry Pournelle, *Lucifer's Hammer* (Del Rey, 1985); Isaac Asimov, *From Earth to Heaven* (Dobson, 1968); Oliver Morton, "In Retrospect: Lucifer's Hammer," *Nature* 453, no. 1184 (2008): 1184.

19. Giggles and Close Calls

1 Milly Alvarez, "Life with a Field Geologist," in *From the Guajira Desert to the Apennines, and from Mediterranean Microplates to the Mexican Killer Asteroid,* ed. Christian Koeberl, Philippe Claeys, and Alessandro Montanar (Geological Society of America, 2022), 7.

2 Erik Conway, Donald K. Yeomans, and Meg Rosenburg, *A History of Near-Earth Objects Research* (NASA Office of Communications, 2022), 87; Walter Alvarez, *T. rex and the Crater of Doom* (Princeton University Press, 2013).

3 Alvarez et al., "Extraterrestrial Cause for the Cretaceous-Tertiary Extinction," *Science* 208, no. 4448 (1980): 1095.

4 Conway, Yeomans, and Rosenburg, *Near-Earth Objects Research,* 87; Alvarez, *Crater of Doom,*34.

5 *A Report of the Discussions at the NASA's Advisory Council's New Directions Symposium,* NASA Advisory Council, September 9, 1980, Tom Gehrels Papers, University of Arizona Special Collections.

6 *Report of a Workshop Held at Snowmass, Colorado, July 13–16, 1981,* Tom Gehrels Papers, University of Arizona Special Collections; Norman C. Rasmussen, *Reactor Safety Study: An Assessment of Accident Risks in U.S. Commercial Nuclear Power Plants* (U.S. Nuclear Regulatory Commission, 1975); K. Zachmann, "Risk in Historical Perspective," in *Risk—A Multidisciplinary Introduction,* ed. C. Klüppelberg, D. Straub, and I. M. Welpe (Springer, 2014), 22.

7 "Collision of Asteroids and Comets with the Earth," Snowmass, Colorado, July 13–16, 1981, University of Arizona Special Collections.

8 "Collision." University of Arizona Special Collections; Conway, Yeomans, and Rosenburg, *Near-Earth Objects Research,* 110.

9 "Asteroid Hits and Near Misses," *Washington Times,* August 24, 1990; James C. Mesco, "Watch the Skies," *Quest* 6, no. 4 (1998); Mark E. Bailey et al., "The 1930 August 13 Brazilian Tunguska event," *The Observatory* 115 (1995): 250; "Collision," University of Arizona Special Collections.

10 Jim Kukowski, "New Asteroid / Comet Nuclei Hazard Studies Announced," NASA News, May 20, 1986, NASA HQ 10003; G. W. Wetherill, "Comments," July 21, 1981, Tom Gehrels Papers, University of Arizona Special Collections.

11 Alex Friedlander to G. W. Wetherill, July 27, 1981, and E. M. Shoemaker, "Draft of Workshop Report," August 19, 1981, Tom Gehrels Papers, University of Arizona Special Collections.

12 Walter Sullivan, "Scientists Ponder Forcing Asteroids into Safe Orbits," *New York Times,* January 4, 1983; "Experts Caution on Space Chunks," *Chicago Tribune,* May 22, 1986.

13 Conway, Yeomans, and Rosenburg, *Near-Earth Objects Research,* 113; Toby Ord, *The Precipice: Existential Risk and the Future of Humanity* (Hachette Books, 2020), 58.

14 James P. Sterba, "Scientists Think Glowing Object Was an Asteroid," *New York Times,* August 26, 1982; "New Large Asteroid Heading Near Earth," *JPL News Clips,* December 9, 1982; Thomas O'Toole, "Space Telescope Finds Thousands of Asteroids," *Washington Post,* December 31, 1983; Brian O'Leary, "Mining the Asteroids," *Aviation Space* (Spring 1984): 56; Delthia Ricks, "Mining of 2 Asteroids Considered Possible," *Daily News,* June 15, 1986; "Scientists Believe Asteroid May Have Started Ice Age," *Washington Times,* March 16, 1987; "Huge Asteroid Striking Earth May Have Spurred Glacial Age," *Washington Post,* July 4, 1988.

15 Conway, Yeomans, and Rosenburg, *Near-Earth Objects Research,* 111.

16 Warren E. Leary, "Big Asteroid Passes Near Earth Unseen in a Rare Close Call," *New York Times,* April 19, 1989; "A Close Call—Cosmically Speaking," *Fairfax Journal,* April 21, 1989; George Johnson, "The Odds of a Great Celestial Smashup," *New York Times,* April 23, 1989; Robert G. Nichols, "Will We Be Ready . . . When Worlds Collide?," *Final Frontier,* March / April 1992; Carolyn S. Shoemaker and Henry E. Holt, "The Palomar Asteroid and Comet Survey (PACS), 1983–1993," in *Asteroids, Comets, Meteors 1993: Proceedings of the 160th Symposium of the International Astronomical Union, Held in Belgirate, Italy, June 14–18, 1993,* ed. A. Milani, M. Di Martino, and A. Cellino (Kluwer Academic, 1994), 269.

17 Lynn W. Heninger to Senator Christopher J Dodd, April 30, 1989; Senator Bob Graham, August 1, 1989; and Martin P. Kress to Senator Alan K. Simpson, August 31, 1990; all in NASA HQ 10003.

18 Conway, Yeomans, and Rosenburg, *Near-Earth Objects Research,* 82; D. Morrison, *The Spaceguard Survey: Report of the NASA International Near-Earth- Object Detection Workshop* (JPL / NASA, 1992), 3, NASA HQ 10003; Edward Tagliaferri, "Dealing with the Threat of an Asteroid Striking the Earth," American Institute of Aeronautics and Astronautics, 1990.

19 Morrison, *The Spaceguard Survey,* 1; "The Big Peril Is Out There," *Washington Times,* July 2, 1991; Blaine Friedlander, "Asteroid Hurtled Near Earth Jan. 18" and "Threat of Asteroids," *Baltimore Sun,* May 15, 1990; "Asteroid Threat Is Real, Quayle Says," *Chicago Tribune,* June 5, 1990; "Asteroids: Clear and Present Danger," *Ad Astra,* 1990; "Tracking the Asteroid Threat," *Baltimore Sun,* April 25, 1992; National Aeronautics and Space Administration Multiyear Authorization Act of 1990, September 26, 1990; "Beep-Beep Veep?," *Seattle Times,* May 19, 1990; John Noble Wilford, "Astronomers Duck as a Tiny Asteroid Passes," *New York Times,* January 25, 1991; Alan R. Hildebrand et al., "Chicxulub Crater: A Possible Cretaceous / Tertiary Boundary Impact Crater on the Yucatan Peninsula, Mexico," *Geology* 19, no. 9 (1991): 867–871.

20 Clark Chapman and David Morrison, *Cosmic Catastrophes* (Springer, 1989), 4.

21 Morrison, *The Spaceguard Survey,* 1; "Scientists Support Building Telescopes to Protect Earth from Asteroids," *Aviation Week & Space Technology,* October 14, 1991; Clark Chapman and David Morrison, "Impacts on the Earth by Asteroids and Comets," *Nature* 367 (1994), 33.

22 Morrison, *The Spaceguard Survey,* vi; Arthur C. Clarke to Alistair Cooke, August 10, 1991, NEO Interception & Detection Workshops, 1991–1992: Questionnaires & Responses, Record Number: 20375, NASA HQ Archives; "Big Bang." *Scientific American* (November 1991).

23 NEO Interception & Detection Workshop 1991–1992, Sessions and Participant List, NASA HQ 20375; Conway, Yeomans, and Rosenburg, *Near-Earth Objects Research,* 115.

24 "Star Dreck," *West,* March 22, 1992; Lowell Wood et al., *Cosmic Bombardment II,* Report no. UCRL-ID-103771, Lawrence Livermore National Laboratory, 1990; Gerrit Verschuur, "This Target Earth," *Air and Space,* October / November 1991.

25 John Rather to Faith Vilas, September 20, 1991, Sessions and Participant List, NASA HQ 20375.

26 John Rather to David Morrison, October 25, 1991, Sessions and Participant List, NASA HQ 20375.

27 Felicity Mellor, "Negotiating Uncertainty: Asteroids, Risk and the Media," *Public Understanding of Science* 19, no. 1 (2010): 19.

28 Verschuur, "This Target Earth."

29 "Important Ideas," Sessions and Participant List, NASA HQ 20375.

30 David Morrison to John Rather, November 7, 1991, and David Morrison to John Rather, November 8, 1991, Sessions and Participant List, NASA HQ 20375.

31 "Steering Group 12-13-91" and Brian Marsden to John Rather, March 4, 1992, Sessions and Participant List, NASA HQ 20375; Gregory H. Canavan, "Future Direction for Research Value of Space Defenses," in *Summary Report of the Near-Earth- Object Interception Workshop,* ed. J. D. G. Rather, J. H. Rahe, and G. Canavan (JPL / NASA, August 31, 1992), 5; "Star Dreck."

32 Rather, Rahe, and Canavan, *Summary Report,* 9, 28, 34; Johndale Solem, "Interceptions of Comets and Asteroids on Collision Course with Earth," R. Hyde et al., "Cosmic Bombardment III," J. G. Hills, "Capturing Asteroids into Bound Orbits around the Earth," and

Anthony Zuppero, "Discovery in Near-Earth Space," all in Rather, Rahe, and Canavan, *Summary Report;* Cory Spencer to John Rather, July 8, 1992, and Gene Shoemaker to John Rather, June 16, 1992, Sessions and Participant List, NASA HQ 20375.

33 Edward Teller, "Why Now?," in Rather, Rahe, and Canavan, *Summary Report,* 36.

34 Brian Marsden to John Rather, March 24, 1992, Steven J. Ostro to John Rather, January 6, 1992, and Tom Gehrels to John Rather, August 19, 1992, Questionnaires & Responses, NASA HQ 20375; Rather, Rahe, and Canavan, *Summary Report,* 38.

35 David Morrison to John Rather April 8, 1992, David Morrison to John Rather, June 10, 1992, Clark Chapman to John Rather et al., and "Critique of Draft Report of NASA NEO Interception Workshop," Sessions and Participant List, NASA HQ 20375; Clark Chapman to John Rather, Jurgen Rahe, and Gregory Canavan, *"Near-Earth-Object Interception Workshop" Report,* June 8, 1992, Sessions and Participant List, NASA HQ 20375.

36 Chapman, "Critique," Eleanor Helin to John Rather, and John Rather to Duncan Steel, June 16, 1992, Sessions and Participant List, NASA HQ 20375.

37 Rather, Rahe, and Canavan, *Summary Report,* 41, 45, 293; Robert L. Park, "Star Warriors on Sky Patrol," *New York Times,* March 25, 1992; "A Rocky Watch for Earthbound Asteroids," *Science* 255 (1992); Bob Davis, "Never Mind the Peace Dividend, the Killer Asteroids Are Coming," *Wall Street Journal,* March 25, 1992; "The Asteroids Are Coming. NOT!," *Washington Times,* April 1, 1992; "Killer Asteroids: The Perfect Peril," *New York Times,* April 6, 1992; "Talk about Star Wars," *Time,* April 6, 1992; Roger Lewin, "How to Destroy the Doomsday Asteroid," *New Scientist,* June 8, 1992.

38 Roger Launius, "Public Opinion Polls and Perceptions of US Human Spaceflight," *Space Policy* 19, no. 3 (2003): 164; Dagomar Degroot, "'A Catastrophe Happening in Front of Our Very Eyes': The Environmental History of a Comet Crash on Jupiter," *Environmental History* 22, no. 1 (2016): 13.

39 "Star Dreck"; Jim Erickson, "Killer Asteroids: Tucsonan Opposes 'Bizarre' Interception Ideas," *Arizona Daily Star,* November 26, 1992; Richard Stone, "Scientists Collide on NASA Comet Report," *Science Scope,* November 13 1992; Jeff Hecht, "Will We Catch a Falling Star?," *New Scientist,* September 7, 1991; Rather, Rahe, and Canavan, *Summary Report,* 38; Carl Sagan, "When Worlds Collide: The Beginnings and Ends of Worlds," *Parade,* March 3, 1991, 4.

40 Edward Teller, "Celestial Search and Deflect," *Washington Times,* May 18, 1992; Kathy Sawyer, "Shooting Back at Space Rocks?," *Washington Post,* March 30, 1992; "Here Are Some Arguments" (1988), "Should the U.S. Spend More on Space Research?" (1992), "We Are Faced with Many Problems" (1991), "Do You Strongly Favor" (1990), and "Are We Spending Too Much" (1990), all available at https://ptn.infobase.com.

41 John Rather, "NASA Asteroid Report," *Science* 259 (1993); Michael Braukus, "NASA Completes Asteroid Workshop Studies," NASA News, March 31, 1992; Richard Truly to George E. Brown, Questionnaires & Responses, NASA HQ 20375; Robert Matthews, "Death in June," *Discover,* June 16, 1992.

42 Leonard David, "Defense Experts Duck Asteroid Threat Hearing," *Space News,* March 29–April 4, 1993.

20. Avoiding Jupiter's Fate

1 David Levy, "Thursday March 25," in *The David Levy Logbooks,* vol. 17, Royal Astronomical Society of Canada; Brian G. Marsden, "IAUC 5800: 1993e," International Astronomical Union

Central Bureau for Astronomical Telegrams, May 22, 1993; Brian G. Marsden, "The Path to Destruction," in *The Great Comet Crash,* ed. John Spencer and Jacqueline Mitton (Cambridge University Press, 1995), 103.

2 Malcom W. Browne, "Astronomers Opening Vigil for Comet's Jupiter Crash," *New York Times,* July 17, 1994, I2; Richard Leach, "Spacecraft System Failures and Anomalies Attributed to the Natural Space Environment," in *Space Programs and Technologies Conference 1995,* 3564, online at https://arc.aiaa.org/doi/book/10.2514/MSPACE95; "Planetary Fireworks: Comet Shoemaker-Levy 9 Impact on Jupiter This Summer," Press Information Note no. 16–94, European Space Agency, June 29, 1994, 1; "Planetary Spacecraft as Observers." *Comet Impact '94: Fact Sheet* (NASA HQ History Division, 1994).

3 Dagomar Degroot, "'A Catastrophe Happening in Front of Our Very Eyes': The Environmental History of a Comet Crash on Jupiter," *Environmental History* 22, no. 1 (2016): 31.

4 Clark R. Chapman, "What If . . . ?," in *The Great Comet Crash: The Collision of Comet Shoemaker-Levy 9 and Jupiter,* ed. John R. Spencer and Jacqueline Mitton (Cambridge University Press, 1995), 105; Jacqueline Mitton, "Introduction," in Spencer and Mitton, *The Great Comet Crash,* 1; Malcolm W. Browne, "Jupiter's Comet Spectacular Ends with a Fiery Flourish," *New York Times,* July 23, 1994, 26.

5 Degroot, "'A Catastrophe,'" 34.

6 Degroot, "'A Catastrophe,'" 34.

7 Degroot, "'A Catastrophe,'" 35; Carl Sagan, "A Warning for Us?," *Parade,* June 5, 1994, 8.

8 Rajiv Chandrasekaran, "Clouds Parted, Jupiter's Drama Plays in Heavens," *Washington Post,* July 21, 1994, B1; Tim Friend, "Jupiter Is Center Stage in Its Collision Course with Comet," *USA Today,* July 12, 1994, 6D; Randy Pakan logbooks and Leo Enright logbooks, Royal Astronomical Society of Canada Archive; Jessica Ellen Sewell and Andrew Johnston, "Material Culture and the Dobsonian Telescope," *Spontaneous Generations* 41 (2010): 155–162.

9 Roylance, "Like a Nuclear Space War," *Baltimore Sun,* January 16 1994, H6; David Levy, "The Collision of Comet Shoemaker-Levy 9 with Jupiter," *Space Science Reviews* 85 (1998): 533; Kathy Sawyer, "Comet Leaves a Trail of Data in Its Wake," *Washington Post,* July 24, 1994, A4; Chandrasekaran, "Clouds Parted"; Malcolm W. Browne, "Comet Pummels Jupiter, Revealing Details," *New York Times,* July 18, 1994, A1.

10 A. R. Hogan, "Deadly, Promising," *Houston Chronicle,* October 1, 1990; William J. Broad, "When Worlds Collide: A Threat to the Earth Is a Joke No Longer," *New York Times,* August 1, 1994; Detjen, "Jupiter Comet Provides Nation Something New to Worry About," *Rome News-Tribune,* July 31, 1994, 12A; Kathy Sawyer, "Comet Barrage Alters Face of Planet Jupiter," *Washington Post,* July 22, 1994, A3; "How Historic Jupiter Comet Impact Led to Planetary Defense," NASA, 2019, available at https://www.nasa.gov/feature/goddard/2019/how-historic-jupiter-comet-impact-led-to-planetary-defense.

11 *Report of the Near-Earth Objects Survey Working Group,* NASA, June 1995, 3, NASA HQ 10002; Erik Conway, Donald K. Yeomans, and Meg Rosenburg, *A History of Near-Earth Objects Research* (NASA Office of Communications, 2022), 148; NASA, *NASA Strategic Plan 1996,* available at http://www.hq.nasa.gov/office/codez/stratplans/1996/Welcome.html; NASA, *NASA Strategic Plan 1995,* available at http://www.hq.nasa.gov/office/codez/stratplans/1995/NSPTOC.html.

12 David Chandler, "Hale-Bopp Will Pass in Spring," *Milwaukee Journal Sentinel,* November 11, 1996; Joe Bauman, "Hale-Bopp: The Common Man's Comet," *Deseret News,* April 10, 1997; Patrick E. Tyler, "Chinese Seek Atom Option to Fend Off Asteroids," *New York Times,*

April 27, 1996; William Broad, "For Killer Asteroids, Respect at Last," *New York Times,* May 14, 1996, C1; C. F. Chyba et al., "Monitoring the Comprehensive Test Ban Treaty," *Geophysical Research Letters* 25 (1998): 191.

13 Conway, Yeomans, and Rosenburg, *Near-Earth Objects Research,* 139.

14 Valerie A. Olson, "Political Ecology in the Extreme," *Anthropological Quarterly* (2012): 1029; Conway, Yeomans, and Rosenburg, *Near-Earth Objects Research,* 231.

15 Ulrike Landfester et al., "Summary Report of the Review of US Human Space Flight Plans Committee," in *Humans in Outer Space—Interdisciplinary Perspectives,* ed. Ulrike Landfester et al. (Springer, 2011), 243–257; Conway, Yeomans, and Rosenburg, *Near-Earth Objects Research,* 282.

16 Conway, Yeomans, and Rosenburg, *Near-Earth Objects Research,* 282, 305; "Discovery Statistics," Center for Near Earth Object Studies, Jet Propulsion Laboratory, available at https://cneos.jpl.nasa.gov/stats/totals.html.

Part V Conclusion

1 Toby Ord, *The Precipice: Existential Risk and the Future of Humanity* (Hachette Books, 2020), 72.

2 Erik Conway, Donald K. Yeomans, and Meg Rosenburg, *A History of Near-Earth Objects Research* (NASA Office of Communications, 2022), 236; Leonard David, "Comet Chasing Makes Deep Impact on Science," *Aerospace America* (2011): 40; Alicia Chang, "NASA Craft Braved Comet Ice Storm during Flyby," *Washington Post,* November 18, 2010; "Deep Impact Comet Encounter," NASA press kit, June 2005, 4, 13, NASA HQ Archives.

3 Conway, Yeomans, and Rosenburg, *Near-Earth Objects Research,* 283; Cristina Thomas et al., "Orbital Period Change of Dimorphos Due to the DART Kinetic Impact," *Nature* (2023): 1.

4 Namrata Goswami and Peter A. Garretson, *Scramble for the Skies* (Lexington Books, 2020); Deganit Paikowsky and Roey Tzezana, "The Politics of Space Mining," *Acta Astronautica* 142 (2018): 10; Asterank, "Asteroid Database and Mining Rankings," www.asterank.com.

Conclusion

1 David H. Levy, *Shoemaker by Levy: The Man Who Made an Impact* (Princeton University Press, 2002), 262.

2 Levy, *Shoemaker,* 264.

3 Levy, *Shoemaker,* 266.

4 Carl Sagan, *Cosmos* (Random House, 1980), 4.

5 Daniel Sage, *How Outer Space Made America: Geography, Organization, and the Cosmic Sublime* (Ashgate, 2014), 3.

6 Gerard K. O'Neill, *The High Frontier: Human Colonies in Space* (1977), 15.

7 Carl Sagan, *Pale Blue Dot: A Vision of the Human Future in Space* (Ballantine Books, 1997), 209.

8 Robert Zubrin, *The Case for Mars* (Simon and Schuster, 2011), xxxvi.

9 Hannah Hunter and Elizabeth Nelson, "Out of Place in Outer Space?," *Environment and Society* 12, no. 1 (2021): 227.

10 Alberto G. Fairén et al., "Searching for Life on Mars before It Is Too Late," *Astrobiology* 17, no. 10 (2017): 962; Frank Tavares et al., "Ethical Exploration and the Role of Planetary Protection in Disrupting Colonial Practices," *arXiv preprint* arXiv: 2010.08344 (2020); Daniel Deudney, *Dark Skies: Space Expansionism, Planetary Geopolitics, and the Ends of Humanity* (Oxford University Press, 2020).

11 O'Neill, *The High Frontier,* 19; Neil M. Maher, *Apollo in the Age of Aquarius* (Harvard University Press, 2017), 38; Patrick McCray, *The Visioneers: How a Group of Elite Scientists Pursued Space Colonies, Nanotechnologies, and a Limitless Future* (Princeton University Press, 2013), 86.

12 O'Neill, *The High Frontier,* 19.

13 McCray, *The Visioneers,* 111.

14 Athanasios Goulas and Ross Friel, "3D Printing with Moondust," *Rapid Prototyping Journal* 22, no. 36 (2016): 864; Harry Jones, "The Recent Large Reduction in Space Launch Cost," 48th International Conference on Environmental Systems, 2018, https://www.aconf.org/conf_130958.html.

15 Donald Goldsmith and Martin Rees, *The End of Astronauts: Why Robots Are the Future of Exploration* (Harvard University Press, 2022).

16 See John Mankins, *The Case for Space Solar Power* (Virginia Edition, 2014).

17 Erica Rodgers et al., *Space-Based Solar Power,* NASA HQ, 2024; L. M. Fraas and M. J. O'Neill, *Low-Cost Solar Electric Power* (Springer, 2014).

18 Keynote Panel, NASA and the Environment Symposium, Georgetown University, September 22, 2022.

19 See Toby Ord, *The Precipice: Existential Risk and the Future of Humanity* (Hachette Books, 2020).

Acknowledgments

The act of expressing gratitude is a form of historical scholarship that can be humbling for the historian. Our job is to identify the causes of historical change, and ideally to formulate an argument for which changes were most important in a particular time and place. Yet it turns out to be nearly impossible to identify every cause—every person worth thanking—responsible even for one's own book, much less to rank contributions by order of significance. In any work of historical scholarship, more is omitted than included. That certainly pertains to these acknowledgments.

My incomplete list of thanks must begin with the Department of History at Georgetown University. I was hired as a historian of climate change, and here I am having written a book about outer space (although, as you have read, there is a lot of climate in it). I am keenly aware that not every department would afford me the freedom to explore my interests, wherever they lead. Nor would every department so generously support a colleague whose methods and sources can seem more at home in the sciences than the humanities.

The most unusual method I used in this book was to follow in the footsteps of my historical protagonists by peering through telescopes. I purchased these telescopes with funds provided by Georgetown University, and again, most universities would not have understood my need to have them. I used telescopes to gain a clear sense of what the observers in this book saw when they gazed at the heavens, and more than that: to develop a visceral understanding of what I call cosmic environments. Yes, the telescope is a scientific tool that allows us to see up close what is far away, but that sterile description obscures its true purpose. It is a device that casts a spell, its magic bending light to unveil the true nature of reality. Telescopes allowed me to experience how small and precarious we are on our little planet as it tumbles through a vast and changeable cosmos, like a mote rushing through a burbling stream. Huddling in parks and courtyards, I was moved to tears by the experience, which shaped how I imagine the influence of the solar system on human history. In many passages of this book, I describe viewing cosmic environments through telescopes. To read more about these observations, you can visit AstroJournals.com.

One of the joys of my professional life has been to work within, and contribute toward, a vibrant community of environmental historians at Georgetown University. John McNeill, the creator of this community, has been a confidant and friend for over a decade. He has been a model of the scholar and mentor that I aspire to be. Tim Newfield has been not only a colleague but a comrade, a partner in teaching and in research. Meredith McKittrick and Kathryn de Luna, historians of Africa who transgress disciplinary boundaries and think deeply about the environment, have educated and inspired me.

I am also fortunate to be affiliated with Georgetown's Earth Commons, our school for the environment. I have enjoyed the friendship and wisdom of Pete Marra, the institute's dean, and its faculty, including Rebecca Helm and Megan Lickley. I have also benefited from grants provided by the Earth Commons, which allowed me to visit many of the archives that informed this book. The grants also allowed me to broaden my mind by hosting major events at Georgetown, on topics such as climate change resilience, NASA's environmental history, and the connection between sustainability and the search for life beyond Earth. I thank my trailblazing colleague Sarah Johnson for helping me manage some of these events and, in the process, educating me on the specifics of astrobiology and planetary science. I also thank Bill Diamond, Amy Hessl, Neil Maher, Teasel Muir-Harmony, and Brian Odom, stars in their respective universes who worked with me to organize different events and, in the process, taught me a great deal that ended up in this book.

I am equally grateful for our Georgetown undergraduate students. I refined the ideas in this book by teaching a popular seminar, "Neighboring Worlds," in which students explore the history of human engagement with Mars, the Moon, and Venus. The insights and enthusiasm of Max Bryant, Benjamin Drake, Grant Draper, Liam Emery Moynihan, Ruby Gilmore, Darren Jian, Ellena Joo, Leila Khan, Clara Ma, Benjamin Manens, Katherine Moar, Justin Potisit, Laura Ratliff, Cody Slutzky, and Jake Wexelblatt, in particular, informed this book's conclusion.

I could not have written this book without the guidance of archivists across North America and Europe. I am especially grateful to Lauren Amundson and Peanut, a cantankerous Chihuahua, at the Lowell Observatory; Tamatha Brumley at the University of Houston–Clear Lake; Colin Fries and Sarah LeClaire at the History Program Office at NASA Headquarters; Benjamin Gross at the Linda Hall Library of Science, Engineering and Technology; Sian Prosser at the Archives of the Royal Astronomical Society; and Randall Rosenfeld at the Royal Astronomical Society of Canada. I am also thankful for Kennedy Delaney and Andrew Ross, brilliant students who helped me explore archival collections at the Library of Congress.

I am profoundly thankful for the uplifting support of my agent, Elise Capron at Sandra Dijkstra Literary Agency, and for the generous guidance of my editors at Harvard University Press and Viking. Rachel Field, in particular, improved this book by offering the most detailed and informative feedback I have received in my career. Her indefatigable reading and rereading of the book yielded a trove of insights that helped me craft stronger arguments and express those arguments with greater clarity. Thanks to Rachel, most of the images in this book are in color.

Three remarkable talents are primarily responsible for the images in this book. I am grateful to the artist Clare Dean for drawing the sketches that introduce the five parts of this book. I thank the planetary scientist and artist William Hartmann for allowing me to reproduce his paintings of solar system environments. And my thanks to the environmental historian and cartographer Geoff Wallace, whose maps of the solar system and climate change elevate this book.

Many colleagues at Georgetown and around the world also deserve thanks, but I am running out of pages to express my gratitude. With apologies to those I do not have space to mention, my profound thanks to Ananya Chakravarti,

David Collins, David Grinspoon, Jacob Haqq Misra, Toshihiro Higuchi, Kathryn Olesko, and Jonathan Wiener. I am also grateful to the organizers of invited lectures across Asia, Europe, and North America, where I tested the ideas in this book in front of diverse and thoughtful audiences.

Lastly, of course, I thank my family. My father, Bas Degroot, nourished my childhood fascination with outer space and human history. Thanks to him, some of my earliest memories involve drawing blue giant stars on our living room floor, checking out armfuls of books from our local libraries, and peering at the ghostly tendrils of the Orion Nebula through a cardboard telescope. This book would not have existed without Bas. I also could not have imagined writing it without my wife, Madeleine Chartrand. A brilliant historian of law, gender, and labor, Madeleine repeatedly read over this book, suggesting improvements and offering encouragement. My cat, Winnie, diligently maintained my work ethic by forcing me to wake up early every morning. Finally, my daughter, Elowyn, and my son, James, are my entire cosmos. I dedicate this book to them, and I hope it does some good to brighten their futures.

Illustration Credits

4 NASA / JPL-Caltec.

26 (*Top*) Reformatted from Dagomar Degroot et al., "The History of Climate and Society: A Review of the Influence of Climate Change on the Human Past," *Environmental Research Letters* 17:10 (2022): figure 2.

26 (*Bottom*) Reformatted from Dagomar Degroot et al., "Towards a Rigorous Understanding of Societal Responses to Climate Change," *Nature* 591 (2021): 539–550, figure 1a.

30 Data sources: D. A. Fordham et al., "PaleoView: A Tool for Generating Continuous Climate Projections Spanning the Last 21,000 Years at Regional and Global Scales," *Ecography* 40:11 (2017): 1348–1358; M. B. Osman et al., "Globally Resolved Surface Temperatures since the Last Glacial Maximum," *Nature* 599 (2021): 239–244; PAGES2k Consortium, "A Global Multiproxy Database for Temperature Reconstructions of the Common Era," *Sci Data* 4, 170088 (2017).

52 NASA / GSFC / SDO.

71 NASA / Alex Gerst.

82 NASA / Rick Guidice.

96 Camille Flammarion, *Les Terres du Ciel* (Paris: C. Marpon et E. Flammarion, 1884), 321.

120 (*Top*) USGS Astrogeology Science Center.

120 (*Bottom*) Maths and Data / Alexis Huet.

122 William K. Hartmann.

123 (*Top*) Reproduced from Technica Molodezhi TM—9 1971.

123 (*Bottom*) NASA.

154 (*Left*) Casimir Marie Gaudibert, "Curious Lunar Mountains," *English Mechanic* 468, March 13, 1874.

154 (*Right*) Franz von Paula Gruithuisen, *Entdeckung vieler deutlichen Spuren der Mondbewohner* (1824), 115.

159 (*Top left*) Johannes Hevelius, *Selenographia,sive Lunae descriptio* (1647).

159 (*Top right*) John Russell, "Lunar Planisphere, Flat Light" (1805).

159 (*Bottom left*) Wilhelm Beer and Johann Heinrich Mädler, *Mappa Selenographica* (Berolini, Carlus Vogel, Simon Schropp, 1837).

159 (*Bottom right*) "LEM-1, Lunar Earthside Hemisphere, 3rd Edition July 1966" (Washington, DC: US Government Printing Office, 1966).

162 (*Top*) A Stanley Williams, "Mare Crisium," Selenographical Club: lunar drawings with notes, 2 vols. (1880–1883), Royal Astronomical Society Archives.

162 (*Bottom*) NARA, AS16-120-19295, View of the Gassendi Crater. National Archives.

168 NASA / JSC.

172 Alan H. Anderson, Jr., "Apparent Activity on Moon's Surface Draws Rising Scientific Interest," *Goddard News,* July 11, 1966. NASA.

178 NASA / JSC / ASU.

208 ESA / DLR / FU Berlin, CC BY-SA 3.0 IGO.

225 Percival Lowell, "Color Drawing of Mars, 1905," Lowell Observatory Archives.

229 ISRO / ISSDC / Emily Lakdawalla.

242 William K. Hartmann and Ron Miller, by permission of William K. Hartmann.

245 "Hello, Mars—This Is the Earth!," *Popular Science,* 1919, 74–75.

246 (*Left*) Arne Nordmann / Wikimedia Commons.

246 (*Top right*) NASA Ames Research Center (NASA-ARC) / Wikimedia Commons.

246 (*Bottom right*) Wikimedia Commons.

276 Museum Rotterdam 11028-AB / CC PDM 1.0.

281 NASA HQ Archives, Record Number: 100020.

288 (*Top*) Steve Jurvetson / Wikimedia Commons / CC BY 2.0.

288 (*Bottom*) National Nuclear Security Administration, Nevada Site Office, "Nevada Test Site Guide," DOE / NV-715, Rev. 1 (March 2005), 59.

291 Stephen Hanley / MIT Museum.

308 Painting by Pamela Lee reproduced from Gerrit Verschuur, "This Target Earth," *Air and Space,* October / November 1991.

315 (*Top*) NASA.

315 (*Bottom*) NASA / STScI.

334 NASA / ARC / Rick Guidice.

Index

References to photos and illustrations are in *italics.*

Timeline of Environmental Change

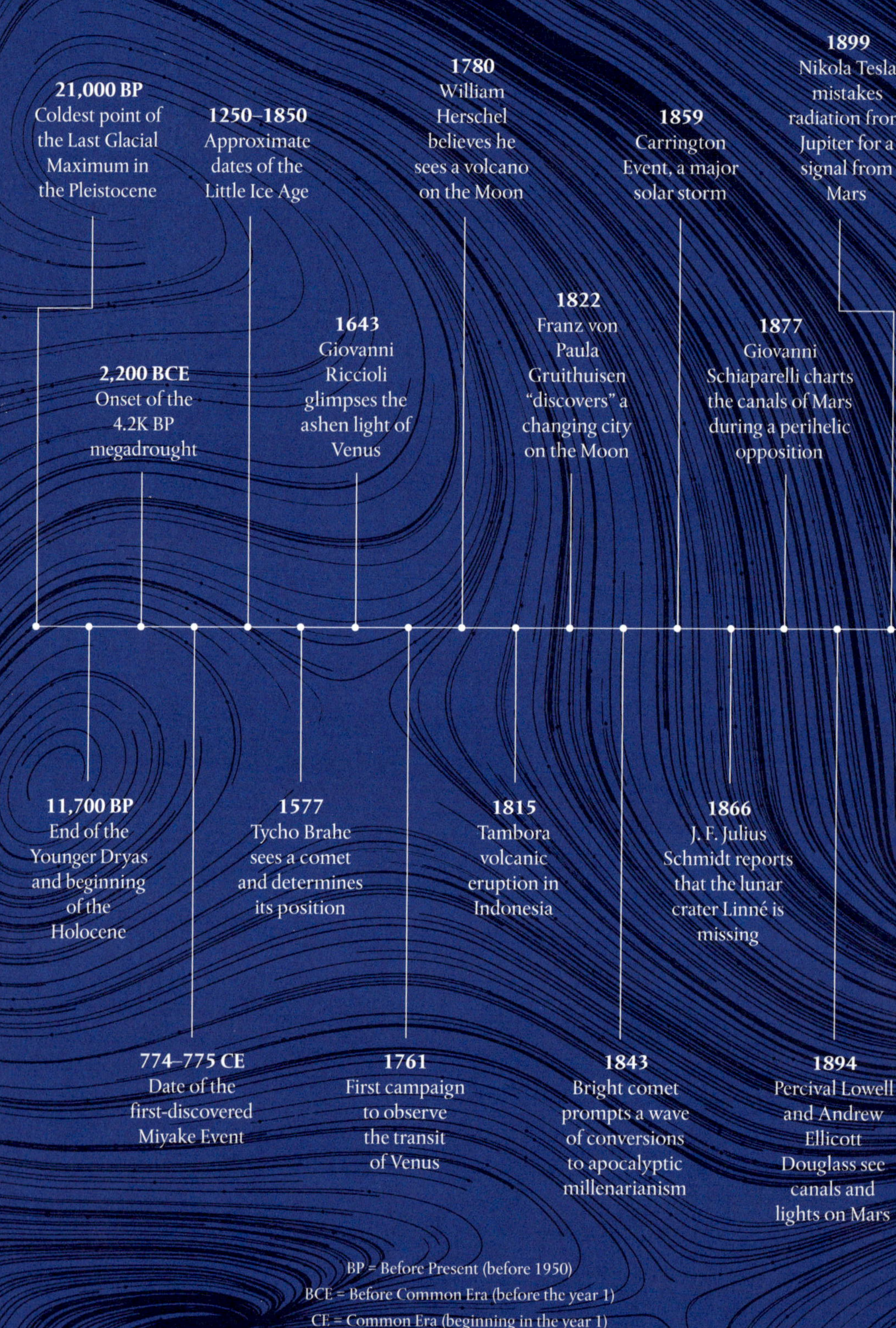